BOTANICAL
ICONS

BOTANICAL ICONS

Critical Practices of
Illustration in the
Premodern Mediterranean

Andrew Griebeler

University of Chicago Press : : Chicago & London

The University of Chicago Press, Chicago 60637
The University of Chicago Press, Ltd., London
© 2024 by The University of Chicago
Published 2024
Printed in China

33 32 31 30 29 28 27 26 25 24   1 2 3 4 5

ISBN-13: 978-0-226-82679-0 (cloth)
ISBN-13: 978-0-226-82680-6 (e-book)
DOI: https://doi.org/10.7208/chicago/9780226826806.001.0001

Publication of this book has been aided by a grant from the Millard Meiss Publication Fund of CAA.

This book was published with the generous assistance of a Book Subvention Award from the Medieval Academy of America.

Library of Congress Cataloging-in-Publication Data

Names: Griebeler, Andrew, author.
Title: Botanical icons : critical practices of illustration in the premodern Mediterranean / Andrew Griebeler.
Description: Chicago : The University of Chicago Press, 2024. | Includes bibliographical references and index.
Identifiers: LCCN 2023006235 | ISBN 9780226826790 (cloth) | ISBN 9780226826806 (e-book)
Subjects: LCSH: Dioscorides Pedanius, of Anazarbos. De materia medica—Illustrations. | Botanical illustration—Mediterranean Region—History—To 1500. | Illumination of books and manuscripts—Mediterranean Region—History—To 1500.
Classification: LCC QK98.16.M43 G75 2024 | DDC 581.022/2—dc23/eng/20230412
LC record available at https://lccn.loc.gov/2023006235

♾ This paper meets the requirements of ANSI/NISO Z39.48-1992 (Permanence of Paper).

For my parents

CONTENTS

FIGURES

INTRODUCTION

A bewildering array of plants inhabit the Mediterranean basin, an area ranging from the Levant in the east to the Iberian Peninsula in the west, from North Africa in the south to Southern Europe in the north. Approximately 24,000 plant species—about 10 percent of the world's known plant species—live in a mere 2.3 million square kilometers. (By contrast, the rest of Europe hosts about 6,000 plant species, or one-quarter the number, in an area more than four times larger.)[1] Nearly 60 percent of the plant species native to the Mediterranean are unique to it, a region encompassing marshy, sandy, and craggy coastlines; high, frosty mountain peaks; and dense woodlands and forested groves, nestled among dry, grassy steppe and thicketed scrub.[2]

The ancient peoples living within this hotspot of botanical endemism and diversity developed a vast botanical lore. They named plants and wove them into their myths, legends, and religions. They collected, compared, described, and depicted them. Plants circulated across Western Eurasia through vast networks of trade for their ritual, medicinal, and culinary uses. Evidence for the systematic descriptive study of plant morphology for medicines appears in Mesopotamia as early as the ninth century BCE, though this tradition may go back to the second millennium BCE.[3] By the fourth century BCE, systematic descriptions of botanical morphology and medicinal properties appear in Greek. Scattered references in that language to a botanical art (*botanikē technē*) or a botanical tradition (*botanikē paradosis*) speak to an ancient awareness of botany as a unique field of inquiry.[4]

This book traces the history of critical practices in the visualization of botanical form within learned medical discourse in the premodern Mediterranean—from the earliest surviving evidence for its existence in the second century BCE to the rise of early modern botanical illustration in the fifteenth and sixteenth centuries. I focus on the illustration of plants in ancient and medieval herbals. George H. M. Lawrence defines an herbal as "a book on plants of real or alleged medicinal properties, which describes the appearance of those plants, and provides information on their medicinal importance and use."[5] Theophrastus of Eresus (d. c. 287 BCE), a pupil of Aristotle (d. c. 322 BCE), has left us perhaps the earliest extant herbal in Greek as part of what is now the final book of his *Historia plantarum* (*Inquiry into Plants*).[6] But while botanical forms proliferated as decoration and emblems in the art of the ancient Mediterranean, illustrated herbals appear relatively late, perhaps only in the second century BCE.[7] This practice of illustrating herbal manuscripts, however, would continue into the modern era. It thus represents one of the longest continuous traditions of secular and scientific image-making in Western Eurasia. I address here evidence for critical practices, namely, *exercises of reasoned or informed judgment* in the visualization and transmission of botanical illustrations. As a result, this book's claims counter prevailing views of premodern botanical art and science as stagnant traditions based on the uncritical copying of earlier manuscripts.

Ancient and medieval botanical illustrations aimed to create and communicate visual knowledge about plants. The art historian John Lowden uses the expression "visual knowledge" to refer to the consistency of iconography in Byzantine religious images, often called icons. He argues this consistency is a product of artists' training and familiarity with the larger tradition of religious painting.[8] Visual knowledge is a matter of knowing what the depiction of a particular subject should look like within a given tradition. In doing so, Lowden makes the case for understanding the transmission and circulation of images on their own terms, independently of the texts (including artists' handbooks) that they might otherwise accompany. This applies to premodern scientific illustrations, too, but with the notable difference that knowledge of a plant's appearance is shaped not just by preexisting imagery, but also by observation and experience. The central idea motivating this book is that an entire body of knowledge about the natural, visible world can be conveyed and transmitted primarily through visual, depictive means, in ways that are complementary and sometimes even contradictory to other, primarily verbal, means of creating botanical knowledge. These plant depictions are truly botanical in the usual narrow sense of the label; they both resulted from and enabled botanical inquiry. In the premodern Mediterranean, image-making was a central way to make and convey knowledge in its own right.

The putative conservatism and rigidity of premodern botanical illustration have typically served as a foil to the vivid, descriptive naturalism of the arts from late medieval and early modern Western Europe. Two studies have played a central role in establishing the main narratives in the early history of botanical illustration. First, Charles Singer's "The Herbal in Antiquity and Its Transmission to Later Ages" (1927) linked various traditions of medieval herbal illustration to their classical forebears, largely by tracing continuities through textual transmission.[9] Despite the preservation of classical traditions of botanical inquiry, Singer nevertheless described medieval herbals in degenerative terms. They are, in his words, "feeble works for feeble minds," products of the "wilting mind of the Dark Ages," a reflection of "the Decay of the Western Intellect."[10] While Singer largely focuses on continuity between antiquity and the Middle Ages, Otto Pächt took up what he called "the other end of the story, the way out of the tunnel of the Middle Ages," in his "Early Italian Nature Studies and the Early Calendar Landscape" (1950).[11] Following Singer, Pächt states that botanical illustration first emerged in the Hellenistic world with a high degree of naturalism, but that it degenerated over the course of the Middle Ages as a result of stylization as well as the successive, uncritical copying of pictures.[12] Accurate illustrations based on the observation of nature only reappear with the invention of novel illustrations for new botanical texts in thirteenth-century Italy.[13] But these illustrations remained schematic, as though "pressed flat—into profile or full frontal views—artificially arranged, prepared for the Herbarium; half picture, half diagram."[14] Fully naturalistic and illusionistic portrayals of plants, what Pächt calls "real life portraits of plants," finally emerge in the fifteenth century.[15]

While subsequent scholars have greatly elaborated upon our view of this history, they have tended to reproduce the familiar narratives of decline and rebirth. Minta Collins's *Medieval Herbals: The Illustrative Traditions* (2000) remains at present the most comprehensive art historical overview of ancient and medieval botanical illustration.[16] Collins provides a valuable synthesis that ultimately fits individual manuscripts into a larger narrative on the emergence of accurate "from life" depictions based on the direct observation of nature. In *Observation and Image-Making in Gothic Art* (2005), Jean Givens offers a closer view of this reemergence of descriptive and naturalistic art in the late Middle Ages. She usefully and incisively distinguishes between illusionism, the extent to which a picture matches the visual field as perceived, and naturalism, the "impression of life-likeness," and descriptiveness, the "rendering of specific factual detail."[17]

The present book offers an alternative to the usual narrative. Rather than trace a linear story of the sequential development of botanical illustration in Western Europe, I attend to the continuous and multivocal development of botanical illustration across the Mediterranean. In doing so, I focus first on the emergence of botanical illustration in the ancient Mediterrnaean and then on its continuous reworking in illustrated medieval versions of the Περὶ ὕλης ἰατρικῆς (*On Medicinal Matter*), often known by its Latin name, *De materia medica*.[18] This pharmacological treatise by the first-century CE pharmacologist Pedanius Dioscorides of Anazarbus would eventually come to serve as the authoritative text on medical botany in the medieval Greek and Arabic Mediterranean. In centering Dioscorides' *De materia medica* in the history of botanical illustration, the present book consequently decenters the medieval Latin botanical tradition, and with it illustrated botanical works such as the *Herbarius* attributed to Apuleius Platonicus or the *Tractatus de herbis*. This shifted emphasis is not to deny the importance of these traditions to the history of botanical illustration. This book posits rather that critical developments in late medieval and early modern botanical illustration emerged not only from linear developments in Western Europe extending from the illustrated *Tractatus de herbis* to the humanist botanical illustrations of the fifteenth century, but also from a vast repertoire of centuries-old critical strategies for botanical visualization from the wider Mediterranean world. These strategies are especially visible in Greek and Arabic illustrated manuscripts firmly rooted in the ancient tradition of the illustrated Dioscorides. The present book argues that ancient and medieval people of the Mediterranean continually corrected, updated, elaborated, and expanded their visual understanding of the botanical world. They aspired to useful and comprehensive visual knowledge of the botanical world. Rather than cataloguing the deficiencies of premodern botanical illustrations, this book positively characterizes what they show, how they show, and what they might tell us about their makers' and original viewers' understanding of plants.

Among the many Greek and Arabic Dioscorides manuscripts discussed in this book, two illustrated Greek manuscripts play an especially prominent role. The first is a famous illustrated copy of Dioscorides from sixth-century Constantinople that is now in the

Austrian National Library (Österreichisches Nationalbibliothek) in Vienna.[19] This manuscript is the earliest extant example of a more or less complete, illustrated herbal. It is therefore perhaps our best witness to antique botanical illustrations, most of which have not survived. The second manuscript that plays an especially prominent role is an illustrated ninth- or tenth-century manuscript of Dioscorides, also from Constantinople, now in the Morgan Library in New York.[20] This manuscript stands as an essential witness to the continuity and development of critical botanical visualization in the early Middle Ages, a time that is otherwise thought to have seen the stagnation of ancient botanical traditions. The Morgan Dioscorides confounds usual narratives with its ample evidence for a variety of critical methods of botanical illustration. That both the Morgan Dioscorides and the Vienna Dioscorides were frequently copied in the later Middle Ages, if not before, also means that they had a direct impact on the subsequent history of botanical illustration. As a result, we can trace a continuous tradition of critical botanical illustration from antiquity to the late Middle Ages through both the Vienna Dioscorides and the Morgan Dioscorides.

The first four chapters of this book cover the emergence of a tradition of botanical illustration in the ancient Mediterranean. Chapter 1, "Rulers and Root-Cutters," surveys the earliest evidence for botanical illustration and puts that evidence in the broader cultural and intellectual context of the Hellenistic Mediterranean. The earliest herbals appear to have been connected to the practice of gathering herbs ("root-cutting," *rhizotomia*). The first illustrated herbals can be further linked to the patronage and intellectual aspirations of Hellenistic and Roman elites. Chapter 2, "Mithridates' Library," takes a closer look at how ancient herbals may have worked, particularly how pictures and descriptions secured an association between a plant name and its medicinal properties. The fact that many ancient illustrated herbals appear to have been illustrated prior to the copying of texts suggests that their primary function was to create and convey knowledge visually. Chapter 3, "Painting, Seeing, and Knowing," considers how ancient pictures carried out this purpose. Drawing on surviving illustrations from the second through the sixth centuries, the chapter further surveys the diversity of ancient botanical illustrations according to their portrayal of space and time, as well as a plant's medicinal properties and names.

Chapter 4, "Illustrating Dioscorides," outlines the formation of illustrated versions of Dioscorides' *De materia medica*. While originally unillustrated and organized according to drug affinities, Dioscorides'

De materia medica was eventually abridged, alphabetically arranged, and illustrated. In the process, pictures were collated from several different sources, and matched to Dioscorides according to synonymous plant names. This new tradition thus linked text and image in new ways. Because word and image were ultimately based on different traditions and sources, each provided information that was otherwise absent in the other. In this way, word and image offered distinct ways of knowing about plants.

The last four chapters of the book consider botanical illustration in medieval and early modern botanical manuscripts in Greek, Arabic, and Latin from across the Mediterranean. Chapter 5, "Medieval Herbals," describes the broad transformations of the botanical tradition in the Mediterranean during this period. It illustrates the vitality of the medieval botanical tradition and how it changed as a result of cultural and institutional shifts, such as the rise of monasteries and medical foundations, including hospitals. The next two chapters survey critical practices of illustration in the medieval Mediterranean. Chapter 6, "The Critical Copy," argues, against the common view that the process of copying can only result in the loss of visual information over time, that medieval copyists employed a variety of critical practices that allowed them to improve and elaborate upon the botanical tradition over time. Chapter 7, "*Ex Novo*," argues against another common view: the idea that medieval botanical illustration was a stagnant tradition based solely on the copying of earlier pictures. The chapter shows instead that medieval illustrators often created new illustrations in order to create more comprehensive herbals. The eighth and final chapter, "Echoes and Reverberations," explores the relationship between critical traditions of botanical illustration in the premodern Mediterranean and the development of scientific botanical illustration in late medieval and early modern Western Europe.

While this book traces a broad narrative arc, it does not provide an exhaustive, comprehensive survey or catalogue of manuscripts. Many such books already exist.[21] Nor does it cover ancient and medieval pharmacology—the powers and properties of roots and herbs—except to the extent that they are relevant for the depiction of plants. Its principal aim is to account for the unique role of illustrations in the production and transmission of botanical knowledge in the premodern Mediterranean, making evident the claim that a picture does not merely document an object in the world, but also actively forms knowledge of it.

Note on Names and Transliterations

This book spans a range of different fields of study that have different conventions for rendering words, names, and titles from other languages. I have chosen to adopt the names and transliteration methods that seem most appropriate to each discipline. Consequently, I use Latinate spellings for classical Greek names and common Latinate versions of the titles of Greek texts if they exist. (In some cases, especially for unedited or lost texts, I simply transliterate the Greek title.) For medieval (post-800 CE) and early modern Greek names, however, I transliterate family names but translate common given names according to their English equivalents. All common Greek words in the text are transliterated into italic Roman letters according to the American Library Association and Library of Congress (ALA-LC) Romanization system for ancient Greek. All Arabic words, names, and titles are transliterated in italics throughout the text according to the system of the *International Journal of Middle East Studies* (*IJMES*). I have also included Hijri dates (AH, that is, *anno hegirae,* "in the year of the Hijra") for authors and rulers of the Islamicate world. I use the terms Islamicate and Latinate to designate broadly inclusive cultural areas. In this book, Islamicate applies to cultures and societies where Arabic and Persian are the primary scholarly languages, while Latinate designates those where Latin is the primary scholarly language. Finally, I have used the names of early modern Western European authors according to their respective modern languages, as opposed to their Latin names (for example, Zsámbóky instead of Sambucus).

Plant names present another set of difficulties. The long history of ancient plants' usage means that they can refer to radically different plants in different contexts and periods. Throughout the body of the text, I have translated ancient plant names with common English names. In their wonderful messiness and imprecision, common names function more like ancient plant names. Still, these common English names should be regarded as rough approximations rather than as one-to-one translations. I provide scientific names in the notes and index whenever possible. I follow each first mention of the plant with the italicized Latin, Greek, or Arabic name given by the source, though spellings often vary even within the same documents.

Several of the manuscripts discussed, illustrated, and cited in this book are referred to by informal designations; for each, the full shelfmark is given on first mention, and the short form is used thereafter. For easy reference, the following key links the informal designations to the manuscript shelfmarks:

Ambrosiana Codex	Milan, Biblioteca Ambrosiana MS C 102 sup.
Ambrosiana Notebook	Milan, Biblioteca Ambrosiana, MS A 95 sup.
Banks Dioscorides	London, Natural History Museum, MS Banks Coll. Dio. 1
Cambridge Dioscorides	Cambridge, University Library, Ee 5.7
Carrara Herbal	London, British Library, MS Egerton 2020
Chigi Dioscorides	Vatican, Biblioteca Apostolica Vaticana, MS Chigi F.VII.159
Egerton 747	London, British Library, MS Egerton 747
Marciana Handbook	Venice, Biblioteca Nazionale Marciana, Cod. gr. XI, 2121 = coll. 453
Morgan Dioscorides	New York, Morgan Library, MS M 652
Munich Dioscorides	Munich, Bayerische Staatsbibliothek, Cod. Clm. 337
Naples Dioscorides	Naples, Biblioteca nazionale, Cod. gr. 1
Old Paris Dioscorides	Paris, Bibliothèque nationale de France, MS gr. 2179
Padua Dioscorides	Padua, Biblioteca del Seminario, Cod. 194
Parchment Arabic Dioscorides	Paris, Bibliothèque nationale de France, MS ar. 4947
Thott Codex	Copenhagen, Kongelige Bibliotek, MS Thott 190 folio
Topkapi Dioscorides	Istanbul, Topkapi Library, MS Sultanahmet III 2127
Vatican Galen	Vatican, Biblioteca Apostolica Vaticana, MS gr. 284
Vienna Atlas	Vienna, Österreichische Nationalbibliothek, Cod. 2277
Vienna Dioscorides	Vienna, Österreichisches Nationalbibliothek, Cod. med. Gr. 1
Vienna *Kitab al-diryāq*	Vienna, Österreichische Nationalbibliothek, Cod. A.F. 10

RULERS AND
ROOT-CUTTERS

Evidence for botanical illustration in antiquity is scant. It appears suddenly in Pliny the Elder's *Natural History*, a comprehensive account of the whole natural world and its products in thirty-seven rich, if occasionally inaccurate, books. Dedicated to Titus in 77 CE, the text was only completed two years later, in 79 CE, when its author died in the eruption of Mount Vesuvius.[1] Concerning botanical illustrations, Pliny complains that

> the Greek authorities that we have already discussed on occasion, among them Crateuas, Dionysius, Metrodorus, produced their work in a most attractive way, but one that demonstrates little more than its difficulties. For they painted pictures of plants [*effigies herbarum*] and then wrote down their effects. A picture with so many colors is truly misleading, especially in the imitation of nature, and the various hazards of copying degenerates them greatly. Moreover, it is not enough for them to be painted at single moments in their lifetime since they change their appearance with the fourfold variations of the year.[2]

Scholars tend to take these comments at face value.[3] We might also be tempted to view Pliny's comments through the lens of distinctly modern experiences: the profusion of affordable, copiously illustrated practical field guides, on the one hand, and a common skepticism about the place of illustrations in serious science, on the other.[4] Much of this book, however, could be read as an attempt to understand Pliny's concerns in light of the longevity and vitality of ancient traditions of botanical illustration in the premodern Mediterranean. This first chapter places ancient botanical illustration in its original social and intellectual context. Ancient illustrated herbals, although initially linked to the practice and profession of root-cutting, appear to have been made for elite consumption, first for Hellenistic rulers, and later for Roman elites as well. Ancient illustrated herbals were relatively rare, treasured objects, unlike today's field guides, but like them were used for a variety of aesthetic, scholarly, and practical purposes.

Pliny the Elder's *Effigies herbarum*

Pliny's comments on illustrated herbals provide some information on how they were made and what factors complicated their use and production. He names several Greek authors who illustrated their herbals and explains how they did it. Despite the attractiveness of these pictures, Pliny questions their utility.[5] They demonstrate "little more" than their "difficulties": their misleading colors, their degradation over time through copying, and their failure to capture the appearance of a plant as it varies over the course of the year.

Ironically, the passage about the degeneration of pictures over time ("the various hazards of copying degenerates them greatly") was itself "corrupted" in transmission. The wording in manuscripts (*fors varia* or *sors varia*) suggest that the degeneration of the pictures is due to "varying hazards" or "varying fate."[6] This baffled editors of the text. Karl Friedrich Theodor Mayhoff replaced these words with *socordia*, or "carelessness," while Carl Friedrich Wilhelm Müller preferred *sollertia*, or "skill."[7] These various readings differ in how they attribute degeneration to the act of copying: whether through fate, carelessness, or skill. But the corruption of the text here further points to the difficulties of copying in a manuscript culture in general, for text as well as for pictures.

We can make sense of Pliny's criticisms in light of his larger project. In other passages in the *Natural History*, he complains about Greek authors and Greek scientific literature.[8] He rails against Greek authors for sacrificing utility for pizzazz.[9] He is dubious of the mutability of Greek medicine and its complicated, inexplicable remedies, which he attributes to the Greeks' supposed proclivity for speculation and professional disagreements.[10] While these complaints point to different conceptions of medicine, they also speak to Pliny's concerns over Roman dependence on Greek expertise.[11] This desire reflects the cultural rivalry between the Latin and Greek worlds, and a consciousness of the elevated place of Greek learning among Roman elites.[12] There were, moreover, clear practical reasons for Pliny, the author of an unillustrated work, to criticize botanical illustration. While other Roman authors, such as Vitruvius and Marcus Varro, had their works illustrated, Pliny did not, presumably because of the practical concerns that he voiced.[13]

Yet Pliny's concerns about botanical illustration were likely shared by some of his contemporaries. Writing about a century later, Galen (c. 129–200/216 CE), for example, criticizes a rival physician for investigating "whether painting was useful to doctors" while failing to define what a disease was.[14] Different physicians doubtless took different positions on the matter, for despite Pliny's critiques, many of his contemporaries continued to produce and read illustrated herbals.[15]

An Alternative Account of Ancient Botanical Illustration

Against Pliny's dim view of ancient botanical illustration, we can place the frontispiece cycle of the Vienna Dioscorides (figs 1.1–1.5), so called after its current location and its principal textual contents.[16] This manuscript was perhaps produced in Constantinople in the early sixth century CE or possibly before.[17] It contains the earliest more or less intact copy of an illustrated herbal, in this case, a version of an encyclopedic account of medicinal substances called the *Peri hylēs iatrikēs* (*On Medicinal Matter*, often called by its Latin name, *De materia medica*) by Pedanius Dioscorides of Anazarbus, a Greek authority in pharmacology and a near contemporary of Pliny. Like Pliny, Dioscorides did not originally illustrate his work. Dioscorides instead emphasized in his preface the need for direct and firsthand observation of actual plants in the field. He does not condemn pictures *per se*, but rather learning about plants only from books. By the sixth century, however, we find versions of Dioscorides' work circulating with illustrations. Medieval authors sometimes assumed that Dioscorides had had his work illustrated.[18] We will return to Dioscorides in later chapters; for now we can look to the Vienna manuscript's frontispieces for an alternative vision of the place of pictures in ancient pharmacology.

Turning over the prefatory folios of the Vienna Dioscorides, the reader finds two scenes of symposia (figs. 1.1 and 1.2) and two author portraits (figs. 1.3 and 1.4). This cycle thus forms a thematically interconnected sequence of portraits referring to the contents of the Vienna

FIGURE 1.1 Symposium of pharmacological authorities, counterclockwise from top: Chiron, Machaōn, Pamphilos, Xenokratēs, Mantias, Hērakleidēs, Nigros (Quintus Sextius Niger). Vienna, Österreichische Nationalbibliothek, Cod. med. Gr. 1 (the "Vienna Dioscorides"), early sixth century CE, fol. 2v. Courtesy of Österreichische Nationalbibliothek.

codex and the historical conditions for its creation. The two symposia scenes show famous pharmacologists in two separate gatherings, each with seven participants. This grouping assimilates the physicians to the fabled Seven Sages of ancient Greece.[19] The two scenes roughly present us with the development of ancient pharmacology from its obscure origins in the mythical figure of the centaur Chiron to its culmination in the figure of Galen. Grasping a mortar and pestle, Chiron acts out the practical origins of medicine. In contrast, the red codex tucked under Galen's arm points to the accumulation of knowledge and medical theory.[20] Pharmacology emerged as a form of practical knowledge communicated orally (Chiron), then through writing on rolls, as carried by most of the symposium participants, and culminating in compendious codices such as the one held by Galen.

Galen turns toward Dioscorides, the author of the codex's main

FIGURE 1.2 Symposium of pharmacological authorities, counterclockwise from top: Galen, Krateuas, Apollōnios, Andreas, Rhouphos (Rufus), Nikandros, Dioskourides. Vienna Dioscorides, fol. 3v. Courtesy of Österreichische Nationalbibliothek.

text and the subject of the next two frontispiece illustrations (figs. 1.3 and 1.4). In the first, Dioscorides sits before a dark building while he gestures toward a mandrake root grasped by a personification of Discovery (*heuresis*). With her left hand Discovery points to a dog. This is a reference to a myth about the extraction of the mandrake root: those wishing to harvest the mandrake could do so by tying a dog to it and having the dog uproot the plant.[21] On the next folio, we find Dioscorides in his study with a helper painting the mandrake, now held up by a personification of *epinoia*. As the literal "thinking on" a thing, *epinoia* encompasses intelligence, thought, invention, design, purpose, or even afterthought or second thought.[22] Dioscorides meanwhile writes into a book that is already illustrated.[23] Botanical illustration is thus presented as a vital part of Dioscorides' authorial practice. The frontispieces in the Vienna Dioscorides tell us nothing

FIGURE 1.3 Dioscorides and the "Discovery" (*heuresis*) of the mandrake root.
Vienna Dioscorides, fol. 4v. Courtesy of Österreichische Nationalbibliothek.

FIGURE 1.4 Dioscorides writing in his study, while an assistant paints a mandrake plant held by a personification of "Invention" (*epinoia*). Vienna Dioscorides, fol. 5v. Courtesy of Österreichische Nationalbibliothek.

about when, why, or how botanical illustrations first appeared. Rather, they depict botanical illustration as an essential, integral part of the ancient pharmacological tradition. The final illustration in the frontispiece cycle shows the presentation of the volume to the Byzantine

FIGURE 1.5 Presentation of a codex to Anicia Juliana by a personification of "Desire for she who loves building" (*pothos tēs philoktistou*), accompanied by personifications of "Magnanimity" (*megalopsychia*), "Prudence" (*phronēsis*), and the "Gratitude of the Arts" (*eucharistia tēs technai*). Vienna Dioscorides, fol. 6v. Courtesy of Österreichische Nationalbibliothek.

princess Anicia Juliana, surrounded by personifications of her virtues (fig. 1.5). I will return to this depiction of the recipient of the codex later on in this chapter. For now, I will take a moment to consider the authors of ancient illustrated herbals more fully.

Authors

Pliny gives us the names of three authors of illustrated herbals: Crateuas, Metrodorus, and Dionysius.[24] Notably, Pliny does not say that any of them invented herbal illustration, or that they were the only authors of herbal texts to have had their works illustrated.[25] The identification of these three authors is difficult. Ancient physicians often adopted similar professional names, which were in turn handed down over generations.[26] Texts were, moreover, frequently misattributed.[27] Admitting these difficulties, we can still identify these authors to a certain degree in order to characterize the authorship of illustrated herbals more generally.

Of the three authors that Pliny mentions, we know the most regarding Crateuas (c. 100–60 BCE).[28] Several fragments attributed to Crateuas survive.[29] In the preface to his *De materia medica*, Dioscorides calls Crateuas a *rhizotomos*, literally a "root-cutter."[30] *Rhizotomoi* are different from the druggists (*pharmakopōlai*) who sold drugs and mixed compounds, although the relationship between the two professions is not well understood.[31] Both are typically associated with traditions of folk medicine that are often contrasted with Hippocratic medicine.[32] A fragment from the *Rhizotomoi*, a lost play by Sophocles, paints a vivid picture of contemporary views of root-cutting in the fifth century BCE.[33] In it, Medea, "naked, shrieking, wild-eyed," uses brazen implements to gather the noxious juice of a plant called deadly carrot [*thapsia*].[34] Theophrastus similarly notes that root-cutting, though sometimes reasonable, is often not.[35] He approves of some practices, such as facing away while gathering noxious plants. Others he regards more skeptically, such as the avoidance of certain birds, the making of offerings, or the drawing of circles in the ground with specific implements.[36] By the Hellenistic period, however, Crateuas and other root-cutters appear to have shed some of the "murkier aspects of classical Greek *rhizotomia*."[37] The development of a putatively more "rational" tradition of *rhizotomia* may go back to the fourth century with the *Rhizotomikon* by Diocles of Carystus (d. c. 300 BCE).[38] Crateuas may have also authored a work of the same name.[39]

In addition to his *Rhizotomikon*, Crateuas may have written other works. The philologist Max Wellmann assumed that Crateuas' *Rhizotomikon* must be different from the illustrated work mentioned by Pliny, because it appears to have involved descriptions of plants, which, Wellmann argues, cannot have been part of the illustrated works described by Pliny, in which only medicinal properties (*effectus*) were mentioned.[40] This illustrated work, Wellmann supposes, was for a more general audience.[41] Wellmann hypothesizes the existence of a second work by Crateuas on metals and spices, following statements by Galen and Dioscorides, because the title of the *Rhizotomikon* would seem to preclude those topics.[42] Yet Wellmann's arguments seem stretched. As Friedrich Ernst Kind has noted, and Alain Touwaide has reiterated, Crateuas could have written a single work on medicinal substances, from which other works were later derived or extracted.[43] That work on medicines may have included herbs, metals, and spices, all under the title *Rhizotomikon*, simply because herbs made up the bulk of the book, or because the title referred generically to medicinal materials. It is also possible that in Pliny's day multiple versions of this work, including an illustrated, perhaps abridged version, may have been in existence. The fact remains that we have only one title associated with Crateuas' name.

Elsewhere in Pliny's *Natural History* we learn that Crateuas named a plant *mithridateia* after Mithridates VI of Pontus (r. 120–63 BCE).[44] From this scholars have supposed that Crateuas may have been in Mithridates' entourage, that he lived at court, or that he was Mithridates' personal physician.[45] Researchers have more recently expressed doubts about such claims.[46] Crateuas' naming of a plant after Mithridates is hardly proof of his presence at Mithridates VI's court.[47] Crateuas may have named the plant after Mithridates simply to honor him or perhaps to refer to his role in its discovery. Pliny tells us of several plants named after their royal discoverers: for example, gentian after the Illyrian king Gentius (r. 180–168 BCE), and artemisia after Artemisia, the sister, wife, co-ruler, and successor of Mausolus (r. 377–353 BCE), of Mausoleum of Halicarnassus fame.[48] In those situations a royal person is supposed to have named the plant after themselves, though other specialists may have simply attributed the discoveries of those plants to those monarchs. In other cases, a specialist might choose to honor someone by giving their name to a plant. Pliny tells us that Juba II of Numidia (r. 30–25 BCE) named euphorbia in honor of his personal physician Euphorbus.[49] In short, it is hard to know the nature of Crateuas' relationship with Mithridates VI on the basis of his having named a plant after him.

Much less is known about the other two authors Pliny mentions. Scholars have proposed a variety of identifications for Pliny's Metrodorus.[50] We find the name Metrodorus several times in Pliny's *Natural History*. Pliny states that Metrodorus was the author of a compendium on herbal remedies called the *Epitomē rhizotomoumenōn*.[51] At another point, Pliny notes that Metrodorus recommends using the herb *peplis* (likely *Euphorbia* sp.) to help remove the afterbirth following childbirth.[52] The identification of Pliny's Dionysius is similarly difficult, due in part to the prevalence of the name.[53] But he may be Cassius Dionysius of Utica, whom Pliny elsewhere calls Cassius Dionysius or simply Dionysius.[54] Cassius Dionysius of Utica is known to have translated a landmark work on agriculture by Mago the Carthaginian from Punic into Greek. The Roman author Columella (d. 70 CE) went so far as to call Mago the father of agriculture.[55] Varro (d. 27 BCE) reports that Dionysius of Utica translated and condensed Mago's twenty-eight books, subtracted eight, and added information from Greek sources.[56] Dionysius dedicated this work to the Roman governor of Africa, the praetor Publius Sextilius in 89 or 88 BCE.[57] Varro adds that Diophanes of Bithynia later abridged Dionysius' text into six useful books for King Deiotarus of Galatia (first century BCE).[58] Dionysius of Utica may have also penned a treatise on medicinal plants called the *Rhizotomika*.[59] It was presumably this work that would have been illustrated.

Many of the titles mentioned above relate to root-cutting. In addition to the *Rhizotomikon* by Diocles of Carystus, Crateuas' *Rhizotomikon*, the *Rhizotomika* by Cassius Dionysius, and Metrodorus' *Epitomē rhizotomoumenōn*, we also know of a work called *Rhizotoumena* or *Peri rhizotomikōn* by Mikion (or Mikkion), the *Rhizotomikon* by Amerias of Macedon, and finally a work of the same name by the little-known Eumachus of Corcyra.[60] By the fourth century CE, Oribasius (d. c. 400 CE), physician to the emperor Julian, refers to such works on root-cutting generally as *rhizotomiai*.[61] Illustrated herbals appear to have emerged in the development and elaboration of these works on root-cutting. But while some *rhizotomika* appear to have been illustrated, others were not. Max Wellmann supposed that early illustrated herbals represented a kind of popular subclass of works on root-cutting, which he distinguished from other, more textually replete and scholarly works within that genre.[62] It is hard to say, though, if these divisions were as clear-cut as Wellmann envisioned. Pliny for one does not allude to different audiences for the illustrated herbals, nor does he criticize their *textual* contents. (Given Pliny's low estimation of herbal illustration, it is unusual that he would forego the opportunity to indulge in yet another criticism.) As will become clearer in the next section, clear-cut divisions between practical and entertaining, as well as between popular and scholarly, are difficult to maintain for early illustrated herbals.

Besides the botanical authors that Pliny mentioned, there were other authors of illustrated medical and scientific texts. Scholars have long supposed that Aristotle's anatomical works were once illustrated with anatomical diagrams.[63] Later Hellenistic anatomists and physicians, such as Apollonius of Citium (fl. c. 60 BCE), also included illustrations in their treatises.[64] Works on toxicology also appear to have been illustrated. Writing in the early third century CE, the Christian author Tertullian (d. c. 230/240 CE) noted that Nicander of Colophon, an author of the second century BCE, "writes and paints [*scribit et pingit*]" of the scorpion's manifold evils.[65] It would seem then that Nicander's verses, or a derivative text, were illustrated, at least by the time Tertullian wrote his *Scorpiace*.[66] Nicander of Colophon wrote two didactic poems—the *Theriaca* and the *Alexipharmaca*—the first on venomous animals, the second on toxic plants and their antidotes.[67] Sometime between the third and fifth centuries CE, a certain rhetor named Eutecnius paraphrased Nicander's poems.[68] Later copies of the paraphrases and the poems are illustrated with depictions not only of animals, but also of plants, depicted either individually or as groups of ingredients in recipes for compound drugs.[69] These illustrations may have been added later.[70] While the exact location and dates of Nicander's literary activity have been debated, he is often placed in the entourage of Attalus III Philometor (r. 138–133 BCE), the last king of Pergamum.[71]

Patrons and Readers

As we have seen, both Nicander and Crateuas may have had connections to the Hellenistic kings Attalus III and Mithridates VI, respectively. We can add that Apollonius of Citium, the author of an illustrated work on the treatment of joints, has been linked to the court of Ptolemy Auletes (r. 80–51 BCE).[72] Illustrated works on astronomy, mathematics, zoology, and medicine, including toxicology, anatomy, and bandaging, may also have been produced in the Hellenistic period.[73] We can surmise that Hellenistic monarchs, aristocrats, and oligarchs played an important role in motivating and patronizing scientific illustration. As Marie-Hélène Marganne has noted, scientific authors orbited the courts of Hellenistic kings interested in the natural sciences.[74] These courts would have been especially fertile ground for the creation and compilation of illustrated herbals due to the presence of skilled artisans and medical specialists, as well as the availability of funding and resources. Hellenistic kings supported the establishment of "research institutions" with vast collections of books and objects, such as the library and *mouseion* in Alexandria and the great library at Pergamum. The period saw the rise of experimentalism, including systematic human dissection (and vivisection) in Alexandria (third century BCE) and the concurrent emergence of opposing medical "schools" or "sects."[75] The rise of medical and scientific illustration in Greek texts also coincides with the elevated status of medical specialists in the Hellenistic world, who were then better able to purchase or commission such works.[76]

Attalus III and Mithridates VI in particular stand out for their purported personal involvement in toxicological research. Both men reportedly tested antidotes and poisons on condemned prisoners, a form of experimentalism perhaps inspired by the vivisections carried out under the Ptolemies.[77] According to Plutarch, Attalus III personally cultivated a number of toxic plants in his royal gardens.[78] The historian Justin (probably second century CE) claims that Attalus III even gifted poisons to his friends.[79] Mithridates VI emulated Attalus III.[80] Galen notes that Mithridates followed the Pergamene king in experimenting on antidotes for toxic sea slugs, scorpion stings, snakebites, and spider bites.[81] The Pontic king enthusiastically patronized research in the medical sciences. He corresponded widely with scholars and physicians. Pliny mentions that Mithridates VI invited Asclepiades of Bithynia, one of the most famous physicians of the day, then in Rome, to join him in Pontus.[82] Asclepiades refused, but sent some written works instead. Similarly, Zachalias of Babylon dedicated several books on gemstones to Mithridates VI.[83] Crateuas may have been among the scholars who were either at the Pontic court or in dialogue with its king. The king himself authored several works, perhaps including recipes for compound drugs.[84] His name would eventually come to be associated with a variety of complicated antidotes, many under the name "mithridatium," sometimes called the "mother of all antidotes."[85]

At the Pergamene and Pontic courts, royal research on pharmacology and botany went hand in hand with the exercise of monarchic power. The kings used their resources and networks to plant gardens and build their collections of books and *materia medica*. Attalus III would have added to the great library already in Pergamum, built by his ancestor Eumenes II (197–160 BCE) in imitation of the Ptolemaic *mouseion* and library in Alexandria.[86] Pliny tells us that Mithridates VI had information on medicinal substances gathered "from all his subjects," and that he compiled a library of such treatises along with a collection of specimens (*exemplaria*).[87] Pliny also reports that Mithridates possessed a collection of gemstones (stored in a *dactyliotheca*), many of which would have been accorded therapeutic properties.[88] I will return to the contents of Mithridates' library and their connections to the function of early herbals in the next chapter. It is worth noting here, however, that in collecting plants, drugs, gems, written treatises, and various reports from his subjects, Mithridates took advantage of a vast network for the acquisition and production of medical knowledge. As Bruno Strasser has noted of early modern and modern collecting,

> establishing this kind of collection, like establishing empires, required the mastery of space. Collectors produced a movement of natural things, which were often dispersed across the world, toward central locations, just as empires produced movements of goods from colonies to metropoles. Unsurprisingly, colonial powers were collecting powers, and colonies constituted rich collecting grounds. The geographical reach of an empire represented an immense field for collecting.[89]

The expansion of Hellenistic collections and resources for pharmacological inquiry were similarly tied to military conquest. According to Pliny, Alexander of Macedon (r. 336–323 BCE) was "inflamed by a desire to know the nature of animals," and had "thousands of persons throughout the whole of Asia and Greece, all those who made their living by hunting, fowling, and fishing and those who were in charge of warrens, herds, apiaries, fishponds and aviaries," answer to the inquiries of Aristotle, "so that he might not fail to be informed about any creature born anywhere."[90] Alexander's interests extended beyond natural history into the art of medicine. The author Plutarch (d. after 119 CE) claims that Aristotle also inspired in Alexander a love for the art of healing (*philiatrein*).[91] Not only was Alexander interested in medical theory, he also prescribed therapies and regimens for his friends. Regardless of their truth, these accounts represent an idealized vision of the relationships among empire, rule, and scientific inquiry, particularly medicine.[92] Alexander of Macedon thus provided Hellenistic monarchs with a model for how to cultivate their own medical interests publicly.

Romans, too, took an active interest in their Greek opponents' research into medicine and botany. Pompey had Mithridates' library transferred to Rome and asked his freedman Lenaeus to translate its works into Latin, so that, as Pliny notes, his victory would "benefit life no less than the state."[93] The Roman Senate similarly mandated that Mago's writings be translated into Latin, although they had already been translated into Greek. Carthaginian libraries, doubtless home to other Punic works on plants and agriculture, passed to the Numidian ruler Juba II, another monarch interested in medicine. As already noted, he authored a now lost treatise on *euphorbia*, named after his personal physician Euphorbus.[94] Botanical research thus continued in the Roman empire and its client states. Pliny's vast project, too, may be viewed as part of this larger transfer and translation of knowledge from the Hellenistic Greek world to Latin.

Besides transferring Mithridates VI's library and collections to Rome, Pompey dedicated to Venus Victrix a Pergamene-style portico in which he planted a large garden with exotic species.[95] Another of Mithridates VI's foes, Lucius Licinius Lucullus, planted a Persian-style pleasure garden (*paradeisos*) on the Pincian Hill in Rome.[96] These large gardens made the reach of empire materially and horticulturally manifest. Later Roman leaders continued to use gardens in these ways. After the fall of Jerusalem in 70 CE during the first Jewish War, Vespasian had the monumental Temple of Peace built in the center of Rome. The temple celebrated Roman imperialism as well as the end of a brief civil war and political turmoil in 69 CE. Elizabeth Ann Pollard has characterized the gardens of this complex as being akin to a colonial botanical garden.[97] The library that was part of this complex, and its proximity to an important spice warehouse, the Horrea Piperataria, as well as to the booksellers on the Vicus Sandalarius, enhanced the broadly scientific air of the complex.[98] In Galen's words, the Temple of Peace complex became "the general meeting-place for all those engaged in learned pursuits."[99]

But gardens in ancient Rome and the Hellenistic world fulfilled a variety of functions; it would be misleading to characterize them purely as expressions of empire, or as research gardens. Gardens—and painted emulations of them—provided Roman elites with spaces for reflection, delight, and the display of prestige.[100] Elite Roman interest in illustrated herbals could be similarly viewed through the twin lenses of practical application and aesthetic pleasure. Roman elites celebrated medical and botanical knowledge within their homes and private lives. Roman interiors from the time of Augustus abound in botanical ornament, as exemplified by the garden room from the villa of the empress Livia at Prima Porta (fig. 1.6). The garden room paintings, dated around 30–20 BCE, present a staggering array of botanical life, including acanthus, cypresses, date palms, ferns, irises, ivies, myrtles, laurels, oleanders, pomegranates, various poppies, quince, rose, spruce, stone pine, strawberry trees, viburnums, violets, and several kinds of oak. Ann Kuttner has linked the botanical specificity of these paintings, and of Augustan art in general, to the circulation of illustrated herbals in the Roman empire.[101] The room's decoration intersects with the interests of the empress Livia (d. 29 CE), known for her cultivation of laurels as well as *liviana* figs, so named in her honor.[102] She likely applied her botanical knowledge to medicine, too: she was celebrated as the inventor of a laxative and of remedies for sore throat, chills, and nervous tension.[103] Moreover, she funded the restoration of the temple of Bona Dea Subsaxana on the Aventine Hill, a healing sanctuary.[104]

The painted walls of the garden room would have created a backdrop for convivial conversation that emphasized the hosts' horticultural interests and botanical erudition. Athenaeus' *Deipnosophistae* (*The Learned Banqueters*), written in the early third century CE, gives us a vivid, if idealized (and exhausting) sense of what such conversations might have sounded like. The last surviving book reports the banqueters' discussions on various topics, including the identities and uses of plants.[105] The interlocutors cite and quote learned treatises by the likes of philosophers, lexicographers, and physicians. Even if idealized, the *Deipnosophistae* portrays elite dinner conversation as a suitable place for the exchange of quotations and displays of botanical knowledge.

The Roman ideal of the supremely erudite elite complicates our familiar division between professionals and experts, on the one hand, and amateurs and dilettantes, on the other. For example, in his *Noctes Atticae* (*Attic Nights*), Aulus Gellius (d. 180 CE) describes one occasion when a Roman aristocrat discovers that he knows more about veins and arteries than a rural physician does.[106] The question of expertise, of scholarly versus popular, or of entertainment versus science is hardly applicable to Roman elites' idealized conceptions of themselves and their learning. All Romans likely had some medical knowhow, especially Roman elites who were responsible for the healthcare of their households and their dependents.[107] Although Livia's and Athenaeus' guests were not professional physicians or horticulturalists in any sense, they certainly aspired to (and presumably succeeded at) a high level of discourse. It is in this broader context that Roman elites purchased, commissioned, and read illustrated herbals.

FIGURE 1.6 "Garden room" in Livia's villa at Prima Porta, now in the Museo Palazzo Massimo, Rome. Photo by author.

While we can only guess that Livia owned or commissioned illustrated herbals, the much later Vienna Dioscorides gives us a concrete example of a Roman aristocrat's ownership of such a volume. The final frontispiece illustration shows Anicia Juliana flanked by personifications of her Magnanimity and Prudence, while the "Gratitude of the Arts" and "Desire for She Who Loves Building" pay homage to her (fig. 1.5). These last two figures are elliptical representations of the people of the Honoratae district of Constantinople who gave Juliana the codex to thank her for her construction of a church in their district.[108] Diliana Angelova has placed the Vienna Dioscorides within a long tradition of elite Roman women publicly cultivating their interest in healthcare, as already seen in Livia.[109] Late Roman imperial women similarly promoted their association with healthcare by visiting healing hot springs, while distributing largesse along the way, and by constructing and refurbishing bathhouses.[110] Within broader Roman society, healthcare was deemed an appropriate female profession.[111] Elite women replicated those roles on a grander scale.

Roman elites could also have used their illustrated herbals for practical domestic purposes. Leslie Brubaker has suggested that the Vienna Dioscorides was an ideal gift for a late Roman aristocratic matron charged with the wellbeing of her household.[112] Centuries before, Livia had employed numerous medical personnel, both enslaved and free, in her household, including a medical superintendent, at least five physicians, a surgeon and eye-doctor, midwives, wet nurses, and male and female orderlies who ran a sickbay for staff.[113] While Livia (or Juliana) might have sought an outside physician for serious illnesses, most healthcare matters were probably dealt with in house, perhaps by resident medics, enslaved and free, who would have relied on simple remedies, such as the herbs depicted in an illustrated herbal. If consulting multiple physicians, Juliana would have needed to choose which recommendations to take. Galen gives us vivid recollections of arguments between physicians at patients' bedsides.[114] The need to arbitrate between different physicians' recommendations, and to find alternative therapies, would have provided a further practical justification for elite use of illustrated herbals.

Fragments of the earliest surviving illustrated herbal also give us a clear view into the context, use, and readership of ancient illustrated herbals. These come down to us in the form of twenty-odd papyrus fragments now housed in different institutions, but originally found in the ancient city of Tebtunis, modern Umm El Baragat, on the southwestern edge of the Fayum basin in Egypt.[115] These fragments all once belonged to a roll or rolls copied in the second century CE.[116]

Kim Ryholt has shown that these fragments likely came from a library deposit at the temple in Tebtunis dedicated to the crocodilian god Sobek, known locally as Soknebtunis, "Sobek, Lord of Tebtunis."[117] The deposit was found in two shallow cellars (only around 1.25–1.30 meters deep) that were, therefore, probably only for storage.[118] The exact ownership of the texts and the circumstances of their deposition cannot yet be established. Peter Van Minnen has characterized these cellars as a kind of private storage for books owned by individual priests.[119] Ryholt, however, suggests that the papyri may simply have been deposited in them at the time of the temple's closure.[120] The vast majority of the papyrus fragments from this temple library were written in Egyptian scripts, although there were also a number of Greek scientific texts, including the illustrated herbal, as well as texts on astrology and medicine. These complemented a sizable collection of Egyptian demotic medical papyri, including a demotic herbal.[121] Many of these medical texts were written rapidly and on reused papyri, "indicating texts for use rather than display."[122] Isabella Andorlini has also noted that many objects related to the preparation and storage of medicine have been found in Tebtunis.[123] That none of the medical texts from this context appears to have been a direct translation from Greek into demotic or vice versa could indicate a comfortably bilingual readership that perhaps appreciated the illustrated herbal as a specifically Greek genre.[124] The illustrated herbal would have been useful in this context for identifying plants by their Greek names.

All of this suggests that the Graeco-Egyptian priests at the temple may have used these medical texts, including the illustrated herbal, to provide medical care.[125] The illustrated herbal could have been used for a variety of practical purposes, ranging from plant identification to consultation with patients. There does not appear to have been a separate profession of herbalists or root-cutters earlier, in Pharaonic Egypt; rather, herb collection fell to the healers themselves.[126] Thus, the priests at the temple may have performed these tasks. Despite its practical use, this illustrated herbal was also a prestige volume. Illustrated herbals were uncommon in Graeco-Roman Egypt: of around 250 Greek medical papyri discovered in Egypt, only two examples— the Tebtunis Roll and a codex from Antinoopolis (see chapters 2 and 3)—were illustrated.[127] This underscores a key difference between the herbal and most of the medical texts from Tebtunis. While many of the texts were cribbed on reused papyrus, the same cannot be said of the illustrated herbal. The Tebtunis illustrated herbal was more than a practical handbook. It may have enhanced social status, as well as the medical expertise and authority of its priestly owners. Ann

Ellis Hanson has further suggested that the Greek medical literature found in Tebtunis may reflect the influence of elites who maintained residences in or near the village, but who were otherwise outsiders.[128] In this view, elite patients made Greek texts available to the priests in order to garner "medical attentions more in keeping with the health care they enjoyed in their official residences" in larger centers, such as Oxyrhynchus and Antinoopolis.[129] At the same time, this privileged clientele would have afforded the priests at Tebtunis more resources with which they could secure an illustrated herbal.

We can see that while ancient illustrated herbals were preeminently for elite consumption, they also circulated between levels of society. While the Tebtunis herbal may have originated in an elite context, we can imagine that it served a broader population once in the hands of the priests of the Soknebtunis temple. It may even have been a gift to priests at the temple from one of the elite residents in Tebtunis. The opposite occurred with the Vienna Dioscorides, which the citizens of Honoratae gave to Anicia Juliana to express their gratitude for her construction of a church in their district. The circulation of books between levels of society repeatedly occurs in the history of the illustrated herbal. While wealthy patrons typically owned and commissioned such books, they also gave them to physicians and charitable foundations, such as temples, and in later times to monasteries and hospitals.

Conclusion

Pliny the Elder's comments on the illustrated herbals of Crateuas, Metrodorus, and Dionysius provide us with the earliest written evidence for botanical illustration in the herbals of the ancient Mediterranean. While Pliny critiques these works—a reflection of his antagonism toward Greek medicine—the fact that he names several authors allows us to consider the authorship and patronage of ancient illustrated herbals and their relationship to other illustrated scientific works, such those by Nicander and Apollonius of Citium. The authors of these texts often had links to Hellenistic courts. Rulers such as Attalus III and Mithridates VI may have supported the writing and illustration of such texts as part of a larger program of medical and pharmacological research that included experimentalism, the planting of gardens, and the acquisition of books and *materia medica*. Roman elite readers emulated these Hellenistic examples. They read and commissioned illustrated herbals as a way to cultivate their interest in botany, medicine, and horticulture. These interests are further indicated by Roman garden culture and the botanical imagery in Roman art.

Roman and Hellenistic elites used botanical culture to express their erudition, their interest in the welfare of their subjects, and their ability to gather and mobilize resources in the exercise of power. Although linked to elite contexts, the Tebtunis Roll and Vienna Dioscorides suggest that illustrated herbals circulated across multiple levels of society, from extremely elite to elite and subelite contexts. Illustrated herbals thus likely served a variety of purposes depending on who read them and why. We can imagine the likes of Livia or Anicia Juliana reading such works for pleasure or for arbitrating a decision about medical care in their household. And while the priestly owners of the Tebtunis Roll likely used it for practical purposes, particularly for gathering herbs, they might also have used it to showcase their social standing and knowledge of Greek medicine in order to enhance their own medical authority.

2
_____

MITHRIDATES' LIBRARY

The earliest illustrated herbals are lost to history. While the previous chapter explored the rise of the ancient illustrated herbal in
its political and social contexts, this chapter addresses its relationship
to the ancient botanical tradition. It aims to reconstruct how the ancient herbal might have been used to refer to plants. In this chapter,
we return to Pliny the Elder's comments on herbals with an eye to
what they might tell us about the purpose and use of the plant depictions in them. In other words, how, according to Pliny, did herbals
refer to plants? And, more specifically, how did ancient botanical illustrations function alongside the herbal texts in which they appeared? I
approach this question by considering illustrated herbals against the
broader landscape of the botanical tradition. I suggest that ancient
herbals constructed a system of reference linking plant names and
medicinal properties to defined kinds of plants. Pictures of plants in
herbals take the place of actual plants as the objects of pharmacological discourse. In line with the concerns of Empiricist traditions of
medicine, illustrated herbals simulated direct visual encounters with
plants.

Referring to Plants

In the famous passage quoted at the beginning of chapter 1, Pliny
complained about the difficulty of using botanical illustrations in the

herbals of his day. He noted that pictures of plants appeared along with texts concerning the plants' effects (*effectus*) or medicinal properties. This juxtaposition linked a plant's name and likeness to a discussion of its properties. As Pliny mentions only *effectus*, it seems that early illustrated herbals may have omitted descriptions of plants altogether.[1] Following this critique of herbal illustration, Pliny continues by contrasting illustrated herbals to unillustrated ones that rely on either verbal descriptions or names alone:

> For this reason the other writers have given verbal accounts only [*sermone eas tradidere*]; some have not even given the shape [*ne effigie*] of the plants, and for the most part have been content with bare names, since they thought it sufficient to point out the properties and nature [*potestates vimque*] of a plant to those willing to look for it [*quaerere volentibus*]. Gaining this knowledge [*cognitio*] is not difficult; I at least was able to examine all but a few plants thanks to the knowledge [*scientia*] of Antonius Castor, the highest authority of our time in the art [of botany]; I used to visit his small garden [*hortulo*], in which he would rear a great number [of plants]. . . .[2]

Pliny here indicates that authors of pharmacological treatises had several options in how they referred to plants. Besides illustrating plants, authors could describe them verbally or simply rely on the name of the plant alone. The last option puts the burden of plant identification on the readers themselves. Max Wellmann has interpreted this passage as indicating that there were three approaches that ancient authors could take in writing about medicinal plants: they could include pictures and properties without descriptions; properties and descriptions without pictures; or names and properties without descriptions or pictures.[3]

Material evidence generally confirms, but also qualifies, Wellmann's interpretation. Surviving fragments of early illustrated herbals tend to lack descriptions of plant morphology. The earliest extant fragments of an illustrated herbal, the second-century Tebtunis Roll, focus on medicinal properties and exclude verbal descriptions.[4] Both the so-called Johnson papyrus, from a fourth- or fifth-century codex from Antinoopolis in Egypt, and a palimpsested parchment leaf dated to the fifth or sixth century, now in an Arabic book at the monastery of St. Catherine's of the Sinai, exclude verbal descriptions of morphology and focus on medicinal properties and preparations.[5] The same observations apply to the textual excerpts of Crateuas' herbal preserved in the Vienna Dioscorides.[6] At the same time, some of these fragments do include information beyond mere *effectus*. For example, some fragments from the Tebtunis Roll refer to a plant's habitat.[7] One fragment (frag. *a*), possibly concerning chondrilla (*chondrilē*), notes the earthy places where it grows.[8] This information may have been useful to the priests of the temple of Soknebtunis, who, as we saw in the last chapter, may have used the herbal to procure herbs.

The illustrated Latin *Herbarius*, perhaps composed in the fourth century CE and falsely attributed to Apuleius Platonicus, follows this pattern. The earliest surviving copy of this text is from sixth-century Italy, but is now in Leiden (fig. 2.1). It typically excludes morphological descriptions except when drawing distinctions between similar plants. It also frequently mentions a plant's habitat and various synonyms under which it is known.[9] The illustrated *Herbarius* thus coordinates between different ways of finding and referring to plants, with slight variations depending on the specific plant in question. In doing so, the *Herbarius* concretizes a broader set of practices involved in botanical inquiry and gathers them all into one text. Users consult the illustrations of plants in tandem with their synonyms, accounts of their habitats, and occasional distinguishing descriptions.

Like many illustrated herbal texts from antiquity, the *Herbarius* is a complex text that was likely composed in several stages from multiple sources. Its compilation may have been similar to that of the illustrated Dioscorides (see chapter 4). As we will see throughout this book, illustrated herbals were regularly added to, abridged, and changed over time, according to the needs of their users. The remnants of the Antinoopolis Codex, for example, show evidence of emendations and corrections by a second hand.[10] We can imagine that subsequent copies based on these specific copies may have contained these emendations and additions as well. Moreover, as we will see in chapter 4, some illustrated herbals eventually came to include descriptions of plant morphology by the sixth century, if not earlier.

These examples suggest that we should not draw too sharply the distinctions between different types of botanical texts. The reality was likely blurrier—and partly driven by the specific uses to which an herbal was put and the general circumstances in which it found itself. Still, surviving fragments tend to confirm a pattern of ancient illustrated herbals excluding or omitting textual descriptions of plant morphology, perhaps on the grounds that the picture was understood to perform that task. These different approaches to writing about plants

point to a central problem: reference. Names, pictures, and verbal descriptions are all different ways of *referring*, of associating specific medicinal properties to a particular plant name.

FIGURE 2.1 Illustration of ironwort (*heraclea*) in the *Herbarius* of Ps.-Apuleius Platonicus. Later marginalia in Latin and Hebrew. Leiden, Universiteitsbibliotheek, MS VLQ 9, sixth century CE, fols. 63v–64r. Courtesy of Universiteitsbibliotheek Leiden.

Naming and Knowing

The simplest way to refer to a plant was by name alone. Pliny expresses a certain faith in the capacity of names to work by themselves—it merely requires familiarity with the plants in question. Confidence in the capacity of names to refer to specific plants may reflect what Andrea Guasparri has called Pliny's tendency to connect a given name ontologically to the properties of its referent.[11] In everyday life, names would have been largely sufficient for procuring herbs from healers, physicians, apothecaries, herbalists, or root-cutters, as well as the various medical practitioners associated with temples and similar foundations.[12]

Ancient plant names were themselves often descriptive in one way or another. As Gavin Hardy and Laurence Totelin note, ancient plant names had many connotations that "evoked the properties, habitats, usages and discoverers of plants."[13] Some plant names closely linked the plant to its purported medicinal properties. For example, *aristo-*

locheia (birthwort), literally "best [*aristo-*] birth [*locheia*]," evokes its use in aiding delivery.[14] Other names suggest a plant's appearance or habitat. The Greek name for parsley, *petroselinon*, suggests the plant's rocky (*petro-*) habitat, while the name for geranium (*geranion*) recalls the resemblance of its seedpods to the heads of cranes (*geranoi*).[15] Ancient plant names also often included epithets and determinatives that made additional distinctions between varieties or subtypes of a given plant.[16] For example, Dioscorides recognized several varieties of birthwort (*aristolocheia*), including *aristolocheia strongylē*, or "round" birthwort, and *aristolocheia makra*, or "long" birthwort.[17] Recognition that many plant names—and medical terms in general—were descriptive led some ancient authors to advocate the study of their etymology.[18] Others, however, were more circumspect. Galen argued, for example, that etymologies often misled.[19] He appears to have written a now lost work, *On the Correctness of Names*, that showed, presumably among other things, that "etymology is an impostor."[20] The title of this lost work recalls Plato's *Cratylus*, a fictional account of a discussion about whether names work by convention or are naturally connected to their objects.[21]

Plant names, however, could present difficulties. They varied greatly by region and language, and they changed over time.[22] Hardy and Totelin have outlined several such complications. Many plants went by multiple names (synonymity).[23] Many names were shared by different plants (homonymity).[24] Sometimes authorities might use false names to conceal the identity of a given plant (pseudonymity).[25] And many plants had no names, especially plants that were uncommon, from other regions, or without known uses (anonymity).[26] Moreover, while many ancient names were descriptive,[27] they could be misleading by not clearly referring to the way in which they pertained to or described the plant.[28] Some names might even apply ironically or antiphrastically.[29] As a result of these difficulties, ancient authorities may have resorted to botanical lexica with lists of synonyms of plant names from different languages and regions.[30] These lists of synonyms were further copied into herbals, as we see in the Leiden Herbarius.

Still, Pliny says that it was easy enough for one to learn about plants by studying them in a garden or by consulting someone knowledgeable. Pliny claims that he learned about plants by visiting Antonius Castor's garden (*hortulus*).[31] Pliny here exalts the ideal of direct observation, of learning by seeing with one's own eyes.[32] Other ancient authors make similar points.[33] Writing at roughly the same time as Pliny, Dioscorides claims in the preface to his *De materia medica*,

I know, on the one hand, from personal observation [*autopsias*] in utmost detail most items, and, on the other hand, . . . I have a thorough understanding of the rest from accounts [*historias*] on which there has been unanimous agreement and previous examination in each case by locals [*epichōriōn*].[34]

Dioscorides here aims to explain his reason for writing his *De materia medica*. Previous authors, he alleges, were less comprehensive, were imprecise, or lacked firsthand knowledge of pharmacology.[35] Dioscorides' emphasis on firsthand knowledge (*autopsia*) and verified textual sources echoes the positions of the so-called Empiricist "school" of medicine. Galen, in fact, notes that the word *autopsia* was a coinage of the Empiricists.[36] Going back to the mid-third century BCE, the Empiricist "school" represents a broader movement and ideological reaction against medical investigation into the causes of diseases and the basing of treatment on them.[37] Empiricists sought rather to provide effective medical treatment through the use of known medicines and therapies. They believed that repeated observation of what worked and what did not—and not theoretical speculation—was the basis of medical experience and medical practice. Experience (*empeiria*) was developed through the repetition of trials and tests (*peirai*). What was beyond individual experience could be gleaned from authoritative texts or rather collective memory of effective treatment (*historiai*).[38] While *historia* has the original and narrower sense of an inquiry, it more generally referred here to the written record. New treatments for unknown or variant cases could be determined through "reasoning by similarity" (*epilogismos*) to known and more familiar cases. *Empeiria* was the practiced summation of this "tripod" of *autopsia*, *historia*, and *epilogismos*.[39] Dioscorides especially echoes Empiricist thought when he criticizes previous authors who neglected to observe drug actions through trials.[40] This is not to say that Dioscorides would have identified himself as an Empiricist, but rather that Empiricist thinking thoroughly saturates his thought.

Dioscorides' use of *autopsia* refers to close and repeated observation in botanical study. He advocates for firsthand knowledge of plants at multiple stages of growth, in multiple areas:

Anyone wishing to have experience [*empeirian*] in these matters must be present [*paratynchanein*] when plants sprout newly from the earth as well as when they are in their prime and past their prime. . . . the person who has come across plants often and in many places will most readily recognize them.[41]

Dioscorides thus further relates his experience to his many travels.[42] Elsewhere in his introduction, Dioscorides notes that it was as a result of his itinerant "soldier's life [*stratiōtikon ton bion*]" that he was able to acquire such broad, firsthand knowledge of medicine.[43] Dioscorides here exposes some limitations to Pliny's learning about plants in the garden of Antonius Castor. The ideal observer sees every plant growing throughout its range and at different times of the year. The fact remains that most premodern medical practitioners would not have had the ability to do so, even within their own country. There were clear limits to the exercise of *autopsia*: the difficulty of access and the normal limits of a single life—surely life is short, but the art of medicine is long.[44] Consequently, even Dioscorides had to draw upon written accounts, but, he assures us, only those that were judged favorably by locals (*epichōriōn*, those who live in the area).

How does the ancient illustrated herbal fit with these broader concerns in the ancient botanical tradition? The principal aim of an ancient herbal, whether illustrated or not, was to connect a specific plant to a name and a list of its medicinal properties. When it came to learning to associate properties and names with particular plants, *autopsia* would undoubtedly always have been the ideal. By contrast, an illustrated herbal, like any other medical text, falls under the category of a *historia*—written texts, the inquiries of others, and collective memory, broadly. *Historia* remained, of course, essential for gaining botanical experience. It provided common ground, a way for specialists to relate their personal experiences to those of others, and to articulate and communicate observations to the wider community. The tension between the necessity of *historia* and the ideal of *autopsia* may go a long way toward explaining the advent of botanical illustration. While a description names and characterizes a plant's features, a picture shows them. It gives viewers a chance to see for themselves what a plant looks like. More than a description, a picture ideally shows what a plant's features look like and how they spatially relate to each other. In other words, a picture offers viewers an opportunity for *autopsia*. In doing so, a picture can stand in for presence of a plant within the text. But while pictures serve these functions to a greater extent than do descriptions, there nevertheless remain important similarities between them that explain why Pliny treats them as similar or even interchangeable.

Descriptions and Depictions

Pictures and descriptions both belong to *historia*. They both perform similar tasks by referring to a plant's outward appearance. This is evident in Pliny's use of similar terms for both depictions and descriptions. While he calls a picture an *effigies*, he uses the term *effigie* to refer to a plant's form as rendered in a verbal description. Pliny's way of referring to the appearance or form of a plant opens itself to ambiguity:

> That plant, however, which the Greeks call *dracontion* has been pointed out to me in *three forms/pictures* [*triplici effigie*]; [the first] has leaves like those of beet, a thyrsus and a purple flower; this is like the *aron*. Others have pointed out [*monstravere*] a kind with a long root, marked as it were and knotted, and with three stems in all. They prescribe a decoction of its leaves in vinegar for the bite of serpents. The third plant pointed out [*demonstratio*] had a leaf larger than that of the cornel, a root like that of a reed, and they say it has as many knots as the plant is years old, the same with the leaves.[45]

At first glance, Pliny seems to complain about three different illustrations.[46] Alternatively, he merely notes that the plant *dracontion* occurs in three different forms. The *triplici effigie* here parallels Pliny's use of *effigie* elsewhere to refer to an herb's outward appearance, form, or shape within a verbal description. The passage's ambiguity concerning *dracontion* is heightened by the words *demonstratio* and *demonstratum*, that is, a showing or pointing out, "as with a finger"—"a vivid delineation," or "picturesque presentation."[47] Pliny's assertion "this is [*hoc est*]" further reinforces the deictic, pointing quality of the passage. All of this pointing suggests a group of experts showing Pliny three different plants, all bearing the same name. In other words, botanically useful descriptions and depictions both point to the distinguishing features that help to define a particular kind of plant.

The author of a description and the maker of a depiction consequently had to choose what features or properties sufficiently defined their subject. Emphasis on similar and distinguishing characters was a central concern in ancient biological definition and classification. Ancient authorities typically regarded an organism as belonging to a general group or class of organisms, an *eidos* or *genos* (*species* or *genus*, respectively, in Latin). These terms can be generally translated as "kinds," "types," or "varieties." The ancient recognition of specific kinds of organisms should not be confused with more modern, fixed conceptions of biological species. Ancient scholars, for example, believed that some plants could change from one kind into another.[48] Others were said to generate spontaneously.[49] Aristotle classed organisms into specific kinds, or *genē*, according to characteristic "differences [*diaphorai*]," now often referred to by the Latin *differentiae*. Aristotle imposed a hierarchy on these divisions, such that a *genos* was composed of different forms (*eidē*). The divisions could, however, telescope in either direction, toward the more generic or toward the more specific.[50] Thus several *genē* could make up another *genos*, while an *eidos* might be composed of other *eidē*. Through a process of successive differentiation, Aristotle aimed to arrive at a single definition of an *eidos* in terms of its *genos* and the final *differentia* that set it off as a distinct group. This final *differentia* thus encompassed many other, earlier *differentiae* that served as only initial or intermediate analytical steps in the process of definition.[51] Aristotle used this method of definition for animals in his *Historia animalium* (*History of Animals*).

Ancient botanical authorities could only apply the Aristotelian system to plants with limited success. They encountered difficulty settling on proper Aristotelian definitions of plant *eidē* according to their final *differentia* and their *genē*. For example, while Aristotle had focused on the most permanent parts in his definitions of animals, Theophrastus recognized that plant parts that were useful for definition were often "indeterminate and constantly changing."[52] Other botanical authorities, such as Dioscorides, employed similar, though less systematic, methods centered on the comparison of plant parts and the identification of differences. While much depended upon the particular plant in question, certain morphological features, such as leaf shape, received far greater attention than others.[53] Conversely, other parts, such as flowers, typically received far less attention. Both Theophrastus and Dioscorides ultimately employed the terms *eidos* and *genos* loosely, interchangeably, and without clear hierarchy.[54] Plant *eidē* were consequently defined simply by salient features that enabled distinctions to be drawn in relation to similar plants. From an Aristotelian point of view, such definitions may ultimately rely on variant or inessential properties, termed "accidents," rather than on final *differentiae* and *genē*. Centuries later, some early modern botanical authorities would ultimately reconcile themselves to established plant kinds according to "accidents" by which plants were typically encountered and distinguished in the field.[55]

Exemplaria

While ancient botanical authorities had several ways to refer to plants—through descriptions, depictions, or just names—there was still another way of doing so: by including actual specimens with the text. Just before Pliny's criticism of illustrated herbals, he tells us that the Pontic king Mithridates VI sought out information on medicinal substances "from all his subjects," and that at his death he left behind "a case [*scrinium*] of these treatises [*commentationum*] along with specimens and [records of their] properties [*exemplaria effectusque*]."[56] Mithridates' medical library included not only treatises, but also *exemplaria* with their (presumably written) properties (*effectus*). *Exemplaria* can refer to written texts, examples, samples, or specimens.[57] That Pliny specifies that the *exemplaria* were accompanied by their *effectus* suggests that the term refers not to written transcripts but rather to actual samples of medicines or perhaps pictures of them.

If *exemplaria* referred to pictures, then this passage could provide some textual support for a theory, proposed by Stavros Lazaris in 2010 and backed by Joshua Thomas in 2019, that the earliest botanical illustrations were on painted panels, whitened wooden boards variously called *pinakes, leukomata*, and *sanides*.[58] Lazaris based this thesis on the "paradigmatic" use of painted panels in ancient lectures and teaching. He further pointed to the layouts in the Vienna Dioscorides and, from the same manuscript, the depiction of the easel painting in the *epinoia* frontispiece, which looks at first glance like a panel painting (fig. 1.4). Thomas further noted the impracticalities of using papyrus rolls for transmitting botanical illustrations. But while painted panels with botanical illustrations may well have existed, especially as a precursor to or in preparation for the illustration of deluxe herbals, there is little direct evidence to support this theory. Teaching medical botany in this way would have required hundreds of panel paintings. Yet no testimony makes unequivocal reference to them. Moreover, the frontispiece from the Vienna Dioscorides shows Dioscorides writing into an already illustrated codex (fig. 1.4), which suggests that the painter's illustration was to be bound into a codex, as in an album or notebook, and not to be circulated independently as a painted panel.[59] The artist in that scene does not seem to work on a prepared panel, but rather a parchment sheet pinned to an easel.

The limitations of papyrus as a support also do not require that botanical illustrations could have circulated only as panel paintings. Fragments of botanical illustrations have in fact survived from both a papyrus roll and a papyrus codex. As David Leith has pointed out, a fragment of the illustrated fifth-century herbal codex from Antinoopolis has a special double-ply papyrus, perhaps to support the illustrations.[60] This suggests that papyrus could, if necessary, have been adapted to make them better suited for the illustrations. It is also possible that early botanical illustrations were improved or elaborated over time, particularly after the adoption of more durable parchment supports.

At the same time, elaborated botanical illustrations could have circulated in parchment codices well before the widespread adoption of the parchment codex for literary works by the fourth century CE. The medical codex on parchment was a special kind of book, and it has a history distinct from that of the literary codex. Galen, for example, counted priceless parchment codices among the valuable books that he lost in the fire of 192 CE.[61] Galen notes that these volumes were created and obtained at great expense—with one pair of volumes costing its previous owner 100 pieces of gold.[62] If we know medical texts circulated in parchment codices already in the second century, it is not much of a stretch to suggest that some of them could have been illustrated. The earlier use of codices as medical notebooks would further explain why illustrated herbal codices look so different from early literary codices in their layout. Herbals tend to have open, relatively heterogeneous layouts with ample spaces for annotations. Due to the relative ease of flipping through a codex and locating discrete items in it, the codex is particularly well suited for herbals used as reference works. The use of medical texts as reference works is further confirmed from an early date by the use of visual separators in early medical texts, including blank lines, indentation, and the *paragraphos* (a line drawn between lines of text).[63] These visual devices aid the reader in skimming through text and locating individual entries and recipes. Illustrations similarly help to divide a text visually. The codex format also more easily accommodated expansion over time in comparison to a roll, as additional folios or quires could simply be sewn in as needed. Simply put, early botanical illustrations could have circulated in the form of a codex. None of this, however, rules out the possibility that there may have been some panel paintings with depictions of botanical illustrations circulating in the ancient Mediterranean; rather, there is currently not enough evidence to argue that they were a primary vector for the transmission of botanical images in the ancient Mediterranean.

Still, if we understand *exemplaria* to be samples and not panel paintings, then we can imagine that they would have been actual drugs or dried plants accompanied by notes on their properties. The rarity of *materia medica* and the need for concrete objects of study would undoubtedly have motivated their collection. The Mediterranean is a large, geographically diverse region. It is simply not possible to grow in one place every plant thought to have medicinal properties. In the case of Mithridates' Pontic kingdom, the cultivation of many Mediterranean plants was especially difficult, given its different climate. Pliny notes, for example, that Mithridates could not grow laurel and myrtle at Panticapaeum (modern Kerch) in Crimea.[64] The other benefit of the collection over the growing of plants in a garden would have been its relative mobility and accessibility for consultation out of season. Plant specimens could have been dried, bundled, and kept in boxes, tins, or earthenware vessels.[65] Roots could be washed and dried, flowers and fragrant plant parts stored in limewood boxes, and seedpods wrapped in papyrus or leaves in order to keep their seeds together.[66] In addition to *exemplaria*, Pliny also reports that Mithridates possessed a collection of gemstones stored in a *dactyliotheca*.[67] While the name literally suggests a chest or case for finger rings, many of these gems were accorded therapeutic or magical properties.[68] A similar word, *pharmakothēkē*, literally "a box for drugs," is attested from the second or third century CE.[69] We can imagine that similar chests or boxes may have been used even earlier for Mithridates' collection of *materia medica*.

That *exemplaria* were paired with *effectus* in Mithridates' library further recalls the pairing of *effigies* with *effectus* in Pliny's description of illustrated herbals. That similarity raises the question: how is an *effigies* analogous to an *exemplar*? Both the *effigies* and the *exemplar* relate a plant to a plant name and set of properties independently of verbal text—that is, without requiring the reader to translate between words and physical experiences. They both enable a kind of *autopsia*, an occasion for an observer to see what this kind of plant looks like. Yet a picture is a representation. As we have seen, it belongs to *historia*: it mediates and constructs an object in the world through the experiences of others. While it simulates *autopsia*, it does not allow direct experience of a plant. In contrast, an *exemplar* is itself the object. It could have been tested to see if it had the same medicinal effect as the *effectus* ascribed to it. But the *exemplar* is selected and labeled by others—it, too, is a product of human intervention, selection, and labeling. In some ways, an *exemplar* thus also belongs to *historia*. The preservation and storage of *exemplaria* result in losing qualities that the plant would possess if encountered alive and still rooted in the earth. A dried plant is typically dull or brittle when compared to a fresh specimen. Unlike an actual *exemplar*, a picture simulates an altogether different occasion of *autopsia*: an encounter with a living plant. Despite these differences, we can see that botanical *effigies* and *exemplaria* respond to similar intellectual needs—the Empiricist impulse for *autopsia*.

The difficulties presented by the study of actual specimens (and hypothetical panel paintings) may have motivated the development of botanical illustration. Potentially cumbersome, fragile, and difficult to store, specimens and panel paintings cannot exist within a text in a fixed way. They are instead linked to texts through names, indices, or labels. In contrast, pictures can be fixed within a text. They appear directly above their *effectus* along with their names. They are thus internal to the herbal text, co-present with writing on the same material support of the papyrus roll. Pictures thus appear in the course of a reading. They visually materialize in the midst of words, while *exemplaria* have to be handled separately from the text and its material supports. The *exemplaria* must be located and brought to the text. In an illustrated herbal, however, the writing already surrounds a picture and is oriented toward it. The picture is where the name, above, and the *effectus*, below, meet. It is a place for the written text to turn and meet its object. The picture is therefore more than just a substitute for the plant, it is also a placeholder for it as the object of pharmacological discourse. In this sense, the image acts as a discrete, concrete place for accumulating botanical knowledge.

Words for Pictures

Herbal illustrations thus may have emerged initially as a way of substituting for an *exemplar* within a text, that is, as a way to fix the plant's form concretely within the text. When present, a picture can perform an important role in a scientific text, particularly when the text omits descriptions of plants. Botanical illustration thus anticipates broader patterns in the history of scientific representation. As Bruno Latour has noted in his study of scientific narratives in a Brazilian soil study:

> The scientific text is different from all other forms of narrative. It speaks of a referent, *present* in the text, in a form other than prose: a chart, diagram, equation, map, or sketch. Mobilizing its own *internal* referent, the scientific text carries within itself its own verification.[70]

Illustrated herbals instantiate their referents within the text through pictures. At the same time, as Latour points out, external referents—the actual plants growing out in the world—only serve to fix the reference. The plant in discourse is otherwise constructed from the "deambulatory" circulation of internally referent pictures within and between scientific texts. The scientific narrative, therefore, does not primarily concern the construction of resemblance between internal and external referents, but rather the transformation of the former. As Latour notes, "the sciences do not speak of the world but, rather, construct representations that seem always to push it away, but also to bring it closer."[71] Nelson Goodman pushes this sentiment even further, stating, "That nature imitates art is too timid a dictum. Nature is a product of art and discourse."[72] Depictions and descriptions both invent their objects.

Material evidence from ancient illustrated herbals also tends to suggest that illustrations functioned as though they were specimens within the text. The earliest examples of herbal illustration tend to depict individual plants as uprooted, without context, in the singular, and with their limbs and leaves flattened and outstretched as though lying against the surface of the page. These features speak to their having been conceived of as being like actual specimens. Moreover, extant illustrated herbals from antiquity up until the ninth century all portray herbs and shrubs, but not trees.[73] They thus depict plants that could easily be uprooted. As noted in chapter 1, an ancient illustrated herbal, a *rhizotomikon*, was based in the ancient profession of root-cutting, *rhizotomia*. Taken literally, the term "root-cutting" presumes plants with roots that can be cut and easily uprooted. As books on root-cutting, ancient herbals would have thus been prototypically concerned with uprootable herbs, although many ancient herbals also dealt with animals, minerals, wines, and oils.[74] Ancient Greek authors in fact used the word "roots" to refer to medicinal herbs in general.[75] The modern English words "herbalism" and "herbal" seemingly imply a similarly limited scope, while both can actually involve substances and plants that are not strictly herbs.

The possibility that botanical illustrations might have substituted for *exemplaria* (whether as separate painted panels or as actual plant specimens) finds further confirmation in the way that ancient illustrated herbals were produced.[76] As Minta Collins has noted, Pliny's comments about illustrated herbals could be read as describing the sequence of book production and illustration.[77] He states that the pictures were executed first, "*and then [atque ita]*" the text was added to them, written down or *under* them (*subscripsere*).[78] Extant early illustrated herbals corroborate this sequence of production.[79] Illustration clearly preceded the copying of the text in several fragments of the Tebtunis Roll.[80] In other early illustrated herbals we find more luxurious margins and spacing, an approach to layouts that could have emerged partly in recognition of the difficulties encountered when illustration preceded the copying of text. Some sixth- and seventh-century illustrated herbals, including the surviving manuscripts of the illustrated Alphabetical Dioscorides, were also illustrated according to this picture-first method of production.[81] These different remains together suggest that ancient herbals were typically first illustrated and then given text.

This sequence of production and illustration is unusual. In the usual order of production, a scribe first copies the text and then sub-contracts other specialists, such as illustrators.[82] This text-first sequence of production emphasizes the accurate transmission of texts, which makes sense for most kinds of literature. By copying the text first, the scribe ensures that there is enough space for it and that any issues with it can be resolved prior to proceeding with the other stages of production. The picture-first sequence, however, places priority on the transmission of pictures. The scribe and the illustrator thereby ensure that the pictures have enough room and are up to their standards before proceeding with the copying of the text. In this system of production, the picture literally precedes; it takes precedence over the text. It is as though the text illustrates the illustration rather than the illustration illustrating the text. In a reversal of the usual meaning of the term "illustration," the text in an illustrated herbal attaches to an illustration as a kind of tag or label, much like the pairings of *exemplaria* with their *effectus* in Mithridates' library.

Conclusion

This chapter reconstructs how ancient illustrated herbals might have worked within the context of the ancient botanical tradition. The primary aim of the ancient herbal, both illustrated and not, was to connect a plant name and a set of medicinal properties to actual plants. The authors of herbals had several different ways to do this. They might rely simply on plant names and their readers' prior knowledge or access to expertise. The ideal way to learn about plants was perhaps always through firsthand observation (*autopsia*), though there were obvious practical limits to living out this ideal. Authors could therefore also describe the plants verbally, or depict them. Descriptions and depictions both primarily worked to set out the morphological properties of the plant, particularly those characteristics that allowed a particular plant kind to be differentiated from others. At the same time, though, pictures in illustrated herbals did not function in the same way as written descriptions. A picture allowed readers to see a plant and directly associate its name and properties to its visual appearance. Contemporaries understood and used these pictures as though they were specimens, such as the ones collected by Mithridates VI. As in illustrated herbals, Mithridates VI had his specimens paired with descriptions of their properties. Both specimens and illustrations simulated *autopsia*. This emphasis on visual knowledge in illustrated herbals further influenced their layout and production.

3

PAINTING, SEEING, AND KNOWING

The earliest botanical illustrations in ancient herbals linked a plant to a name and a list of properties. As we saw in the last chapter, these illustrations were analogous to descriptions of plant morphology. But unlike descriptions, they also simulated firsthand experience (*autopsia*) and concretized visual reference to plants within the text. This chapter explores the ways that early extant botanical illustrations depict plants by showing their general appearance or habit, as well as distinguishing features. Early illustrations demonstrate a shift toward the increasing articulation of plant parts that more fully shows how each part connects to the whole. They also increasingly treat the blank background less as depth than as a shallow fictive surface against which the individual plant rests. This flattening or "pinning" of the plant further reinforces the rendering of the plant's distinguishing features, including even "mnemonic" references to their names and medicinal properties.

Determining the pictorial contents of the earliest illustrated herbals is challenging. The earliest surviving botanical illustrations include the extant fragments of the second-century Tebtunis Roll and the fifth-century Antinoopolis Codex, as well as the palimpsested fragment of a fifth- or sixth-century herbal now in the monastery of St. Catherine's of the Sinai. In each case, we have at best two illustrations from what were much larger books. The earliest surviving, more or less complete examples of illustrated herbals are the Vienna and Naples

Dioscorides manuscripts, of the early sixth and late sixth or early seventh centuries, respectively. Scholars have long supposed, however, that many of the illustrations in these manuscripts were based on much earlier illustrations, many of which were of Hellenistic origin.[1] It is on these grounds that I consider them in this chapter. But the Vienna and Naples Dioscorides manuscripts also have a staggering variety of pictures. It is not possible here to document the content of every illustration; rather, I characterize the main strategies in order to outline the diversity of the tradition. My focus here is not on the accuracy, naturalism, or illusionism of the pictures, but simply on what properties they show a given kind of plant to have.

Pictures for Learning

Before considering the pictorial content of ancient botanical illustrations, we should first look at the broader relationship between the visual and scientific content of a picture. Researchers have long supposed that ancient botanical illustrations served instructional or didactic purposes. In a posthumous 1945 publication, the scholar Erich Bethe argued that the origins of book illustration were to be found in the instructional pictures (*Lehrbilder*) of scientific treatises, rather than the "decorative" illustrations (*Schmuckbilder*) of literary texts.[2] He reasoned that pictures and diagrams are necessary for comprehending scientific texts. The art historian Otto Pächt elaborated on the didactic visual qualities of premodern botanical illustrations. In his 1950 essay on Italian nature studies, Pächt distinguished between premodern botanical illustrations that were "pressed flat—into profile or full frontal views—artificially arranged, prepared for the Herbarium; half picture, half diagram," and the more naturalistic and illusionistic botanical illustrations of fifteenth-century Italy.[3] More recently, Stavros Lazaris has argued that ancient scientific illustrations emerged within the context of ancient medical instruction, where they fulfilled a "paradigmatic" function in teaching.[4]

Given their didactic quality, it may be tempting to view these pictures as diagrams. Pächt's phrase, "half picture, half diagram," highlights the possibility that premodern botanical illustrations functioned as diagrams.[5] Pictures of plants that present their subjects flattened and didactically arranged can certainly look like diagrams. But while they may share some visual features with diagrams, they do not function diagrammatically. They do not "allow imperceptible theoretical objects to become visible and tactile" as diagrams do.[6] Nor do they coordinate between different representational and cognitive systems through, for example, verbal discursive features, as emblematized by the frequent presence of writing in diagrams.[7] Beyond the plant's name, premodern botanical illustrations typically lack text, and they do not typically carry out a synthesis of text and image within themselves.[8] Even though some botanical illustrations occasionally have diagrammatic features, the larger tradition is primarily based on their use as pictures. The point is that Pächt's comparison to diagrams is more about the optical flattening of the plant depicted, its simplification, and its didactic arrangement of parts.[9] Consequently, more relevant here is Pächt's characterization that premodern botanical illustrations present plants "as if they had been pulled up by the roots and taken to the herbalist's studio, pinned down and neatly arranged for didactic purposes."[10] The premodern botanical illustration prototypically presents its botanical subject in a highly mediated and artificial way, with its component parts flattened and fully legible.

In another article, published several decades later, Pächt further elaborated on the differences between early modern Italian "portraits" of plants and premodern botanical illustrations.[11] For Pächt, tedious pictorial conventions subservient to didactic and instructional aims overburdened earlier modes of botanical illustration.[12] These forces led artists to abandon conceiving of depicted plants as objects in the world (Gegenstandsvorstellung).[13] As he notes, "In short, what the ancient illustrated herbarium offers is usually a manipulated nature, an empiricism aimed at recognizability and objective determinability, never the subjective impression of the thing as spontaneously perceived."[14] By contrast, the emphasis on plasticity, modeling, and natural patterns of growth in fifteenth-century Italian botanical illustrations makes them appear as genuine re-creations (Nachschaffen). In the eyes of the viewer, they "awaken back to life" (Wieder-zum-Leben-Erwecken) the plant as a natural object in the world.[15] Pächt's principal concern here is essentially the formal elements that contribute to the naturalism and illusionism of a picture. But these concerns seem to stand apart from the question of use or utility. In a sense, all botanical illustrations are instructional or didactic, manipulated, and aimed at recognizability, insofar as they are botanical, that is, useful for botanical inquiry. The principal differences between ancient and early modern modes of illustration are not their basic didactic aims, but rather the pictorial conventions they employ.

Minta Collins in *Medieval Herbals: The Illustrative Traditions* (2000) uses terms of analysis from Pächt's studies.[16] She categorizes ancient and medieval botanical illustrations on a spectrum, ranging from schematic, with little to no resemblance to the plant, to recognizable, and finally, to naturalistic or lifelike, pictures that show both "artistic accomplishment" and "botanical expertise."[17] Collins further distinguishes between decorative but inaccurate plant illustrations and more accurate plant portraits that aim to record or instruct.[18] Collins acknowledges that plant illustrations and portraits are not mutually exclusive categories. While her study often correlates utility with depictive accuracy, the nature of those connections often remains unclear. As Hardy and Totelin note, "schematic plant illustrations can at times be more useful to students of botany than more 'artistic' portraits."[19] Conversely, as Collins recognizes, it is difficult to determine function on the basis of form, whenever a form admits to multiple uses.

More recent research has bracketed out considerations of lifelikeness and illusionism in premodern botanical illustration. In *Observation and Image-Making in Gothic Art* (2005), Jean Givens distinguishes illusionism (the extent to which a picture matches the visual field as perceived) from both naturalism (the "impression of life-likeness") and descriptiveness (the "rendering of specific factual detail").[20] Givens restricts her definition of descriptive images to those that "visually communicate information concerning the external and sometimes, internal physical structure of real-world objects and phenomena," noting that "they need not be lifelike."[21] But pictures are always necessarily descriptive of the things they depict in one way or another. All pictures that represent something can be said to be descriptive and more or less accurate to a particular understanding or view of the object depicted. Moreover, while an illustration's accuracy can shed light on an artist's grasp of the plant in question, and therefore their working methods, it does not account fully for an illustration's role within premodern systems of knowledge. A picture actively forms knowledge of the object it represents. Learning about plants from premodern illustrations means seeing plants according to how they were known and understood by people in the past.

Selection and Aspect

How, then, do these botanical illustrations typically present plants? Pächt's observation that premodern botanical illustrations appear "as if pinned down and neatly arranged for didactic purposes" remains incisive.[22] It provides us an opening for rethinking how premodern botanical illustrations "describe" plants. Pinning, a central practice in the preparation and display of zoological specimens, involves the manipulation and fixing of a specimen into a specific configuration in order to show aspects of its morphology. Central to both pinning and botanical illustration is the act of selecting what features to show and deciding how to show them. We have seen in chapter 2 that selection is involved in the description of plants. But selection operates differently in a depiction. Ernst Gombrich long ago noted that selectivity is a necessary feature of depiction, in general: artists must pick and choose what to depict because pictures are restricted in scope whereas the "visual world is incalculably large."[23] Pictures represent a subject selectively with only some of its properties.

Dominic Lopes develops Gombrich's idea of selectivity further in terms of a picture's "aspectivity."[24] According to Lopes, a picture can depict an object as having or not having a property, or as portraying another property that explicitly precludes inclusion of other properties. For example, the portrayal of a curving, bending leaf can obscure its shape.[25] Showing a leaf's top (adaxial surface) can preclude a view of its underside (abaxial surface). A picture can also be inexplicit or "inexplicitly noncommittal" about the possession of a property if it is vague about the presence of a given detail. This often occurs when a botanical illustration simplifies or omits aspects of a plant's anatomy. Moreover, by Lopes's definition pictures *must* represent spatial properties. Pictures characterize spatial qualities as they show how properties appear in relation to each other. Consequently, pictures show objects as being "spatially unified." While Lopes's terminology here resonates with Givens's definition of descriptive images as those that correlate with the structure of real-world objects, he avoids casting this spatial structure in terms that invite explicit comparison to a verbally discursive act such as description. Descriptions, while able to capture an object's nonvisible properties, such as taste and scent, cannot capture its overall spatial structure, particularly its internal spatial relationality. While words allow us to construct a mental image, a picture presents visual content directly in a denser and more replete way. Hence, as discussed in chapter 2, a picture readily simulates the visual experience of *autopsia*. According to Lopes, spatial unification in a picture does not necessarily mean that depicted features share the same viewpoint (as in linear perspective), but merely means that the picture indicates spatial relations between distinct parts of a given scene.[26] In other words, pictures have a spatial unity that relates different parts to each other, regardless of how that unity is ultimately mapped. The representation of a plant's spatial properties further requires that pictures make a number of explicit noncommitments, due to the impossibility of fully representing a three-dimensional subject within a two-dimensional picture.[27]

Aspectivity is relevant to botanical illustrations not only insofar as they are depictions, but also as they are a means of creating and communicating visual knowledge about the plant portrayed. By attending to those "distinct qualities or properties" that designate particular kinds of plants, selectivity in depiction works analogously to the identification of distinguishing characteristics shared by members of a particular kind of plant. At the same time, the aspective arrangement and spatial unifications of the subject depicted work together to convey the plant's general appearance or growth habit, that is, the tendency of a plant to grow a particular way (for example, as an upright shrub, a low, spreading tree, or a prostrate herb). Both concerns—visual distinction and habit—shaped the early history of botanical illustration.

Abstracting Habits

Botanical artists must choose what details to include in a picture and how to show them. These choices require artists to exercise critical judgment. Too many details or poorly chosen ones undermine the clarity and utility of an illustration for visually defining a kind of plant. This was no simple matter in the past. As we saw in chapter 2, the ancient world did not possess a single way of systematically defining plants. What parts were necessary, from what angles, and at what stages of growth?

The scene of Dioscorides at work in his study in the Vienna Dioscorides (fig. 1.4) illustrates this process of critical judgment. It portrays a personification of *epinoia* holding up a mandrake to an artist. As we saw in chapter 1, the word *epinoia* encompasses a wide range of meanings, including "thought," "invention," "design," or "purpose." These signal the active process by which the mandrake is invented as an object of verbal and pictorial discourse. In philosophical and theological texts of the time, *epinoia* narrowly designates the conceptual existence of a thing in contrast to its concrete reality.[28] In his discussion of the theological meaning of the term, Lewis Ayres glosses *epinoia* simply as "the activity of reflecting on and identifying the distinct qualities or properties of something."[29] Also in discussing the theological sense of the word, E. C. E. Owen writes that "it denotes not direct perception or conception, but reflexion on a percept or concept already formed. In the case of a science or art it takes a percept given to it by sensation, 'refines on it,' explains, and analyses it."[30] These different senses of the term *epinoia* undercut our understanding of the scene as portraying an artist closely copying a model directly from life. Dioscorides and the anonymous painter attend not to an individual mandrake, but rather to the mandrake as a concept or class. Ancient botanical depictions do not aim at portraiture; their focus is the reliable and essential definition of a plant as a distinct kind. This process of taking from observed reality, analyzing it, and refining it is abstraction—literally "drawing" (*-trahere*) "away from" (*abs-*) perception. The abstraction of essential features from perception within an ancient botanical illustration is thus not a mere consequence of depicting without observing; rather, it is an important part of how ancient people conceived of ancient botanical illustration as a critical and scientific art form.[31]

Abstraction, however, can take many forms. Ancient botanical artists often simplified and abstracted their illustrations by pruning them down to their most essential features. The abstraction of botanical detail may not hinder the use of such pictures for identification, comparison, and the training of the eye. Ute Mauch has shown that ancient plant pictures could emphasize a plant's general appearance or habit.[32] A generalizing or abstracting approach to illustration can usefully convey a plant's habit. Such an illustration need not attend to the specific morphology of a plant, as long as it adequately captures its general shape and features. Focusing on a plant's habit can get in the way of showing its features in detail. Many modern botanical illustrations tend to sacrifice the portrayal of habit in favor of botanical detail (see chapter 8). At the same time, premodern botanical artists typically eschewed perspectival portrayal of growth habit. An attempt to show three-dimensional habit would require adopting depictive strategies, such as foreshortening, that would break with the desire to depict a plant as though it had been clearly pinned out.[33] With some exceptions, premodern botanical artists generally favored clarity and general habit over specific detail.

Habit often assumes a prominent place in non-Linnaean botanical systems. Ever since Carl von Linné (d. 1778) published his *Species plantarum* (*The Species of Plants*) in 1753, the anatomy of flowers and fruits has served an essential role in the delineation of plant species. But there are other ways to do botany. In an interview with Emmanuele Coccia, the botanist Francis Hallé described his astonishment on learning that the chief of a Baule village in Côte d'Ivoire was able to identify trees to species simply on the basis of their architecture or growth habit. This conversation led Hallé to an epiphany: "You could be a botanist without being a Linnean."[34] Botanical inquiry in the ancient and medieval Mediterranean likely worked in a similar way. As we will see throughout this book, flowers were rarely used for classification and identification. A flower was thought to be merely "a beautiful appendage."[35] Bees were thought to "steal" nectar from flowers.[36] And sexual reproduction in plants was only recognized in a few cases, most notably in the date palm.[37] Theophrastus did not define the function of flowers, although he does distinguish them on the basis of their anatomy in several different ways.[38] And Dioscorides typically pays flowers little attention in his descriptions of plants.[39] In practical terms, too, flowers are often unavailable for identifying a plant for most of the year. Ancient herbalists, moreover, tended to think that powerful medicinal plants were flowerless and of a wild, masculine, or "harsh" character.[40] As a result, ancient botanists chose to focus on other features of botanical life, such as habit and leaf shape.

The simplification and abstraction of plant parts may also have facilitated the comparison and classification of plants according to their similarities and differences. For example, leaves with margins that are crenate, crenulate, erose, erosulate, biserrate, dentate, denticulate, and so forth, can appear in ancient botanical illustrations as simply smooth, dentate, or serrate. Although detail is lost, the herbal's user can more easily group plants on the basis of these broader types. In a similar way, flowers are often reduced to a few basic types, such as composite, labiate (with "lips," as in a sage flower), umbellate (with clustered inflorescences characteristic of carrots), papilionaceous (butterfly-like, that is, like a pea flower), or simple unopened buds. This mode of abstraction may have allowed for a simplified, and thus easier, approach to the classification, comparison, and memorization of plants according to similar morphology. This mode of simplification is essentially a form of abstraction by way of a generalized typology of each plant part, and is thus different from abstraction according to habit.

These modes of abstraction and patternmaking are distinct from mistakes or decorative treatments of illustrations.[41] Ancient botanical artists abstracted, ornamented, and stylized plant depictions for many other reasons. They adapted illustrations to fit broader stylistic norms, to follow artistic conventions, or to enhance aesthetically pleasing visual effects, such as scrolling vines and interlacing branches. These approaches could introduce visual properties into a depiction that did not provide information on what the plant was thought to look like. For example, a thick outline typically does not shed much light on the features or habit that a plant was thought to have. In some cases, adaptations in a picture could result in outright errors. For example, the number of petals on a flower in ancient botanical illustrations is often inaccurate. These simple errors, which can be relatively easily overlooked in a non-Linnaean context, could have been the result of aesthetic considerations, or could reflect the dominance of certain morphological patterns, plants, or families in the botanical tradition or the broader visual culture. Such mistakes could occur not only when copying an illustration but also when depicting a plant from direct observation. A subsequent copyist might later choose to rectify these errors or remove them from an illustration by simplifying it further.

The Search Image

Emphasis on habit further suggests that early botanical illustrations worked by training visual recognition of plants. To train viewers to recognize a plant successfully, an illustration should provide a basis for the formation of a "search image."[42] For example, as Robin Wall Kimmerer recalls of her visit to tide pools on the Olympic Peninsula in Washington state:

> I knew from poring over field guides in anticipation of the trip that we "should" see starfish in the tidepools and this would be my first. . . . As I looked among the mussels and the limpets, I saw none. . . . Disappointed, I straightened up from the pools to relieve the growing stiffness in my back, and suddenly—I saw one. Bright orange and clinging to a rock right before my eyes. And then it was as if a curtain had been pulled away and I saw them everywhere. . . . The sensation of sudden visual awareness is produced in part by the formation of a "search image" in the brain. In a complex visual landscape, the brain initially registers all the incoming data, without critical evaluation. Five orange arms in a starlike pattern, smooth black rock, light and shadow. All this is input, but the brain does not immediately interpret the data and convey their meaning to the conscious mind. Not until the pattern is repeated, with feedback from the conscious mind, do we know what we are seeing. It is in this way that animals become skilled detectors of their prey, by differentiating complex visual patterns into the particular configuration that means food.[43]

Kimmerer learned initially what to look for from field guides. This planted an initial image that she honed into a search image through repetition and reflection in the field. The resulting search image is abstracted from select distinguishing features that flexibly and reliably signal the desired object. The eye catches a familiar pattern that disrupts the usual patterns of a landscape and habitat. Much like the field guide in this anecdote, a functional botanical illustration helps to implant an initial image that can then be trained into a reliable search image. It need not be a highly detailed picture, but rather can be a simple one that attends to a plant's general habit and any simple, especially pattern-like, distinguishing features. A complex, detailed illustration may consequently not work as well as a simple one for the purposes of training a viewer's search image.

The formation of a search image takes place within a particular set of contexts. We learn to recognize the patterns of a desired object in the field against a dense backdrop of other patterns and textures. In some cases, a search image that is learned in one habitat or environment may prove to be useless in another. Perhaps you learn to recognize a plant on the sole basis of its red leaves. Such a search image might work well where there are no other plants with red leaves. But where there are many plants with red leaves, a different search image is needed to pick out the desired object. Ultimately, a search image's efficacy is greatly dependent on context.

The use of a picture for developing a search image does not require much intentionality on the part of either the artist or the viewer. A viewer simply acquires a search image through habitual exposure, reflection, and practice applying it in the field. An artist does not necessarily intend to initiate or sharpen a viewer's search image. They rather simplify a picture to the extent that it is still useful. Yet simplifying and patternmaking can be critical practices in illustration to the extent that the artist must decide which features are necessary and which are expendable.

Jacob von Uexküll, who coined the term "search image" in 1934, further notes that "the search image obliterates the perception image."[44] As Uexküll explains, he had been accustomed to drink water from an earthenware pitcher at his meals. When the pitcher broke in his absence and was replaced at mealtime by a clear glass carafe, he failed to register that there was any water on the table at all. His search image based on the earthenware pitcher canceled out his perception of the clear glass carafe. It follows that people may see a plant more according to their search image of it than as it appears to their perception within a specific context. Similarly, people may tend to see an illustration more according to the search image than as it appears to perception. If so, the presence of errors or distortions in premodern botanical illustrations may have mattered little in terms of an illustration's basic utility for shaping viewers' search images, as long as distortions did not lead the viewer to look for an entirely different plant. Errors can be overlooked up to a point.

My own experiences learning about plants from ancient illustrated herbals echo these observations on search images. I found it easy to recognize a plant already known from life in an illustrated herbal even though it meant overlooking obvious errors. Conversely, as I became more familiar with illustrations in early herbals, I began to see actual plants more as they appear in the herbals. It did not matter that many of the illustrations were inaccurate. It did not even matter that resemblance between a picture and a plant might only be slight. The dominance of the search image may thus help to explain how errors or "drift" in copying can occur without necessarily affecting the perceived utility of an ancient botanical illustration. Distortions or errors need not mean a degraded botanical practice. Nor do they mean that a picture was botanically unusable. A simplified or inaccurate picture can still serve the basic reference functions outlined in chapter 2. As long as viewers are acquainted with a plant and have a reliable search image of it, they do not require detailed and accurate pictures of it. Any corrections or additions that occurred in the process of copying a picture are thus indicative not that the picture was used, but that it was actively compared against an actual plant or the image-maker's personal experience and familiarity with the plant.

Elaboration and abstraction are two contrasting approaches to depiction that had to be balanced in a constant tug-of-war throughout the history of botanical illustration. Ultimately, how an artist illustrated a plant came down to their familiarity with a particular plant, its context, and the ways an illustrated herbal was used. As we will see, some artists favored more details; others, less. While abstraction could enhance a picture's reference to a plant and make it easier to train recognition of it, it could also dampen the utility of a picture as an opportunity for *autopsia*, at least when the viewer used it for an extended study of plant morphology.

While this section has largely concerned the use of pictures for visually learning, more detailed and accurate pictures could have assumed a more direct role in botanical inquiry. A user could appeal to such a picture to identify an unknown plant or to distinguish it from another. These uses demand greater attention to distinguishing visual details and their correspondence to the features of an actual plant.

Depth and Ground

While ancient botanical illustrations vary in their treatment of morphological detail, all surviving examples portray plants at a shallow depth against a blank background. These two conventions have remained remarkably consistent throughout the early history of botanical illustration. The earliest surviving examples of botanical illustration, the second-century fragments of the Tebtunis Roll (figs. 3.1 and 3.2), show their subjects hovering against a blank ground. As Daniela Bleichmar has noted of eighteenth- and early nineteenth-century botanical illustrations, the blank ground in botanical illustrations can signal the erasure of a botanical subject's specific locality within its usual habitat.[45] The blank ground thus extracts and transforms the botanical subject into an object of a universalizing, metropolitan science. The blank further marks the closure of temporal, geographical, and physical distances between an actual botanical subject and its presentation within a scientific discourse. A botanical illustration presents the image of a plant in the time and space of the viewer. To go a step further, the blank of the botanical illustration echoes the operations of empire in expropriating and moving resources, peoples, and things from one place to another. Similar observations hold for the blank grounds of early illustrated herbals. As we saw in chapter 1, the earliest known illustrated herbals arose within the context of Hellenistic empire building and the control of diverse territories.

Although the remnant paintings of the Tebtunis herbal lack complex modeling, such as highlights and form shadows, they nevertheless create an impression of pictorial depth through blank spaces and the pooling of colors.[46] The artist created a sense of atmospheric or aerial perspective, that is, when more distant elements appear lighter and less clearly defined. When viewing this depiction, we must "fill in" the gaps the painter left between the various plant parts.[47] In doing so, we essentially follow *Gestalt* principles of perception, whereby we conceive of the figure as an interconnected whole first, and then infer its details. As a result, the background is ambiguous. The blank space is alternately a gap between plant parts to be mentally bridged by the viewer, or it is a depth into which the form recedes. So while many premodern botanical illustrations recall specimens lying against the surface of the page as though pinned, pressed, or flattened, this may not have been the case for the earliest illustrated herbals.

A consequence of this "impressionistic" approach in the Tebtunis Roll is a striking loss of visual detail. The pictures omit leaf shape, margin, and venation, as well as petioles and pedicels. Do the flowers grow directly from the shoot (sessile), or do they sit on supporting stalks or pedicels (pedicellate)? The picture does not say. As we have seen, the picture's inattention to floral anatomy reflects ancient botanical thinking.[48] The pictures of the Tebtunis Roll would have still worked well to convey botanical habit, to provide a basis for the formation of

FIGURE 3.1 Illustration of false dittany (*pseudo-diktamnon*) from fragment *e* of the Tebtunis Roll. University of California, Berkeley, Tebtunis Center, P. Tebt. II 679, second century CE. Courtesy of Tebtunis Center, Berkeley.

FIGURE 3.2 Fragment *f* of the Tebtunis Roll. University of California, Berkeley, Tebtunis Center, P. Tebt. II 679, second century CE. Courtesy of Tebtunis Center, Berkeley.

a search image, and to serve as memory aids for plants already known. The Tebtunis Roll did not set out to provide an exhaustive account of plant morphology. The priests from the Soknebtunis temple that used this herbal may have used its pictures to train or affirm their recognition of plants in the field, or to remind them of plants already known.

The Antinoopolis Codex presents a different approach to the rendering of depth and space (figs. 3.3 and 3.4). It echoes the Tebtunis illustrations in their emphasis on growth habit. And, like the Tebtunis

FIGURE 3.3 Illustration of a comfrey (*symphyton*) on a fragment of the Antinoopolis Codex, the "Johnson Papyrus," recto/Side A. London, Wellcome Collection, MS 5753, fifth century CE. Courtesy of Wellcome Images.

Roll, the Antinoopolis Codex shows us plants represented as single uprooted individuals lying against a blank ground. But compared to the Tebtunis Roll, the Antinoopolis fragments employ a greater variety of techniques, involving more complex modeling and layering of colors. For example, the leaves of a mullein (*phlomos*, spelled *phlommos* on the papyrus) on Side B are either light green with thick outlines, or dark green without outlines, evidently indicating differences between the upper and lower surfaces of leaves (fig. 3.4). The

FIGURE 3.4 Illustration of mullein (*phlommos*) on a fragment of the Antinoopolis Codex, the "Johnson Papyrus," verso/Side B. Courtesy of Wellcome Images.

artist apparently outlined only the lighter leaves so as to better define them against the blank ground. Most leaves lie parallel to the picture plane, though several curve in profile. The plants here are more clearly articulated: all of their parts are connected to each other. As a result, the blank ground serves only as a background. Thus, while the blank ground in the Tebtunis Roll indicated a background, depth, and the gaps between parts, the blank ground in the Antinoopolis Codex is simply a shallow surface behind the plant. Full articulation over a less ambiguous background may further allow viewers to see the plant's potential three-dimensionality, perhaps by way of mental rotation.[49] The articulation of plant parts over a shallow, blank background becomes one of the most defining and widely conserved conventions in botanical illustrations over time. At the same time, however, these early illustrations show a tension between the flattening of plant parts and the suggestion of depth.[50]

Although palimpsested, the remains of two illustrations in the Sinai Fragment seem to follow the Antinoopolis Codex in their rendering of space and the full articulation of plant parts. Both illustrations seem to show plants with rather rigid sprays, radiating outward from the base of the plant and a large root mass (fig. 3.5). That none of the plant's leaves appear in profile gives them a radically flattened appearance. The straightened limbs and flattening of the plant's shoots and leaves all emphasize the legibility and clarity of leaf shape, structure, and arrangement, while obscuring the plant's three-dimensionality.

We can extend this consideration of space in ancient botanical illustrations by turning to two illustrated manuscripts of Dioscorides: the early sixth-century Vienna Dioscorides and the late sixth- or early seventh-century Naples Dioscorides. Although the manuscripts present their own idiosyncrasies and unique concerns, which are the subject of the next chapter, both have a shared source and reflect the same broader tradition.[51] The pictures in both manuscripts that match tend to agree in their details with some minor differences. The pictures in the Vienna Dioscorides tend to be more modeled, and detailed.[52] The illustrations in the Naples Dioscorides are more stylistically uniform and have thick black and yellow outlines throughout. The Vienna and Naples Dioscorides manuscripts follow earlier illustrations in their emphasis on single individual specimens and plant habit. Like the Antinoopolis and Sinai illustrations, the illustrations in these codices eschew the impressionistic quality of the Tebtunis Roll illustrations in favor of fully articulated plants with a shallower, more flattened spatial aspect. They combine thick outlines and modeling that present the plant both as a round figure in space, and as flattened out against a

FIGURE 3.5 Palimpsest with trace illustration of a kind of St. John's wort (*philetairion*), undertext dated to fifth or sixth century CE, St. Catherine's of the Sinai, Arabic "New Finds" NF 8, fols. 16v–17r. Courtesy of sinai.library.ucla.edu, a publication of St. Catherine's Monastery of the Sinai in collaboration with Early Manuscripts Electronic Library (EMEL) and University of California—Los Angeles (UCLA). Photo: Keith Knox.

shallow, blank ground. Kurt Weitzmann noted that this quality in the illustration of a violet (*ion porphyroun*, fig. 3.6) aimed at providing a clear and "advantageous viewpoint" free of confusing overlaps.[53] The same observations apply equally well for most of the illustrations in the Vienna and Naples Dioscorides manuscripts.[54] While we can see some of these qualities in the Antinoopolis and Sinai herbals, it is

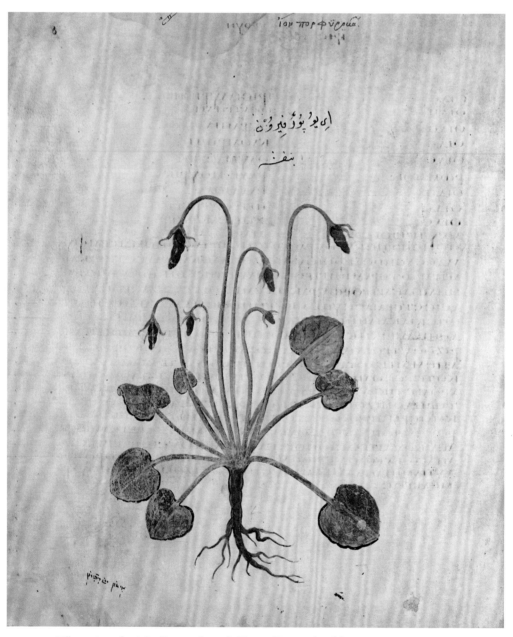

FIGURE 3.6 Illustration of a violet (*ion porphyroun*). Vienna Dioscorides, fol. 148v. Courtesy of Österreichische Nationalbibliothek.

only with these manuscripts that we arrive at a conception of space and composition fully in line with Pächt's pinned "museum exhibits." The painting of the violets indicates the plant's growth habit, leaf shape, margin, and even venation. As in earlier illustrations, however, it avoids depicting floral anatomy by opting instead to show the plant with unopened buds.[55]

In contrast to earlier extant illustrations, those in the Dioscorides manuscripts clearly give greater attention to specific distinguishing features. So although there were common conventions in botanical illustration, such as the presentation of uprooted individuals on a blank ground, an emphasis on habit, and the flattening and "pinning" of botanical form, the illustrations nevertheless vary tremendously in how they approach morphological detail. Each picture consequently has its own content, its own inaccuracies, and hedgings of morphological detail.

Time and Timing

Space is inextricably entwined with time. By isolating the botanical subject from its context, the blank ground presents its subject in an abstracted, seemingly eternal space, ever present to the viewer. Yet, as Theophrastus observed, many of a plant's most characteristic features, such as flowers or fruits, only appear for a brief time.[56] The selective presentation of these features in a picture consequently delimits the apparent temporality of the botanical subject. Concern over the temporality of plants leads to three different pictorial strategies: the painter could emphasize temporal change by showing plant parts at different stages of their life cycle, exclude temporal variance entirely by emphasizing only the most permanent characters, or represent only a single moment in a plant's life.

Examples of the first approach to time can be seen in illustrations of blackberry bush (*batos*) in the Dioscorides manuscripts. These stand out for their virtuosic portrayal of multiple stages of growth across different plant parts (fig. 3.7).[57] This illustration provides flowers and fruit at different stages of maturity, including unopened floral buds, blossoming flowers, flowers dropping their petals (top left), as well as mature and immature fruit. We also find new (apical) growth, and the broken end of a dead cane. Dead, broken branches, typically on the lower right side of the plant, appear throughout the codex. These evidently signal the effects of senescence or the bare appearance of the plant in winter or when dead.[58] The illustration of the blackberry further shows the tendency of canes to root at nodes, which

FIGURE 3.7 Illustration of a blackberry (*batos*). Vienna Dioscorides, fol. 83r. Courtesy of Österreichische Nationalbibliothek.

allows the plant to grow into dense thickets. The leaves appear largely flattened and tilted toward the viewer, though many curl or bend to show their lighter undersides. The illustration gives the impression of having been designed to show as many plant parts at as many different stages of growth as possible.

Other illustrations portray petals falling from the flower head, such as the illustration of the anemone (*anemōnē hē phoinikē*, fig. 3.8).[59] The "petals" (actually tepals) appear frozen in time, as though they had been captured in a photographic snapshot. While seeming to depict a single instant or moment, the petals refer to the transition from one plant part to another, that is, from flower to fruit. In the larger composition, the petals clarify the temporal or sequential relationship between the bud, the flower head, and the seedpod. In this case, since the falling petals are related to the seedpod, the viewer can infer that the other closed forms are unopened flower buds. The petals thus clarify the temporal aspect and transformation of the plant from one life stage to another. This sensitivity to temporal change and variation in particular plant parts, moreover, parallels Pliny's complaint that a plant needed to be depicted according to its "fourfold" variation over the course of the year.[60] Such depictions can be termed "diachronic" in their attention to a plant's changes through time.[61] By depicting discrete stages in the life of the plant, the illustrations demonstrate ancient scholars' interest in annual processes, especially in fruiting and the development of seeds. Aristotle and Theophrastus both treat the formation of fruit as the *telos*, or purpose, of the plant.[62] In alluding to these processes, ancient botanical illustrations may be seen as aspiring to the portrayal of causes and principles of growth in botanical life, that is, to more speculative philosophical discourses within the ancient botanical tradition.

Still, many plant depictions in the Naples and Vienna Dioscorides do not depict flowers at all. For example, the depictions of comfrey (*symphyton*, spelled *synphyton* here) and milk thistle (*sillybon*) in the Naples Dioscorides (fig. 3.9) lack their inflorescences entirely.[63] By excluding the flowers, the painter seems to have focused on only the most permanent characters of the plant, such as its leaves and roots, both of which were accorded medicinal properties. The decision to exclude flowers could further reflect the thinking that they were simply less available for identification and thus less suitable for identifying and defining a particular kind of plant. The absence of flowers, however, may also indicate the ideal time for harvesting the plant for medicinal use, that is, before flowering. In the first case, the illustration would present the most permanent features of the plant, arguably its

FIGURE 3.8 Illustrations of two anemones. Vienna Dioscorides, fols. 25v–26r.
Courtesy of Österreichische Nationalbibliothek.

ΑΝ€ΜωΝΗ ΦΟΙΝΙΚΙΟ

ΟΙΔΑ€	ΗΜΙΟΝΙΟΝ
ΟΙΔΑ€	ΜΙΚωΝΙΟΝ
ΟΙΔΑ€	ΤΡΑΓΟΚΕΡωϹ
ΟΙΔΑ€	ΓΗϹΤΑΡΙΗΗ
ΟΙΔΑ€	ΒΑΡΕΥΑΗ
ΟϹΘΑΛΗϹ	ΒΗΡΥΛΛΙΟϹ
ΟΜΟΙωϹΟΡΗΟϹ	ΚΕΡΑΗΙΟϹ
ΠΥΘΑΓΟΡΑϹ	ΛΛΡΑΚΤΥΛΙϹ
ΠΡΟΦΗΤΑΙ	ΚΝΗΚΟϹΑΓΡΙΑ
ΡωΜΑΙΟΙ	ΟΡΚΙΤΟΥΠΙΚΑΙ
ΑΦΡΟΙ	ΧΟΥΦΦΟΙϹΤ

Η Μ€ΝΤΙϹΑΓΡΙΑΗΔ€ΙΜ€ΡΟϹΚΑΙΤΗϹΙΜ€ΡΟΥΗΜ€ΝΤΙϹΦΟΙΗ
ΚΑΙΑΦ€ΡΙΤΜΙΑΗΟΗΗΔ€ΥΠΟΛ€ΥΚΑΓΑΛΛΑΚΤΙΖΟΗΤΑΗΓΟΡ
ΦΥΡΑΦΥΛΛΑΔ€Ϲ€ΧΙΚΟΡΙΟ€ΙΑΗΛ€ΠΤΟϹΧΙΔ€Ϲ€ΡΑΤΠΡΟϹ
ΤΗΓΗΚΑΥΛΙΑΧΗΟΦΑΗΛ€ΠΓΑΥΤΤΡΟΦΩΗΤΑΛΛΟΗΟϹΤ€ΡΜΗ
ΚΦΗΟϹΚΑΙΜ€ϹΑΚ€ΦΑΛΛΙΑΜ€ΛΑΠΑΗΚΥΔΗΙΖΟΗΤΑΡΙΖΑΗΚΑ
ΤΑΜ€Γ€ΘΟϹ€ΑΛΙΑϹΗΜΙΖΟΗ ΟΙΟΗ€ΙΓΟΗΑϹΙΝΑΙ€ΙΛΗΜ
Μ€ΝΟΗΔΥΝΑΜΙΗΔ€€ΧΟΥϹΙΝΑΡΙΜ€ΛΙΑΜΦΟΤ€ΡΑΙΟΘ€Η
ΟΧΥΛΟϹΤΗϹΡΙΖΗϹΑΥΤωΗΡΙΗ€ΧΥΤΗϹΤΡΟϹΚ€ΦΑΛΗϹ
ΚΑΘΑΡϹΙΗΚΑΙΜΑϹΘΟΙϹΑΛ€ΗΡΙΖΑΓΙΦΛ€ΓΜΑΤΑϹ€ΝΘ€ΙϹΑ
Δ€€ΝΓΛΥΚ€ΙΚΑΙΚΑΤΑΤΓΛΑϹϹΟΜ€ΝΟΦΘΑΛΜωΗΦΛ€ΓΜ
ΝΑϹΙΑΤΑΗΚΑΙΤΑϹ€ΝΟΦΘΑΛΜΟΙϹΟΥΛΑϹΚΛΑΜΒΑΟΤΤΙΑϹ
ΑΤΤΟϹΜΑ ΑΝΑΚΛΘΑΙΡΙΔΑϹΚΑΙΤΑΡΥΤΤΑΡΑΤωΗ€ΛΚΦΗΤΑΔ€
ΦΥΛΛΑΚΑΙΟΙΚΑΥΛΟΙϹΥΝ€ΨΗΘ€ΗΤ€ϹϹΥΠ€ΤΤΙϹΗΗΚΑΙ
€ϹΘΙΟΜ€ΝΑΓΑΛΛΑΚΑΤΑϹΤΓΑ€ΠΤΡΟϹΘ€Τωॏ€ΜΜΗΗΠΔΑΓΙ
ΚΑΤΑΤΓΛΑϹϹΘ€ΗΤΑΔ€Λ€ΠΤΡΑϹΑΦΙϹΤΗϹΙ :

ΑΝ€ΜωΝΗ ΗΑ ... ΑΓΡΙΑ€ΜΕΡΟϹ

ΚΡΑΤΕΥΑϹ
ΑΝ€ΜΟΗΗ ΦΟΙΝΙΚΗ
ΑΝ€ΜΟΗΗ ΑΥΗΛΑΜΠ€ΧΑΡΙΜΙΑΝΟΘΕΗΟΚΥΛΟϹΤΗϹΡΙΖΗϹΑΥΤΟΥ
ΓΙΝ€ΤΑΙΟΗΧΥΤΟϹΠΡΟϹΚ€ΦΑΛΗϹΚΑΘΑΡϹΙΗΚΑΙΜΑϹΘΟΙϹΑΛ€ΗΡΙ
ΔΙΠΦΛ€ΓΜΑ€ΨΗΘΟΙϹΑΛ€ΗΓΛΥΚ€ΙΚΑΙΚΑΤΑΠΛΑϹϹΟΜ€ΝΟΦΘΑ
ΜωΝΦΛ€ΓΜΟΝΑϹΛ€ΠΤΕΙΟΜΟΙωϹΚΑΙΤΑϹΟΥΛΑϹΑΠΟϹΜΗΧ€Ι ΤΑ
Λ€ΦΥΛΛΑΚΑΙΟΙΚΑΥΛΟΙϹΥΝ€ΨΗΘ€ΗΤΑΠΤΙϹΑΗΗΚΑΙ€ϹΘΙΟΜ€ΝΑ
ΓΑΛΛΑΚΑΤΑϹΠΓΑ€ΠΤΡΟϹΘ€ΤωΔ€Ϲ€ΜΜΗΗΠΔΑΓ€ΙΚΑΤΑΠΛΑϹϹΟΙϹΛΔ€
Λ€ΠΡΑϹΑΦΙϹΤΗϹΙΗ:

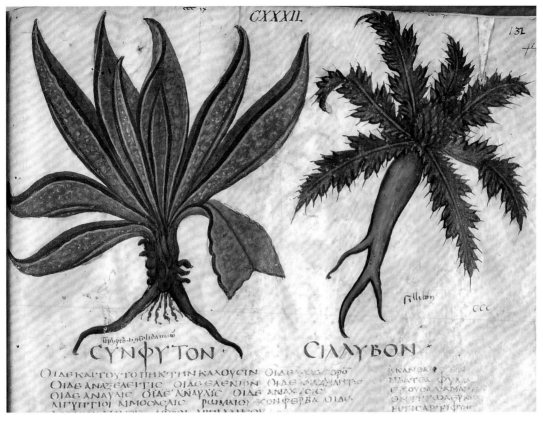

FIGURE 3.9 Illustration of comfrey (*synphyton*), left, and milk thistle (*sillybon*), right. Naples, Biblioteca Nazionale, Cod. gr. 1 (the "Naples Dioscorides"), late sixth or early seventh century CE, fol. 132r. By permission of Ministry of Culture (Ministero della Cultura) © Biblioteca Nazionale di Napoli.

most eternal visual aspect. In the second case, the illustration would refer to a specific moment: the timing of the plant's harvesting. In the absence of a clear label or text specifying the illustration's temporality, either explanation could be true.

Names and the Powers of Roots

Ancient botanical illustration could also attend to a plant's name, an essential aspect of its identity but not one easily captured through illustration. The falling "petals" in the illustration of anemone, for example, may also illustrate the name of the plant, taken from *anemos*, or "wind." As Ovid explains, "The winds from which it takes its name shake off the flower."[64] Such an inclusion obviously enhanced the picture's use as a mnemonic device.[65] This detail also highlights an aspect of the plant's lifecycle, namely the brevity of its flower.

We find a more extreme illustration of a plant's name and properties in the picture of an eryngo, specifically a sea holly (*ēryngion*).[66] In the Naples Dioscorides, the plant's long taproot terminates in a Medusa head (fig. 3.10).[67] The title below the picture names the plant as *ēryngion* or *gorgonion*. The Medusa-shaped root evidently relates to this alternate name, *gorgonion*: Medusa and her sisters were the Gorgons, that is, the "grim" ones. The unusual name and picture recall the Medusa head as a familiar apotropaic emblem that was ubiquitous in the ancient world. It is perhaps in this vein that Dioscorides notes that the root destroys growths when suspended around the patient's neck.[68] We can find a more likely explanation in Plutarch's treatise, *Maxime cum principibus philosopho esse disserendum* (*That a Philosopher Ought to Converse Especially with Men in Power*):

> Of the plant *eryngium* they say that if one goat take it in its mouth, first that goat itself then the entire herd stands still until the herdsman comes and takes the plant out. Such pungency, like a fire spreading over everything near it and scattering itself abroad, is possessed by its potent emanations.[69]

Here Plutarch reports that the root petrifies goats, and that this effect is transmissible from one goat to many.[70] This putative property of the plant is perhaps the source for the name *gorgonion* as well as the peculiar depiction of its root. The inclusion of a Gorgon head in the illustration likely also reinforced the association of this eryngo with its other name and its purported petrifying properties. In this sense, the Gorgon head also operates as a mnemonic device. There is reason, however, to think that this lore about the sea holly may eventually have been supplanted by a belief that the eryngo did indeed have a Gorgon-like head at the end of its long taproot.[71] This was perhaps due in part to a literal understanding of its illustration.

The eryngo's Gorgon-headed root may also allude to a preoccupation with roots in ancient herbal medicine. The term "root" (*rhiza*) was commonly used to refer to a medicinal herb in general. Despite the significance of the root to herbalism in the ancient Mediterranean, however, most roots depicted in surviving ancient herbals look rather generic. The most notable exceptions tend to be plants with unusual or potent roots, such as the sea holly or the mandrake, or plants with rhizomes or prominent underground stems. Some illustrations in Dioscorides exclude roots entirely, because they depict either species that lack roots or plants that were believed to be rootless.[72] For example, the rootless depiction of a snapdragon (*kynokephalion*) may reflect the belief that it did not have roots (fig. 3.11).[73]

FIGURE 3.10 Illustration of groundsel (*ērigerōn*), left, and eryngium (*ēryngion* or *gorgonion*), right.
Naples Dioscorides, fol. 78r. By permission of Ministry of Culture (Ministero della Cultura)
© Biblioteca Nazionale di Napoli.

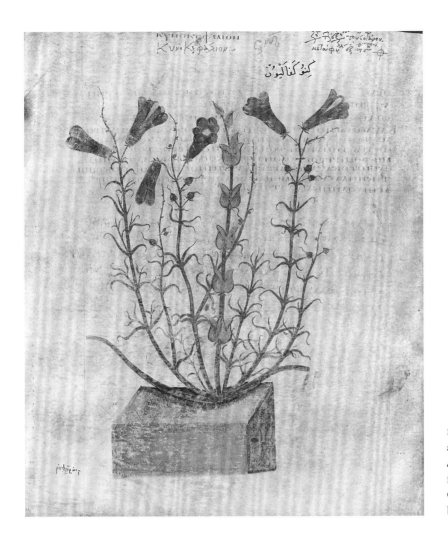

FIGURE 3.11 Illustration of a snapdragon (*kynokeph-alion*). Vienna Dioscorides, fol. 159v. Courtesy of Österreichische National-bibliothek.

Finding Aids

A few illustrations in Dioscorides manuscripts also include references to the terrain or habitat where a plant can usually be found. The most elaborated reference to a "plant" habitat occurs in the illustration of coral from the Vienna Dioscorides (fig. 3.12).[74] The illustration here accompanies not Dioscorides' text, but rather the otherwise unillustrated *Carmen de viribus herbarum* (*Song on the Powers of Herbs*), a series of verses on remedies associated with different deities. (While we now recognize coral to be an animal, ancient authorities were uncertain.[75] Some considered it to be a plant, hence its inclusion here in a text on the powers of herbs.) The illustration shows a complex coral rising from a turbulent sea inhabited by various sea creatures. Nearby sits a half-nude female figure crowned with crab claws. She rests a paddle on her shoulder and, leaning back, sets her elbow against

FIGURE 3.12 Illustration of a coral (*enalydrys*). Vienna Dioscorides, fol. 391v.
Courtesy of Österreichische Nationalbibliothek.

an obsequious-looking dog-faced sea monster (a *kētos*).[76] Her attributes suggest she is a personification of the sea, Thalassa, or perhaps a Nereid or sea goddess, such as Amphitrite or, less likely, Thetis.[77] Whatever the specific identification, she is suggestive of the marine habitat in which coral can be found.[78] (The poem does not directly refer to this female figure, though it does mention Poseidon. This led a fifteenth-century reader to write above the figure, "Poseidon, so they say." Later copies of this miniature consequently adapt the figure into a beardless Poseidon, for example, fig. 8.8.)[79] Beyond indicating the sea, this female figure gazes toward the coral and points to it with an open right hand, as though presenting it to us. Her inclusion further recalls the fact that ancient herbal texts often noted where plants could be located.

The significance of this one human figure for the broader history of premodern illustrated herbals has been missed by earlier scholars. Most scholars tend to view the inclusion of subsidiary figures as a later innovation.[80] That such a figure appears already in the sixth-century Vienna Dioscorides suggests that they may have been more widespread in ancient herbals than is otherwise indicated by extant fragments. Heide Grape-Albers has similarly dated the figurative scenes in the Latin herbal attributed to Ps.-Apuleius Platonicus to the period of late antiquity and not the Middle Ages.[81] Subsidiary figures certainly become more common in later Arabic and Latin illustrated herbals, though they remain relatively rare in Byzantine illustrated herbals. These figures typically draw attention to particular plants or plant properties. At the most basic level, they act as finding aids both for the plant within a given habitat and for the plant within the herbal, like a tab or an arrow.[82]

Conclusion

All plant depictions are invariably "descriptive" of the plants they purport to show. They simulate the visual appearance of plants according to their overall habit and the specific features thought necessary to know a plant visually. They helped to train visual recognition of a plant by providing users with a starting point for learning a search image. Despite their diversity, ancient botanical illustrations generally share several conventions: they all use blank backgrounds and portray plants as single specimens, usually uprooted. But ancient botanical illustrations varied greatly in their treatment of details, space, and time. The second-century Tebtunis Roll demonstrates aerial perspective and an "impressionistic" handling of color that emphasized proximate surfaces, and tended to result in the omission of morphological details. Such depictions work well to indicate a plant's general habit. They do not tell the viewer much about specific morphology. Such illustrations would have served best as memory aids and as training for visual recognition. Later illustrations instead articulate plant parts fully and treat the blank background as a shallow surface against which the plant is "flattened" or "pinned." By the sixth century, though perhaps earlier, we find increased attention to a plant's distinctive or characteristic features. Illustrations attend with varying degrees of detail to the morphology of leaves, flowers, and fruits; to the plant's name; and to its medicinal and special properties. In some cases, illustrations show diverse plant parts at different stages of development, thus echoing Pliny's concerns about the ability of a picture to show a plant's changes over the course of the year. We also find illustrators taking the opposite approach: either the portrayal of a plant at a single stage of its life cycle, or the elimination of variable morphological features. The most elaborated illustrations considered in this chapter come from illustrated manuscripts of Dioscorides. The next chapter takes up the formation of that tradition.

4

ILLUSTRATING
DIOSCORIDES

The earliest surviving illustrated copies of Dioscorides' *De materia medica* were radically different from Dioscorides' original work in their scope and organization. They were shorter, illustrated, and alphabetically arranged. The fact that many, if not most, pictures in this new Dioscorides came from earlier herbals meant that they portrayed botanical morphology and lore absent from Dioscorides' *De materia medica*. At the same time, however, this new version retained Dioscorides' descriptions, thus combining three different ways of referring to plants—names, pictures, and descriptions—into a single herbal. While mistakes arose through this process of compilation and rearrangement, it also led to the expansion and transformation of the ancient botanical tradition. This chapter begins by tracing the transformation of Dioscorides' *De materia medica* from an unillustrated treatise arranged by drug action into a fully illustrated Alphabetical Herbarium. The chapter then turns to the illustration of the original version of Dioscorides' text first evident in an eighth- or ninth-century manuscript now in Paris.

The Ordering of Plants

Writing an herbal, whether illustrated or not, required authors to organize the botanical world, to present it sequentially within a discursive text. Ancient authors used a variety of methods to organize their herbals. Ancient herbals could be arranged alphabetically, by drug action, astrologically, or by body part affected (head to toe).[1] Some herbals, however, do not adhere to any obvious ordering principle, as seen in the Latin *Herbarius* of ps.-Apuleius Platonicus.[2] Most writers of herbals, however, seem to have favored alphabetical organization, likely on account of its utility for reference.[3]

Dioscorides followed a two-tiered approach in the "original" version of his *De materia medica*. (In what follows, any mention of the "original" Dioscorides refers to Wellmann's three-volume reconstruction of the original text and thus not the exact text that Dioscorides produced.)[4] First, Dioscorides arranged his *De materia medica* into five books: Book I, on aromatics, oils, salves, trees, and shrubs; Book II, on animal parts and products, cereals, pot herbs, and sharp herbs; Book III, on roots, juices, herbs, and seeds; Book IV, on more herbs and roots; and Book V, on wines and minerals.[5] In doing so, he followed a quasi-Theophrastian system of organization. Theophrastus had divided plants into four main categories: tree (*dendron*), shrub (*thamnos*), undershrub (*phryganon*), and herb (*poa*). Theophrastus attempted to apply Aristotelian division by similarities and differences into groups of specific kinds (*eidē*), and he further subdivided kinds into yet more subtypes (also *eidē*).[6] This system ran into some difficulties when moving from broader kinds to subtypes. It did not easily lend itself to multiple levels of intermediate groupings with clearly defined relationships, as found in modern systematics. Theophrastus had to adopt a variety of different categories, such as usage (*kata tēn chreian*).[7] Thus he placed plants that are useful for making chaplets and wreaths (*stephanōmatika*) into a subcategory between undershrubs and herbs.[8]

Dioscorides organized the chapters, each addressed to a single medicinal substance, within each book by drug actions.[9] Though complex, Dioscorides' system had some benefits: it was easier to memorize with a view to medical practice, and it enabled users to find substitutions.[10] He thus eschewed alphabetical organization within each of the books in his *De materia medica*. Despite its prevalence, Dioscorides complained that alphabetization separated materials with similar properties and, as a result, made them more difficult to memorize.[11] Nevertheless, later authors, including Galen, Oribasius, Aetius of Amida, and Paul of Aegina, all organized their pharmacological writings alphabetically.[12] Dioscorides' *De materia medica*, too, was eventually subjected to alphabetization.

These different systems of organization presumed different uses. Dioscorides' readers seem to have been more prepared to carry their pharmacological knowledge with them in their heads; while the readers of alphabetical texts likely required regular access to their pharmacological reference works. The former may have been more active and itinerant; the latter more stationary and dependent on libraries. These generalizations fit what we know about these physicians: Dioscorides claimed to have lived an itinerant "soldier's life," while Galen had an impressive collection of books, which he stored in various locations.

Making the Alphabetical Herbarium

Several different versions of Dioscorides' *De materia medica* appeared between the first and sixth centuries. The two earliest surviving and more or less complete manuscripts of Dioscorides—the early sixth-century Vienna Dioscorides and the late sixth-century or early seventh-century Naples Dioscorides—are much shorter than the original, and are focused mainly on herbs, omitting the chapters on trees, minerals, wines, oils, and animal products. The remaining chapters are arranged alphabetically.[13] (The original preface, wherein Dioscorides lays out his method of organization by drug action, is also missing.) Both manuscripts also include lists of plant synonyms, as well as illustrations. The two manuscripts contain slightly different versions of the text, though they ultimately descend from a common source, perhaps compiled as early as the second or third century CE in Italy, perhaps Rome.[14] This version of the text has been called the Alphabetical Dioscorides, on account of its organization, or the Alphabetical Herbarium, on account of both its organization as well as its limited scope.

The original goal of the Alphabetical Herbarium may have been to create a comprehensive and fully illustrated herbal. Instead of illustrations, Dioscorides had originally used descriptions alone, largely comparative in nature.[15] This approach requires the reader to have some prior knowledge of common plants, or access to other sources or authorities that could clarify the text. By rearranging Dioscorides' text alphabetically and abridging it to focus only on herbs, the compilers of the Alphabetical Herbarium transformed Dioscorides' *De materia medica* into an herbal more in line with earlier illustrated books on root-cutting.[16] Yet, unlike older illustrated *rhizotomika*, the Alphabetical Dioscorides now contained both depictions and verbal descriptions of plants. As a result, word and image now stood as separate elements that might mutually reinforce or contradict each other. In doing so, the illustrated Dioscorides expanded the reference capacities of the work.

Scholars have long tried to discern the various sources for the illustrations in the Alphabetical Herbarium. Early scholars, including Wellmann and Charles Singer, supposed that the compilers took the illustrations from the illustrated herbal by Crateuas.[17] In fact, quotations of text attributed to Crateuas appear alongside several illustrations in the Vienna Dioscorides.[18] It remains unclear, however, which illustrations in the Alphabetical Dioscorides, if any, can actually be attributed to Crateuas' herbal. Curiously, the original sixth-century index at the front of the Vienna Dioscorides names only about half of the plants contained in that manuscript.[19] (The striking insufficiency of this index led scholar John Chortasmenos, working in the first years of the fifteenth century, to add a lengthy index to the codex.)[20] Some scholars speculate that this old index may reflect an earlier herbal with more naturalistic images that had perhaps served as a source for many of the illustrations in the Alphabetical Dioscorides.[21] But the reality may be more complicated. Heide Grape-Albers follows earlier scholars in distinguishing the plant pictures in the old index from the others in the Vienna codex, yet she suggests dividing the illustrations into three groups on the basis of their relative naturalism: first, the oldest and most naturalistic group of pictures, which she calls "Hellenistic"; second, an intermediate group, some of which are associated with the old index; and finally, a third and more recent "late antique" group that is the least naturalistic.[22] Giulia Orofino adds that the intermediate group dominates in the old index beginning with the letter *delta*, which Premerstein similarly noted.[23] John Riddle and Minta Collins, however, question these reconstructions, especially the association of the most naturalistic pictures with Crateuas and the Hellenistic herbals.[24] Both note that naturalistic illustrations appear throughout the codex irrespective of their association with quotations of Crateuas. At present, it seems doubtful that the old index and stylistic analysis are enough to fully elucidate the ultimate sourcing of the illustrations in the Alphabetical Herbarium. It is unclear how we might distinguish stylistic changes made in copying the manuscripts from the style of their putative archetypes. Another complicating factor is that we simply do not know what the pictures inside Crateuas' herbal looked like. In the end, we can merely hypothesize that the Alphabetical Herbarium drew upon multiple sources; some of these sources themselves may have been compilations of yet earlier sources. At the same time, it remains possible that some illustrations in the Alphabetical Herbarium may have been made specifically for this text, perhaps even on the basis of Dioscorides' descriptions.

Despite their similarities, the Vienna and Naples codices diverge in their layouts. The Naples Dioscorides typically reserves the top part of each folio for pictures, with the text below (see, e.g., figs. 3.10, 4.3). Content is generally restricted to the recto of each folio, though sometimes text spills over onto the verso. This approach to layout echoes what we see in earlier herbals, such as the Tebtunis Roll and the Antinoopolis Codex, and may ultimately mean that the makers of the Naples Dioscorides simply imitated layouts from a papyrus roll (or perhaps a codex that followed an earlier system of layouts). At the same time, however, this particular approach to page design, whereby any excess writing flows onto the otherwise blank verso, ensures relatively consistent formatting across the manuscript. In contrast, the Vienna Dioscorides typically allocates a full folio to each picture with the accompanying text on the facing folio (see, e.g., figs. 3.8, 6.9). This approach to page design seems to be a logical extension of what is seen in the Sinai palimpsest (fig. 3.5), in which the pictures of plants were enlarged to full-page size, with much shorter texts occupying a limited space around the roots of each plant. Moreover, the Alphabetical Herbaria in both the Viennese and Neapolitan codices appear to have been illustrated prior to the copying of text.[25] As we saw in chapter 2, this approach also appears in earlier herbals.

The two manuscripts also diverge in their contents. While most illustrations go back to a common source, some do not.[26] When individual pictures in the Naples and Vienna Dioscorides manuscripts differ, those in the Naples tend to be more accurate to Dioscorides' text.[27] For example, the picture of bugloss (*bouglōsson*) in the Naples Dioscorides (fig. 4.1) is closer to the actual plant that Dioscorides describes as resembling mullein than it is to the thistle-like plant found in the Vienna Dioscorides under the same name (fig. 4.2).[28] For this reason some scholars view the Naples Dioscorides as a more faithful reflection of an earlier version of the Alphabetical Herbarium.[29] But it is also possible that the Naples Dioscorides reflects an improved and edited version of the Alphabetical Herbarium.[30] Pictures in the Vienna Dioscorides, however, are often more accurate, or have more details that might reflect either earlier versions of the Alphabetical Herbarium, their having been improved in transmission, or perhaps even based on direct observation.[31] If the Naples Dioscorides appears to be more true to the "archetype" of the Alphabetical Herbarium, then the Vienna Dioscorides, by contrast, may represent a reworked or "expanded" version of that archetype.[32] As we have seen, the makers of the Vienna Dioscorides added a number of quotations in smaller script from other authorities—not only Crateuas, but also Galen.[33] That these quotations are largely concentrated at the beginning of the

FIGURE 4.1 Illustration of a bugloss (*bouglōson*). Naples Dioscorides, fol. 28r. By permission of Ministry of Culture (Ministero della Cultura) © Biblioteca Nazionale di Napoli.

volume suggests that the Vienna Dioscorides remains unfinished. The large margins, additional quotations, and open layout make the entire manuscript look like a work in progress—as though it had originally been conceived of as a deluxe heirloom notebook, similar perhaps to the parchment codices that Galen reported owning.[34]

Exactly how Dioscorides' *De materia medica* became the Alphabetical Herbarium remains unclear, in part because we know so little about the early circulation of the text.[35] Multiple versions may have circulated even in Dioscorides' lifetime. Galen, for example, famously complained that his writings were "subject to all kinds of mutilations, whereby people from all over circulate texts under their own names, with all sorts of cuts, additions, and alterations."[36] By the fourth century, we find Oribasius (d. after 395/6), physician to the emperor Julian, including alphabetically arranged and condensed excerpts from Dioscorides' *De materia medica* in his vast compilation of earlier medical authorities.[37] Max Wellmann supposed Oribasius' alphabetized Dioscorides might have emerged in the third or maybe

FIGURE 4.2 Illustration of a thistle labeled "bugloss" (*bouglōsson*). Vienna Dioscorides, fol. 76v.
Courtesy of Österreichische Nationalbibliothek.

fourth century.[38] But, as John Riddle has pointed out, Oribasius alphabetically arranges only large chunks of excerpts from Dioscorides concerning plants and metals.[39] Oribasius' excerpts of Dioscorides' discussions of wines generally follow Dioscorides' original organization by drug action.[40] Riddle concludes that we cannot know when Dioscorides' work was alphabetically rearranged on the basis of this evidence.[41] More recently, Marie Cronier has found no relationship between the excerpts in Oribasius and the Alphabetical Herbarium.[42] She has suggested that the compilers of the Alphabetical Herbarium may have worked by adapting a nonalphabetized version of the *De materia medica* to earlier alphabetized and illustrated herbals.

Lists of synonyms were also added to most, but not all, chapters in the Alphabetical Herbarium.[43] These typically appear at the beginning of each chapter, either as continuous text or divided into two columns. Synonyms within the same language are merely set off by the expression "but others . . . [*hoi de*]," short for "but others call it." While the lists focus on Greek and Latin synonyms, they also occasionally give synonyms purportedly used by twenty-six different ethnic groups, including Africans (*aphroi*), Dacians, Dardanians, Egyptians, Etruscans (*thouskoi*), Gauls, Spaniards, and Syrians, as well as "marginal" authorities, such as Democritus, Pythagoras, Zoroaster, Osthanes, and "the prophets" (that is, the Magi of ancient Persia).[44] We should take these synonyms with more than a grain of salt.[45] Max Wellmann argued that they were taken from the work of the first-century Alexandrian lexicographer Pamphilus.[46] While this may be true, Hardy and Totelin cautiously note that there may have been other sources.[47] This way of handling synonyms appears in other ancient herbals. For example, lists of synonyms in the earliest copy of the Latin *Herbarius* by Ps.-Apuleius Platonicus, dated to the sixth century, essentially follow the same pattern as found in the Vienna Dioscorides (compare fig. 2.1 and fig. 3.8).[48] Such lists may have been fairly common in ancient illustrated herbals.

Not only does the Alphabetical Dioscorides remove chapters—particularly the chapters on trees, minerals, spices, wines, and oils—it also adds chapters originally excluded from Dioscorides' *De materia medica*: chapters on hoary stock, dog's cabbage, hound's tongue, two kinds of delphinium, large and small varieties of hawkweed, and an unidentified plant, perhaps cudweed or marigold, called *zōonychon*.[49] While these are likely from an earlier source, it remains difficult to say where this interpolated material came from. These chapters provide different kinds of information. For example, some interpolated chapters follow the rest of the Alphabetical Herbarium in providing synonyms, a brief description of the plant morphology, and thera-peutic uses.[50] Others, however, include only a list of synonyms and medicinal properties, much in line with the illustrated herbaria de-scribed by Pliny and seen in the *Herbarius* of Ps.-Apuleius Platonicus, or simply a list of synonyms.[51] Most of these interpolated chapters appear in the lacunose table of contents at the beginning of the Vien-na Dioscorides.[52]

The interpolated and abridged contents overshadow deeper re-workings of Dioscorides' *De materia medica* in the division and des-ignation of plant kinds.[53] For example, while the original version of Dioscorides' *De materia medica* and the Alphabetical Herbarium both mention three different kinds of succulents called *aeizōon*, they do not refer to the same kinds of plants. Dioscorides mentions a big *aeizōon*, a small *aeizōon*, and a "third kind of *aeizōon* [*triton eidos aeizōou*]." In the Vienna Dioscorides, we find that the third *aeizōon* is labeled "the narrowleaf *aeizōon* [*aeizoon to leptophyllon*]," but the text correspond-ing to it is actually a shortened version of the same chapter on the small *aeizōon*. What the Alphabetical Herbarium apparently shows is two subtypes of small *aeizoa*: a small *aeizoon* and a small narrowleaf *aeizōon*, evidently confirmed by the fact that "small *aeizōon* [*aeizōon to mikron*]" is listed as a synonym for the narrowleaf *aeizoon*.[54] It is as though the makers of the Alphabetical Herbarium had pictures of two "small" *aeizōa* from another source and simply adapted Dioscorides' chapter on the small *aeizōon* to fit both. If we want to find the text for Dioscorides' "third kind of *aeizōon*" in the Vienna Dioscorides, we must actually look to the chapter on wild purslane (*andrachnē agria*), where we find "wild *aeizōon*" in the list of synonyms (*aeizōn [sic] agri-on*).[55] All of this indicates that Dioscorides' text was here adapted to fit the sources of the illustrations.

Similar observations can be made throughout the Alphabetical Herbarium. In some cases, the number of subtypes given in the Al-phabetical Dioscorides differs from that in the original Dioscorides. For example, the original Dioscorides notes three kinds of *artemis-ia* (wormwoods, mugworts): a small, fragrant, single-stalked inland herb, and two coastal shrubs, one with large leaves, and the other smaller and foul-smelling, with white flowers.[56] In contrast, the Al-phabetical Herbarium gives only two: the single-branched *artemisia* (*artemisia monoklōnos*) and the many-branched *artemisia* (*artemisia hetera polyklōnos*).[57] The Alphabetical Herbarium apparently desig-nates the small inland herb as the single-branched *artemisia*, while it conflates the two shrubby coastal *artemisia* types with each other under a single name, the many-branched *artemisia*. The Alphabetical

Herbarium thus clearly differs from the original Dioscorides in how it designates plants and how it groups them into kinds. While some of these differences could be regarded as mistakes, or as merely the result of collating different texts, we could also see them as reflections of different ideas about what names designate what kinds of plants.[58]

The Naples and Vienna Dioscorides manuscripts also differ in their handling of the relationship between text and picture for related kinds of plants. While the *aeizōa* in the Vienna Dioscorides involve the duplication and abridging of text, more commonly different pictures, each with their own names and synonyms, were grouped alongside the same text from Dioscorides. The Naples Dioscorides consistently places similar plants with different titles and synonyms side by side, while they share the same text below, as seen in the entry for the two *artemisia* types (fig. 4.3). The Vienna Dioscorides, by contrast, uses three different approaches. In the case of the two *artemisia*, it puts them on separate folios with separate titles and lists of synonyms, even though they ultimately share the same text.[59] A second approach involves rendering a related plant at a much smaller scale. We see this in the inset illustration of one anemone below the main text, with a much larger illustration of an anemone on the facing folio (fig. 3.8).[60] Finally, related kinds sometimes appear side by side on the same folio, where they share a text despite having distinct names and lists of synonyms.[61] While this variable emphasis could have been used to make useful distinctions between related plants, to compare them, or to assert the greater significance of one plant over another, it remains unclear why the illustrators chose one layout over another at any given moment.

Recognition that the Alphabetical Herbarium may ultimately draw upon several earlier illustrated herbals raises the question of how the compilers matched Dioscorides to the pictures.[62] In some cases, the compilers could have matched the names of the plants in Dioscorides to those in the illustrated texts. In other cases, the variability of plant names would have required the compilers to rely on synonymy and textual descriptions as well. We have already seen that this may have occurred when the makers associated Dioscorides' third *aeizōon* not with the other *aeizōa* in the manuscript but rather with the wild purslane. It may also have occurred when they linked the Gorgon-rooted depiction of sea holly to Dioscorides' text.[63] Dioscorides makes no mention of the head in his description.[64] In the Alphabetical Herbarium, however, *gorgoneion* (or *gorgonion*) appears as an alternate name for eryngo.[65] The original version of Dioscorides' *De materia medica* does not mention this alternate name. We can

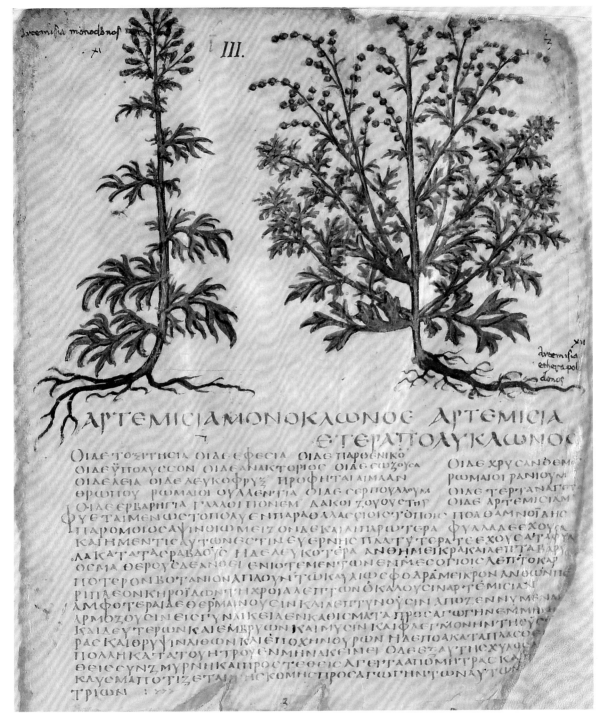

ΑΡΤΕΜΙϹΙΑΜΟΝΟΚΛΩΝΟϹ ΑΡΤΕΜΙϹΙΑ
ΕΤΕΡΑΠΟΛΥΚΛΩΝΟϹ

ΟΙΔΕ ΤΟ ΕΙΠΝΟΙΑ ΟΙΑΕ ΕΦΕϹΙΑ ΟΙΑΕ ΠΑΡΘΕΝΙΚΩ ΟΙΑΕ ΧΡΥϹΑΝΘΕΜΟ
ΟΙΑΕ ΥΠΟΛΥϹϹΟΝ ΟΙΑΕ ΑΝΑΚΤΟΡΙΟϹ ΟΙΑΕ ϹΩΖΟΥϹΑ ΡΩΜΑΙΟΙ ΡΑΝΙΟΥΝΙ
ΟΙΑΕ ΛΕΙΑ ΟΙΑΕ ΛΕΥΚΟΦΡΥϹ ΠΡΟΦΗΤΑΙ ΑΙΜΑΛΛΑΝ ΟΙΑΕ ΤΕΡΓΑΝΑΓΕΤ
ΘΡΩΠΙΟΥ ΡΩΜΑΙΟΙ ΟΥΛΛΕΝΓΙΑ ΟΙΑϹ ϹΕΡΠΟΥΛΛΟΥΜ ΟΙΑΕ ΑΡΤΕΜΙϹΙΑΜ
ΟΙΑϹ ΕΡΒΑΡΗΓΙΑ ΓΑΛΛΟΙ ΠΟΝΕΜ ΔΑΚΟΙ ΖΟΥΟΥϹΤΙΡ ΠΟΛ ΘΑΜΝΟ ΤΑΠΟ
ΦΥΕΤΑΙ ΜΕΝ ΩϹ ΤΟ ΠΟΛΥ ΕΝ ΠΑΡΑΘΑΛΑϹϹΙΟΙϹ · ΤΟΠΟΙϹ ΦΥΛΛΑ ΔΕ ΕΧΟΥϹΑ
ΚΑΙ ΗΜΕΝ ΤΙϹ ΑΥΤΩΝ ΕϹΤΙΝ ΕΥΕΡΝΗϹ ΠΛΑΤΥΤΕΡΑ ΤΕ ΕΧΟΥϹΑ ΤΑΤΩΝ
ΛΛΑ ΚΑΤΑ ΤΑϹ ΡΑΒΔΟΥϹ Η ΑΞ ΛΕΥΚΟΤΕΡΑ ΑΝΘΗ ΜΕΙΚΡΑ ΚΑΙ ΛΕΠΤΑ ΒΑΡΥ
ΟϹΜΑ ΘΕΡΟΥϹ ΛΕΑΝΘΕΙ ΕΝΙΟΤΕ ΜΕΝ ΤΩΝ ΕΝ ΜΕϹΟ ΤΟΠΟΙϹ ΛΕΠΤΟΚΑΡ
ΠΟΤΕΡΟΝ ΒΟΤΑΝΙΟΝ ΑΠΛΟΥΝ ΤΩ ΚΑΥΛΙΩ ϹΦΟΔΡΑ ΜΕΙΚΡΟΝ ΑΝΘΩΝ ΠΕ
ΡΙΠΤΑ ΕΟΝ ΚΗΡΟΙ ΑϹΘΝΤΗ ΧΡΟΙΑ ΛΕΠΤΩΝ Ο ΚΑΛΟΥϹΙΝ ΑΡΤΕΜΙϹΙΑΝ
ΑΜΦΟΤΕΡΑΙ ΔΕ ΘΕΡΜΑΙΝΟΥϹΙΝ ΚΑΙ ΛΕΠΤΥΝΟΥϹΙΝ ΑΠΟ ΖΕΝΝΥΜΕΝΑΙ
ΑΡΜΟΖΟΥϹΙΝ ΕΙϹ ΓΥΝΑΙΚΕΙΑ ΕΝ ΚΑΘΙϹΜΑΤΑ ΠΡΟϹΑΓΩΓΗΝ ΕΜΜΗΝ
ΚΑΙ ΔΕΥΤΕΡΩΝ ΚΑΙ ΕΜΒΡΥΩΝ ΚΑΙ ΜΥϹΙΝ ΚΑΙ ΦΛΕΓΜΟΝΗΝ ΤΗ ΟΥ
ΡΑϹ ΚΑΙ ΟΡΥ ΓΙΝΑΙ ΩΘΩΝ ΚΑΙ ΕΠΟΧΗΝ ΟΥΡΩΝ Η ΑΕ ΠΟΑ ΚΑΤΑ ΠΛΑϹϹΟ
ΠΟΛΛΗ ΚΑΤΑ ΓΟΥ ΗΤΡΟΥ ΕΝ ΜΗΝΑΚΕΙΜΕΙ ΟΛΑ ΕΞ ΑΥΤΗϹ ΧΥΛΩΘΕ
ΘΕΙϹ ϹΥΝ ΖΜΥΡΝΗ ΚΑΙ ΠΡΟϹ ΤΕΘΕΙϹ ΑΓΕΤΤΑ ΑΠΟ ΜΗΤΡΑϹ ΚΑΙ
ΚΛΥϹΕΜΑΤΟ ΤΙΖΕΤΑΙ ·ΗϹ ΚΟΜΗ ΠΡΟϹΑΓΩΓΗΝ ΤΩΝ ΑΥΤΩΝ
ΤΡΙΩΝ : ⁘

FIGURE 4.3 Illustrations of two artemisias. Naples Dioscorides, fol. 3r. By permission of Ministry of Culture (Ministero della Cultura) © Biblioteca Nazionale di Napoli.

similarly explain the rootlessness of the snapdragon (*kynokephalion*) through reference to a tradition not preserved in Dioscorides, according to which a synonymous plant (*antirrhinon*) was believed to lack roots.[66] Again, the compilers evidently matched the picture to the text by way of a synonym.

Relying on synonyms for matching plant pictures to Dioscorides' text came with complications. We can see this in the multiplication of illustrations of hart's tongue fern in the Alphabetical Dioscorides under names such as bracken fern (or "female fern," that is, *thēlypteris*, fig. 4.4), male fern (or "another fern," *pteris hetera*), and, in the Vienna

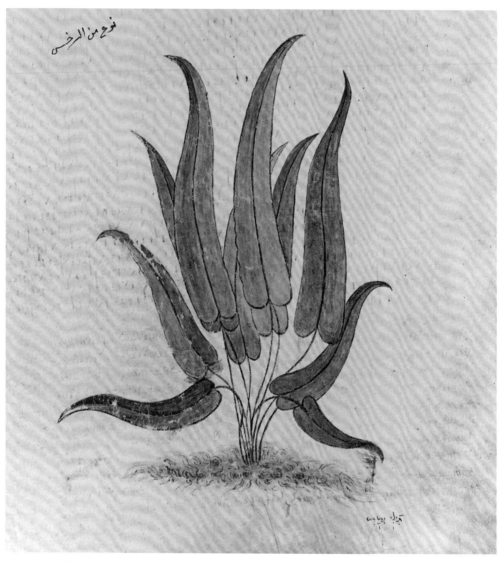

FIGURE 4.4 Illustration of a bracken fern (literally "female fern," *thēlypteris*). Vienna Dioscorides, fol. 142r. Courtesy of Österreichische Nationalbibliothek.

Dioscorides, *nymphaia* (fig. 4.5), which usually designates a water-lily, as seen in the Naples Dioscorides (fig. 4.6).[67] The association of hart's tongue with the name *nymphaia* in the Vienna Dioscorides may go back to the name *nymphaia pteris*, which Dioscorides gives as a synonym for bracken fern.[68] This picture of hart's tongue may have been linked with *thēlypteris* due to the Latin name *lingua cervina*, "hart's tongue," that appears as a synonym in the Alphabetical Herbarium.[69] (The apparent weight accorded this Latin synonym might provide further support for Cronier's hypothesis that the Alphabetical Dioscorides originated in Italy.)[70] As *thelypteris*, which is in the

FIGURE 4.5 Illustration of a fern labeled *nymphaia*. Vienna Dioscorides, fol. 239r. Courtesy of Österreichische Nationalbibliothek.

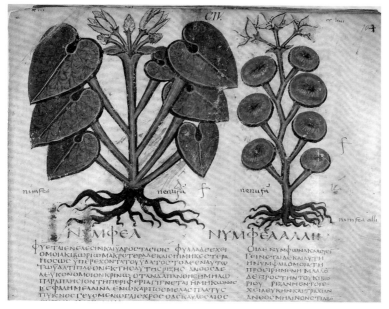

FIGURE 4.6 Illustration of waterlilies (*nymphea* and *nymphea allē*). Naples Dioscorides, fol. 104r. By permission of Ministry of Culture (Ministero della Cultura) © Biblioteca Nazionale di Napoli.

old index, is related to the images associated with *pteris hetera* and *nymphaia*, which are not in the old index, the manuscript sources for these illustrations already seem to reflect overlapping, underlying traditions, hence the difficulty of using the old index to reconstruct some hypothetical earlier source. That these duplicated images are related to each other but stylistically distinct also indicates the difficulty of isolating hypothetical sources for the pictures according to their stylistic differences. The Vienna Dioscorides is the product of a long, complicated manuscript tradition. While this multiplication of pictures under different names seems to have arisen from errors in collation, it is worth noting that hart's tongue ferns are particularly variable. Dioscorides' descriptions of ferns also tend to be rather vague. As a result, the duplications in the Vienna Dioscorides might simply reflect uncertainty or disagreement among ancient botanical authorities on the identification and naming of ferns.

The compilatory nature of the illustration of the Alphabetical Dioscorides means that text and image often provide different kinds of information. This is particularly evident in the example of the sea holly. We saw in chapter 3 that the Gorgon-headed root of the sea holly actually reflects botanical lore that Dioscorides excluded: namely, the belief that if a goat took the root of the plant into its mouth, its entire herd would come to a sudden stand-still.[71] The absence of roots in the illustration of snapdragon likewise refers to a tradition absent from Dioscorides. These examples demonstrate that text and image represent different strands of the larger botanical tradition. In this way, the compilation of the illustrated Alphabetical Herbarium enabled the confluence of different botanical ideas, knowledge, and lore.

The ability of images to include earlier or omitted knowledges may have gone hand in hand with the generally archaizing flavor of earlier illustrated herbals. Max Wellmann has noted that some plant names in the Alphabetical Herbarium had an old-timey ring to them.[72] We can see similar archaism already in the Tebtunis Roll, where Ann Ellis Hanson has pointed to the ancient use of the word *petalon* for "leaf," as opposed to its then more common narrower usage for "petal."[73] Whatever pretense the Alphabetical Dioscorides might have had to represent an expansive, venerable, and ancient botanical tradition, the compilatory nature of its illustration also led to novel expansions of that tradition. In some cases, these expansions were outright errors, as we have seen in the multiplication of hart's tongue illustrations in the Alphabetical Herbarium. In the case of the small and the narrowleaf *aeizōon*, however, the makers arguably improved on Dioscorides by recognizing more varieties of *aeizōon*.

The Alphabetical Herbarium was probably not the only herbal to undergo this process of textual and pictorial compilation. The herbal attributed to Ps.-Apuleius Platonicus is similarly heterogeneous in its textual contents. On the basis of its comparison to her stylistic groupings of the pictures in the Vienna Dioscorides, Grape-Albers has argued that this fourth-century textual compilation likely copied many of its illustrations from earlier templates originally from around the third century.[74] This process of adding pictures to texts is thus indicative of much larger patterns. Yet as far as I know the Alphabetical Herbarium was novel in its pairing of Dioscorides' descriptions with earlier depictions.

Illustrating the Original *De materia medica*

The original version of Dioscorides, arranged in five books according to drug action, was also eventually illustrated. The earliest surviving illustrated copy of the original *De materia medica* can be found in a parchment codex now in Paris.[75] Despite some disagreement, scholars tend to place the manuscript in Syria or Palestine toward the end of the eighth or the beginning of the ninth century.[76] The text is of a higher quality than those of the other illustrated herbals that we have seen so far.[77] The illustrations tend to appear on the right side of the text column (see fig. 4.7).[78] Sometimes they appear in the margins (fig. 4.8). And sometimes they are rotated so as to fit in the allotted space (figs. 4.7 and 4.8).[79] As with earlier illustrated herbals, this Old Paris Dioscorides typically shows one picture per plant named in the text.[80] Unlike the plant pictures in the Naples and Vienna Dioscorides manuscripts, the illustrations in the Old Paris Dioscorides were copied after the copying of the text.[81]

The pictures in the Old Paris Dioscorides differ stylistically from those in the Vienna and Naples Dioscorides manuscripts. The pictures are streamlined, as though any extraneous detail had been removed. Branches are often symmetrical along a central axis, typically coinciding with the main axis of the plant. Although stylistically distinct

FIGURE 4.7 Illustrations of fumitory (*isopyron*), above, and below, a violet (*ion*). Paris, Bibliothèque nationale de France, MS gr. 2179 (the "Old Paris Dioscorides"), eighth or ninth century CE, fol. 119r. Photo: Bibliothèque nationale de France.

FIGURE 4.8 Folio with illustrations of *zea* and *tragos* in right margin, and oat (*bromos*) below. Old Paris Dioscorides, fol. 98r. Photo: Bibliothèque nationale de France.

ΦΑΝ ΤΙΒΟΣ ... ΤΙ ΚΟ ... ΕΥΦΟΡ ...
ΚΑΙΟΠΑ ... ΦΑΓΓΙ ... ΕΤΑΙ ... ΜΕΧ ... ΕΝ ...
ΚΑΙ ΤΟ ΚΑΛΟΥΜΕΝΟΝ ΔΕ ΚΟ ... ΡΜ ... ΕΚ ...
ΜΕΝΟΝ Δ ... ΕΤ ... ΚΗ ... ΚΑΙ ΑΝΤΙ ...
ΝΟΥ ... ΜΑ ... ΝΤ ...
ΚΑΛΓΓΟΡ ... ΤΙΝΕ ... ΚΑΛΕΟ ΥΜΕΝΕ ...

... ΓΑΛΛΙ ... ΚΑΝ ... ΠΑΣΑ ... ΠΟ ...

ΚΡΙ ... ΜΝ ...

ΚΑΙ ... Γ ΑΔΕΣ ...
ΚΑ ... ΤΙ ...

... ΠΕΡ ...
... Γ ...

ΒΡΟΜΟΣ ...

Βρομος

ΑΚΥ ...

from the Vienna and Naples Dioscorides manuscripts, the pictures in the Old Paris Dioscorides work in similar ways that were noted in the last chapter. A handful of pictures in the Old Paris Dioscorides allude to medicinal properties as well as to harvest and collection methods.[82] The illustration of aloe (*aloē*, fig. 4.9), for example, shows liquid pouring from the plant.[83] While most manuscripts containing the Alphabetical Herbarium avoid depicting extraction and harvest, such scenes do appear in later illustrated copies of Dioscorides.[84] The vast majority of pictures in the Old Paris Dioscorides, however, show plants rather simply, as though flattened in a single plane, with usually only one side visible and no overt visual references to properties or harvesting.

Despite notable differences between the Old Paris Dioscorides and the Alphabetical Herbarium, some of the illustrations in the Old Paris Dioscorides appear to be descended from the Alphabetical Herbarium, or rather from a common pictorial source, as is evident in the illustration of violets (fig. 4.7, compare to fig. 3.6).[85] While previous scholars have commented on the possible connection between the Old Paris Dioscorides and the Alphabetical Herbarium, there are also many dissimilarities.[86] Any similarities, moreover, do not necessarily indicate the direct transmission of images from the Alphabetical Herbarium but could perhaps point back to a common source for the illustrations in both the original Dioscorides and the Alphabetical Herbarium. As we will see in the next chapter, Marie Cronier has

FIGURE 4.9 Illustration of aloe (*aloē*). Old Paris Dioscorides, fol. 16r. Photo: Bibliothèque nationale de France.

even raised the possibility that some of these illustrations may have come from an early Arabic translation of Dioscorides, given that the manuscript contains a number of Arabic annotations in cursive Greek letters that seem to function as instructions to the illustrator.[87]

The degree of change that can occur in the course of copying certainly complicates our attempts to understand the sourcing of the illustrations in the Old Paris Dioscorides. Some illustrations in the Old Paris Dioscorides may have come from the Alphabetical Herbarium or one of its sources, but any clear, visible relationship has been lost through successive changes over time. The illustration of greater celandine (*chelidonion*) in the Old Paris Dioscorides gives a sense of how images change through the process of copying (fig. 4.10).[88] The pale red ink of the underdrawing is plainly visible. It is clear that in the

FIGURE 4.10 Illustration of greater celandine (*chelidonion*). Old Paris Dioscorides, fol. 3v. Photo: Bibliothèque nationale de France.

FIGURE 4.11 Illustration of "mouse ear" plant (*myos ōta*), and an unknown plant, possibly dock, labeled as "woad" (*isatis*). Old Paris Dioscorides, fol. 5r. Photo: Bibliothèque nationale de France.

initial underdrawing the basal leaves of the plant were bent as seen in the illustration of woad (*isatis*, fig. 4.11).[89] When painting the greater celandine, however, the painter opted to depict only one side of the basal leaves. As a result, the flexed basal leaf of the greater celandine has become unflexed.

Some illustrations in the Old Paris Dioscorides were clearly based on Dioscorides' text and not the Alphabetical Herbarium. This is especially evident in the picture of a plant called *lonchitis* (fig. 4.12), probably a kind of orchid.[90] The picture closely adheres to all of the major points in Dioscorides' description:

> *Lonchitis* has leaves like a sliced leek but wider and reddish; very many of them are near the root, bending on the ground, as it were; it also has a few around the stem upon which there are flowers resembling little felt hats, shaped like gaping comic masks; they are black and there is something white that protrudes from their opening toward the lower lip, as if it were a little tongue. The seed is like a spearhead [*lonchē*], triangular, and encapsulated, whence it earned its name.[91]

The illustrator either worked from an incomplete or lacunose illustrated manuscript, or perhaps rejected earlier pictures because they did not sufficiently echo the textual description.[92]

The illustrator based another plant depiction in the Old Paris Dioscorides not on Dioscorides' description, but rather on the plant's name. The illustration of *tragos* (fig. 4.8) shows a feathery plant with small goat-heads popping up out of the plant's axils. Dioscorides' description is rather terse.[93] *Tragos* may have originally referred to a grain product, perhaps made of spelt.[94] The makers of the manuscript (or its source) evidently did not know this, but assumed *tragos* was a plant. The curious goat-heads in the illustration reflect the fact that the Greek name *tragos* also refers to a he-goat. The matching of pictures to text here speaks in another way to this edition's fidelity to Dioscorides' text, at least to the extent that it was understood. This is less apparent in the Alphabetical Herbarium, where divergences between text and image, particularly in the form of additional information, spoke to the autonomy of the pictures and of the need to bend the text to the pictures available.

The Old Paris Dioscorides includes six illustrations of human figures interacting with pictures of plants concentrated at what is now the beginning of the codex (see, for example, figs. 4.10 and 4.11).[95] We can suppose, however, that more figures may once have populated

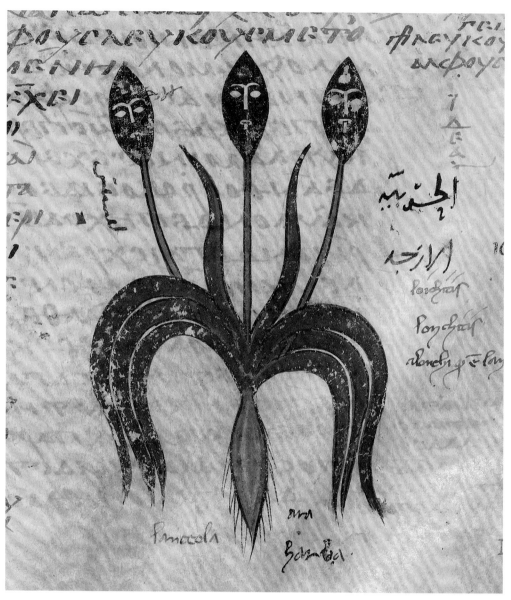

FIGURE 4.12 Illustration of *lonchitis*. Old Paris Dioscorides, fol. 65r.
Photo: Bibliothèque nationale de France.

the codex, as it is currently missing the first book and most of the
second.[96] The loose, bold, and somewhat blocky coloring of the figures departs from the clear linearity of the botanical illustrations. The
human figures instead seem closer to those in older books of the sixth
and seventh centuries. This could suggest that they represent a kind
of archaism entirely consistent with the elements of classical iconography present, such as the figures' bare feet, fluttering drapery, and
pastoral costume.[97] Kurt Weitzmann and John Riddle have suggested

that the figures in the Paris codex have a didactic function, while Alain Touwaide agrees but adds that they also have a decorative role.[98] The three authors cite the figure beside the now unidentified "mouse ear" plant (*myos ōta*, fig. 4.11).[99] Holding his hand to his face, the figure apparently refers to the root's purported ability to cure eye ulcers or lachrymal fistulae.[100] Minta Collins considers this figure the exception to the rule: most figures, she argues, "add no information to the plant illustration."[101] But it seems likely that the figures in the Paris codex probably carry out a variety of functions by directing viewers' attentions and by mediating between text and image, even if some of these functions are no longer obvious today.[102] As we saw in chapter 3, the earliest surviving example of a gesturing, didactic figure appears in the illustration of coral from the Vienna Dioscorides, where the figure helps to indicate the plant's habitat.

A ninth-century fragment of an illustrated copy of the original Dioscorides was also preserved as a flyleaf to an Armenian manuscript in Yerevan (fig. 4.13).[103] The text concerns ground pine (*chamaipitys*) and two kinds of St John's wort (*androsaimon* and *koris*).[104] Only the illustration of the *koris* St John's wort survives. It departs radically from the illustrations that we have seen so far. It omits the plant's roots, making the plant appear as though it were a branch. There is little modeling, and the leaves appear to be flattened and are visible from only one side. With the possible exception of leaf shape and arrangement, the picture generally lacks differentiating details. This approach

FIGURE 4.13 Illustration of St. John's wort (*koris*), Yerevan Fragment, ninth century CE. Yerevan, Matenadaran, MS arm. 141. Oxford, Bodleian Library, MS gr. class. E. 19. Courtesy of Bodleian Libraries, Oxford. Photo by F. C. Conybeare.

to illustration again appears in some medieval herbals, including the Morgan Dioscorides of the late ninth or early tenth centuries.[105]

Conclusion

The earliest surviving illustrated copies of Dioscorides consist of an alphabetically arranged herbarium, radically different in organization and scope from the original *De materia medica*. Dioscorides' text was abridged—made to focus exclusively on uprootable herbs—and alphabetically arranged, while synonyms, illustrations, and even whole chapters from other works were added to it. The compilers appear to have matched pictures from earlier illustrated herbals to Dioscorides on the basis of plant names and synonymy, often with mixed results. In doing so, however, the compilers confronted disjunctions between Dioscorides and the rest of the ancient botanical tradition. As a result, they found plants not mentioned by Dioscorides and plants in Dioscorides that had not been illustrated elsewhere. The fact that many pictures in the Alphabetical Herbarium came from earlier herbals meant that they often portrayed botanical morphology and lore absent in Dioscorides' *De materia medica*. Some of these changes made Dioscorides' *De materia medica* align with other illustrated books on root-cutting. Yet the Alphabetical Herbarium diverged from these earlier works in that it retained Dioscorides' verbal descriptions of plants, so that pictures and text could now be directly compared against each other. In time, the earlier original version of Dioscorides was also illustrated, perhaps due to the success of the Alphabetical Herbarium. Early surviving manuscripts of the illustrated original Dioscorides themselves provide evidence for the spread of several innovations in the treatment of Dioscorides' text, including illustration on the basis of text and the addition of subsidiary explanatory human figures to the illustrations of Dioscorides' text. Late antiquity and the early Middle Ages saw the proliferation of these different illustrated versions and redactions of Dioscorides. The next three chapters examine this expansion of the ancient tradition in the Middle Ages.

5

MEDIEVAL HERBALS

In the half-millennium spanning from the fourth to the ninth century, the Mediterranean world changed entirely. Christianity became the principal faith of the Roman empire. The empire splintered into smaller kingdoms in Western Europe, while a new Rome, Byzantium, carried on in the Eastern Mediterranean. Islam rose in the seventh century, ushering in its wake widespread transformations in Central, Western, and South Asia, North Africa, and Iberia.

The medieval Latin-, Greek-, and Arabic-speaking peoples of the Mediterranean world collectively inherited, adapted, and innovated upon the ancient botanical traditions of the Roman and Hellenistic world. The physicians of the Mediterranean world accepted the Galenic synthesis, nearly unanimously, although they simultaneously focused on empirical medical practices in their day-to-day work. Medical botany emerged as a major focus of medieval medicine. At the same time, the rise of monasteries, as well as new charitable foundations and hospitals in the Islamicate world, provided new institutional spaces for the promotion of a medical culture deeply invested in the ancient botanical tradition, including the writing and reading of illustrated herbals.

This chapter looks at the remarkable persistence of herbal illustration in a radically transforming world. It surveys developments in the botanical tradition in the medieval Mediterranean and their implications for botanical illustration. Despite the botanical tradition's many transformations during this time, illustrated botanical works continued to be produced, read, and disseminated. Medieval people valued botanical illustrations as carriers of ancient visual knowledge about plants, and as a storied, but above all useful way to learn about them. But while the ancient tradition of illustrating herbals continued into the Middle Ages, much changed. Illustrated herbals were produced by new makers in new ways for new patrons and readers in new institutional contexts. The illustrated herbal adapted to these new circumstances to become one of the longest unbroken traditions of image-making in Western Eurasia.

Medieval Medical Botany

By the sixth century, Dioscorides' writings on herbs (and various texts attributed to him) had become central to the study of pharmacy in the Mediterranean. At the Vivarium monastery in Calabria, Italy, for example, the monks were exhorted in the sixth century to read Dioscorides before any other medical work. The monastery's founder, Cassiodorus, a prominent statesman and scholar who had earlier served at the Ostrogothic court, tells the monks of his monastery:

> Even if the eloquence of Greek letters is unknown to you, you have first of all the *Herbal* by Dioscorides, who has discussed and depicted the herbs of the fields with remarkable accuracy. After this read the Latin translations of Hippocrates and Galen (that is, Galen's *Therapeutics*, addressed to the Philosopher Glaucon) and a certain anonymous work compiled from various authors. Then read Caelius Aurelius' *On Medicine* and Hippocrates' *On Herbs and Cures* and various other works written on the art of medicine that, with the Lord's aid, I have left to you in the recesses of our library.[1]

Dioscorides' writings served as the foundation for the monks' medical studies.[2] While illustrated herbals had existed earlier, Cassiodorus gives us the earliest example of a scholar recommending a medical text in part because of its illustrations. As we have seen, Dioscorides never intended his work to be illustrated. But Cassiodorus believes, on the contrary, that Dioscorides had his herbal illustrated, in accordance with the frontispiece cycle from the Vienna Dioscorides (figs. 1.1–1.5). Cassiodorus' thoughts on medicine are moreover entirely rooted in Christianity. Before proposing his course of medical study, Cassiodorus praises the compassion of those brothers who provide medical care. They will reap eternal rewards for their attention to the earthly sufferings of their neighbors, though, he cautions, no one should place their hope only in medicine, as the Christian God alone gives life.

The medical tradition, too, transformed during this time.[3] The great medical authorities of the late antique world—Oribasius of Pergamum (d. 400), Aetius of Amida (fl. 530), and Paul of Aegina (d. c. 630)—focused much of their attention on massive encyclopedic compilations of earlier medical texts. Moreover, the fracturing of the Roman world and the erosion of its institutions forced medical practitioners to adapt to new social and cultural realities.[4] As Vivian Nutton notes, the loss of ancient institutions that had earlier sustained the Greek medical tradition, particularly in the Latin-speaking Western Mediterranean, led to the emergence of a medical self-help culture.[5] Medical works focused on herbs, such as Dioscorides, would have had obvious appeal in such a context. The period likewise witnessed a widening gulf between medical theory, dominated by Galenism, and actual medical practice, based on a narrower range of therapies that were known to work and thus more in line with the Empiricist medical tradition.

Some of these transformations are certainly evident by the late ninth century, as demonstrated by a letter written in the early 870s from Photios, the once and future patriarch of Constantinople, to Zacharias, the metropolitan of Chalcedon. In it Photios advises:

> I think bloodletting would be advantageous for your bodily condition even though it is summer. But if that is against the expectations of the physicians that are now in fashion, that is not at all unexpected: Well, let's not mention their other errors in the art [of medicine]: horsetail is deemed to be mare's tail; hartwort, Hercules' woundwort; crowfoot is judged to be swallowwort; spurge [is taken for] purslane; sea lavender for pondweed; white pine thistle for black chameleon thistle; *argemōnē* instead of anemone; and the natures of countless other plants are enlisted at strange times, for [strange] uses and under [strange] names (clearly these things [are] obvious, and were previously never among the things befitting controversy in the ancient halls of medicine): what a marvelous thing then, that bloodletting, being necessary for you, would seem strange to them? But if you accept my advice, God the Savior permitting, you will find both the refutation of them with evidence as well as your reward.[6]

Photios' brief letter paints for us a vivid picture of the state of medicine in ninth-century Byzantium. While Photios recommends Zacharias pursue a regimen of therapeutic bloodletting, the larger part of his letter enumerates the various errors of contemporary physicians. Photios portrays a medical field dominated by "fashionable physicians" preoccupied with herbal medicine—particularly the correct identification and usage of plants. By the time Photios wrote this letter to Zacharias, nearly eight centuries separated him from Dioscorides. Plant names, of course, had varied tremendously in the ancient Mediterranean, but centuries of linguistic change, the vagaries of manuscript transmission, and disruptions and changes in the trade of *materia medica* compounded whatever difficulties had existed earlier. It should come as no surprise that disputes over the identification of plant names would have preoccupied physicians in the ninth century to such a degree, or, conversely, that Photios should be so dubious of their pharmacological knowledge. Besides, bloodletting had much to recommend it: it was certainly simpler, enjoyed widespread acceptance, and found staunch support in Galen's writings.[7]

The fashionable physicians of Photios' day would have learned pharmacology primarily from several different versions of Dioscorides' *De materia medica* and other works on pharmacy, such as those by Galen and those excerpted in late antique compilations. Dioscorides' comprehensiveness and direct language, however, made his work especially suited to be the essential reference on pharmacology in the early medieval Mediterranean. The main Greek versions of Dioscorides' *De materia medica* essentially reflected two different approaches to the text, which we have seen in chapter 4: the first was to maintain Dioscorides' original organization according to drug affinities, while the second was to reorganize it alphabetically as a kind of reference work. It is uncertain how well Dioscorides' system of drug affinities fared in the Middle Ages, due to the rise of Galenic pharmacy and the prevalence of alphabetized versions of Dioscorides' text.[8] Still, original versions of Dioscorides continued to circulate. Photios' litany of errors curiously hints at the survival of that organization: most of the faulty identifications that he mentions involve plants that appear near each other in the original *De materia medica*.[9] At the same time, these mixups also involve humorous lexical juxtapositions: the opposition between *potamogeitōn* and *leimōnion*, for example, suggests that these fashionable physicians cannot tell the difference between a river (*potamos*) and a meadow (*leimōn*).[10] Whether the proximities of these plants within the original *De materia medica* speak to the compositional shortcuts that Photios might have taken simply for rhetorical effect, or to actual debates among contemporaries about appropriate substitutions following Dioscorides' system of drug affinities, we cannot say. At the same time, alphabetized and illustrated versions of Dioscorides clearly continued to circulate in the early Middle Ages. That the Naples Dioscorides was produced in the seventh century and the Old Paris Dioscorides in the eighth or early ninth century speaks to the ongoing production of illustrated herbals at a high level in the midst of a "dark" period often characterized by its relatively low literary output.[11] This is all to say that Dioscoridean pharmacy was alive and well in various forms in at least some parts of the early medieval Mediterranean.[12]

Medieval interest in Dioscorides' *De materia medica*, however, extended beyond its practical applications in medicine. In his *Bibliotheca*, a collection of book reviews, Photios notes that Dioscorides' *De materia medica* is "useful not only for medical practice but also for speculations in philosophy and natural science."[13] Such sentiments reflect the development of an attitude already apparent in the sixth-century Vienna Dioscorides, in which the Alphabetical Herbarium had been accompanied by paraphrased works on toxicology, herpetology, ornithology, and ichthyology—a summary distillation of the natural world into the domains of land, air, and water. We can imagine that interest in Dioscorides as a vehicle for ruminating on natural philosophy would have had an impact on contemporary readership of Dioscorides, including the viewership of any of the text's illustrations.

By the end of the ninth century, Dioscorides' *De materia medica* had also been translated into Syriac and Arabic.[14] According to medieval sources, the famous translator Ḥunayn ibn Isḥāq (d. AH 260/873 CE) translated Dioscorides into Syriac for a member of the prominent Bukhtīshūʿ family who served as a physician at the court of the Abbasid caliph al-Mutawakkil (r. AH 232–47/847–61 CE).[15] Syrian physicians dominated the medical field in Syria and Mesopotamia during the early Abbasid period.[16] There were also already Syriac translations of Greek pharmacological literature, such as the translation of Galen's *De simplicium medicamentum facultatibus* by Sergius of Rēsh ʿAynā (d. 536).[17] Ḥunayn's Syriac translation was soon followed by a translation into Arabic by Iṣṭifan ibn Bāsīl for Muḥammad ibn Mūsā, an official at al-Mutawakkil's court.[18] Ḥunayn checked and authorized Iṣṭifan's translation. Many plant names entered Arabic at this time through Syriac or were transliterated directly from Greek, either because the plant was unknown or perhaps in recognition of the meaning of the Greek names in the original text.[19] It should be noted, too, that there was an earlier botanical tradition in Arabic, as represented by the now fragmentary *Kitāb al-nabāt* (*Book of Plants*) by

Abū Ḥanīfa al-Dīnāwarī (d. AH 282/896 CE).[20] Besides Iṣṭifan's and Ḥunayn's translations, medieval sources also mention a thirteenth-century translation of Dioscorides for the Artūqid ruler Najm al-Dīn by Abū Sālim al-Malṭī, which does not survive, and another for the same patron by Mihrān ibn Manṣur.[21] We now know, however, that the translation and redaction of Dioscorides in Arabic was much more complicated than might be expected from the narratives provided by medieval sources.[22] Manfred Ullmann has demonstrated that there was already an Arabic translation of Dioscorides based on an interpolated version of the Greek text before the translation by Iṣṭifan.[23] George Saliba and Linda Komaroff have also pointed to the differences between illustrated and unillustrated versions of the text, as well as the existence of many reworkings, rectifications, and abridgments of the text.[24]

The translation of Dioscorides into Arabic laid the foundations for a complex, expansive botanical tradition in Arabic. Ibn al-Nadīm (d. AH 385/995 CE), Ibn al-Qifṭī (d. AH 646/1248 CE), and Ibn Abī Uṣaybiʿa (d. AH 668/1270 CE) mention around one hundred different Arabic writers who wrote on pharmacology.[25] These authors greatly expanded the botanical tradition: the immense *Kitāb al-jāmiʿ li-mufradāt al-adwiya wa-l-aghdhiya* (*Compendium on Simples and Foods*) by Ibn al-Bayṭār (d. AH 646/1248 CE) includes more than 400 different drugs unknown to Dioscorides.[26] Arabic scholars further expanded the tradition through commentaries on Dioscorides.[27] Greek and Arabic botanical traditions remained in dialogue over the course of the Middle Ages. By perhaps as early as the late ninth century, Arabic medical and pharmacological texts were translated into Greek, including most notably *Zād al-musāfir* (*Provisions for the Traveler*, known in Greek as the *Ephodia tou apodēmountos*) by Ibn al-Jazzār (d. AH 400/1004 CE), as well as a treatise on smallpox (Arabic: *Kitāb al-judarī wa-l-ḥaṣba*, Greek: *Peri loimikēs*) by Abū Bakr al-Rāzī (d. AH 313 or 323/925 or 935 CE).[28]

The paths that texts took from one language into another over the course of the Middle Ages could be quite complex. For example, the *De plantis* (*On Plants*) attributed to Nicolaus of Damascus (fl. first century BCE) passed from Greek into Syriac and Arabic by the ninth century CE, then Latin by about 1200, and back into Greek around 1300.[29] The topic of Greek translations of Arabic remains an area of active research, complicated by the fact that many translations were misattributed or remain anonymous. Beyond the translation of entire works, translation also took place on a smaller, ad hoc basis as is evident in many brief, marginal, and now fragmentary passages and recipes that were added to medical texts by physicians and scholars with varying degrees of bilingualism.[30] Traces of this dialogue (or at least manifest interest in it) can also be seen in the proliferation of botanical lexica during the Middle Ages.[31]

The practice of pharmacology and medical botany in the Latinate Mediterranean proceeded differently from what we have seen in the Islamicate and Byzantine worlds. The *Herbarius* of Ps.-Apuleius Platonicus enjoyed especially wide circulation. There were also classical and late antique Latin texts, including Pliny the Elder's *Naturalis historia*, the *Medicinii Plinii* (*Medical Pliny*), the book by Marcellus Empiricus (fourth to fifth century CE), and the works associated with Gargilius Martialis (fl. third century CE). There were also medical entries in the *Etymologiae* (*Etymologies*) by Isidore of Seville (d. 636).[32] An early anonymous translation of Dioscorides existed in Latin, too, though it appears to have had rather limited circulation.[33] Dioscorides' writings also influenced some early medieval illustrated Latin herbals, such as the *Liber medicinae ex herbis femininis* (*Book of Medicine from Feminine Herbs*) and the *Curae herbarum* (*Cures from Herbs*), though the origins of these texts (and their illustrations) remain unclear.[34]

The robust medical culture, a "school" (in the broadest sense), in the Southern Italian city of Salerno between the eleventh and thirteenth centuries, however, transformed Latin medicine and pharmacology. It developed organically from the multilingual medical culture already present in Southern Italy, particularly in Campania and Calabria, sustained by monasteries such as Montecassino as well as by Arabic, Greek, and Jewish communities that maintained contacts across the wider Mediterranean world.[35] The region also saw increasing connectivity with the Islamicate world through trade connections and the establishment of Islamic polities in Sicily and Apulia.[36] By the eleventh century, we find Salernitan practitioners and teachers authoring therapeutic handbooks, commentaries, and compilations, such as the *Passionarius* (*Book of Diseases*) by Gariopontus of Salerno, which had been drawn from a variety of classical and late antique sources.[37] A monk named Constantine the African (d. after 1098 or 1099) at Montecassino also made a number of important Latin translations of Arabic medical texts, most notably the *Pantegni Practica* adapted from the medical encyclopedia by 'Alī ibn al-'Abbās al-Majūsī.[38] The interests and writings of Salernitan masters became increasingly theoretical and "scholastic" in the twelfth century.[39] Of particular significance among Salernitan works on pharmacology was the *Circa Instans* written by Matthaeus Platearius in the twelfth century.[40] This alphabetically arranged herbal focused on the therapeutic properties and humoral qualities of around 270 substances. In the thirteenth century, a version of this text called the *Tractatus de herbis* was extensively illustrated, with many pictures created *ex novo* that demonstrate close observation of nature (see chapter 7). Other important pharmacological Latin works from the later Middle Ages include the *Antidotarium Nicolai*, the herbals ascribed to Macer Floridus, and translations of an Arabic work attributed to Serapion the Younger, which was itself a compilation from several Arabic pharmacological texts.[41] Much later, a translation of Serapion's text from Latin into Paduan Italian was extensively illustrated in a deluxe illustrated herbal for Francesco Carrara the Younger between 1390 and 1404 (see chapter 8).[42]

By the thirteenth century, medical education in Salerno became more formalized, with the adoption of a central core curriculum through the *Articella* "textbook" and by the implementation of regulations and certifications for medical professionals. An actual university, however, was not formally organized until much later, by which time Salerno's preeminence in Latin medicine had waned.[43] Still, the emergence of a robust medical culture of Salerno foreshadowed the subsequent development of medical schools in the thirteenth century at universities, such as those at Padua and Montpellier, which would come to have a greater impact on the development of medical botany in late medieval and early modern times.

Botanical Visual Knowledge in the Middle Ages

Throughout this period, illustrated herbals continued to be produced. As in antiquity, these illustrated texts were rare; most herbals were unillustrated.[44] Illustrated herbals were expensive and difficult to produce—yet they were still sought out and copied. Medieval botanical illustrations may have fulfilled a variety of functions, ranging from offering practical assistance in the identification and study of plants to providing a visual means for philosophical reflection and speculation about the natural world. The idea that all illustrated medieval herbals were simply luxury objects for book collectors and were therefore not studied for their pictures is indefensible. Already in the sixth century, we find Cassiodorus recommending that monks study Dioscorides in part because of its illustrations.

The medieval Mediterranean world remained interested in the cultivation of visual knowledge. In Byzantium, the eighth and ninth centuries had seen an intensification of debates over the capacity and necessity of images for creating and conveying knowledge. The victory of iconophile factions in favor of religious icons in 843 could have resulted in greater confidence in pictures as a basis for knowledge. Iconophile authors justified their use of religious images on the grounds that they were necessary for thinking and learning about the divine. In his first oration on images, John of Damascus notes:

> If, therefore, the Word of God, in providing for our every need, always presents to us what is intangible by clothing it with form, does it not accomplish this by making an image using what is common to nature and so brings within our reach that for which we long but are unable to see? A certain perception takes place in the brain, prompted by the bodily senses, which is then transmitted to the faculties of discernment, and adds to the treasury of knowledge something that was not there before.[45]

A little later, Theodore the Studite (d. 826) added that all representative thought necessarily occurred through images.[46] In other words, the mind needs images. Such arguments could work just as well for the justification for botanical icons, as they did for religious ones. But it remains difficult to say how the iconophile triumph in 843 would have impacted scientific illustration. There is no reason to suppose that iconoclasts would have opposed scientific image-making in the first place. Paul Magdalino has argued that iconophile humanism tended to downplay the natural sciences, particularly astronomy and astrology—fields of study that iconoclast intellectuals had actively pursued.[47] If anything, iconophiles may have been more uncomfortable with the depiction of nature than iconoclasts were. Henry Maguire has shown that iconophile victories resulted in the elimination of natural imagery from Byzantine religious artworks, because iconophiles had associated nature with idolatry and had opposed iconoclastic espousals of symbolic and allegorical imagery drawn from nature.[48]

And yet surviving illustrated herbals from Byzantium after 843 exhibit their makers' careful attention to botanical form and its diversity. The late ninth- or early tenth-century manuscript of Dioscorides now in the Morgan Library in New York contains hundreds of botanical illustrations created and compiled from different sources.[49] This manuscript, which I cover in the next two chapters, demonstrates contemporaries' interest in detailed and accurate botanical illustrations, some of which may have been created through the observation of nature. Byzantine botanical illustration was therefore clearly not subjected to the same strictures and discomforts that followed depictions of the natural world in religious art. Medicine, too, may have been especially favored by iconophiles. We have, for example, seen the iconophile patriarch Photios' espousal of Dioscorides for speculations on "philosophy and natural science," and his display of botanical erudition to Zacharias.[50]

It is hard to say how these developments in Byzantium correlate to attitudes in the Islamicate world. Scant evidence for botanical illustration in Arabic manuscripts survives prior to the eleventh century. This situation is not unique to botanical texts; very few manuscripts survive from the earliest centuries of the Arabic book in general.[51] This is perhaps due not only to the fragility of paper and papyrus, and the vicissitudes of transmission, but also to a broader book culture in which new copies and interpersonal oral transmission were especially prized. Early scientific treatises in Arabic were, of course, illustrated, as demonstrated by the occasional and rare survival of fragments from learned treatises.[52] The illustration of scientific and philosophical texts may even have provided a major stimulus for the development of illustration in Arabic books at this time.[53] And like Cassiodorus, some Arabic authors clearly assumed that Dioscorides had himself illustrated his writings. In his *Fihrist* (*The Book Catalog*), Ibn al-Nadīm quotes the historian Yaḥya al-Naḥwi's praises Dioscorides' illustration of plants.[54] Ibn Abī Uṣaybiʿa similarly states in his *ʿUyūn al-anbāʾ fī ṭabaqāt al-aṭibbāʾ* (*The Best Accounts of the Classes of Physicians*) that Dioscorides illustrated plants in the course of his botanical inquiries.[55] From the perspective of these authors, botanical illustrations were clearly thought to have been a part of Dioscorides' original work.

We should also keep in mind that scholars in the Islamicate world could have used illustrated scientific books in other languages, notably Greek and Syriac, especially before the translations of the Abbasid period. The Old Paris Dioscorides contains many annotations in early Greek cursive minuscule letters that transliterate Arabic plant names (see, for example, figs. 4.7 and 4.11).[56] These annotations put the Old Paris Dioscorides in a ninth-century Graeco-Arabic context.[57] Marie Cronier has suggested that the annotator may have been a Greek-speaker with an advanced command of Arabic.[58] The annotations appear to have preceded the illustration of the codex, which could suggest that they played a role in illustration, and were perhaps made by the same individual who copied the text.[59] Later annotations in Arabic script both transliterate the Greek plant names into Arabic letters and give the Arabic plant names.[60] That these later annotations are often concentrated around the illustrations rather than the text testifies to their maker's regard for the pictures as a focal point and reference for the text. This pattern of annotating illustrated Greek

Dioscorides texts in other languages occurred many times over the course of the Middle Ages and into the early modern period.[61] Conversely, we also occasionally find Greek annotations in Arabic translations of Dioscorides, as seen in the Greek plant names that were added to an Arabic Dioscorides now in Leiden (fig. 5.1). The colophon to this manuscript, dated to 1083 (AH 475), notes that it was a copy of a rectification of the Arabic Dioscorides by al-Nātilī, who not only reworked it in 990 (AH 380) but also executed its illustrations.[62] The Greek annotations were evidently added perhaps as late as the fifteenth century.[63]

Medieval Arabic scholars recognized that ancient botanical illustrations could convey useful knowledge. The polymath Ibn Waḥshiyya (d. AH 318/930–1 CE) reports in his *Kitāb al-filāḥa al-nabaṭiyya* (*The Book of the Nabataean Agriculture*) that there were some 118

FIGURE 5.1 Illustrations of fruit trees with Greek marginalia in a copy of al-Nātilī's rectification of Iṣṭifan's Arabic translation of Dioscorides. Leiden, Universiteitsbibliotheek, MS Or. 289, colophon dated 1083, fols. 46v–47r. Courtesy of Universiteitsbibliotheek Leiden.

ancient pictures among the remains of a book that once had a thousand pictures, each having "curious and useful meanings in many (fields of) knowledge."[64] For example, a picture of a "vine of convalescence" shows "many secrets and hidden things which everyone should know":

> He [Dawānāy, also called the "Drawer" (al-Muṣawwir)] drew a large and wide vine with many branches which have become entangled with each other so as to become circles, 49 in all which is seven times seven. In each of these circles, there is a picture of bunches of grapes hanging from the branches of the vine. . . . On the upper side of the vine he drew a fire and below it the earth, in front of it on the right he drew air and to the left water. In each circle he drew the picture of some animal which is harmful to the vine and is its enemy. With this he gave us the useful piece of information, firstly, that the vine has 49 creeping animals which are its enemies and harmful to it, O people. He also drew in it the picture of farmers and all that is needed in caring for the vine. In their hands he drew the tools with which they work in the vineyard and all other things which they need.[65]

Ibn Waḥshiyya clearly believed that ancient pictures could convey varied information. While the picture described may be apocryphal, Ibn Waḥshiyya's description recalls late antique floor mosaics showing inhabited grapevines, as seen at the mid-sixth-century Church of Lot and Procopius at Khirbat al-Mukhayyat in Jordan (fig. 5.2). It is tempting to think that Ibn Waḥshiyya based his account on a similar mosaic. This could suggest in turn that many different images from antiquity could have been analyzed specifically for their scientific content. Ibn Waḥshiyya may have been further inspired by botanical illustrations, such as those in the Old Paris Dioscorides, that show the harvesting and use of plant products (see fig. 4.9).

Ibn Waḥshiyya's writings also highlight the place of horticulture and agriculture in medieval botanical traditions. Medieval rulers across the Mediterranean devoted considerable resources to the planting and maintenance of large garden estates. For example, the Umayyad emir of al-Andalus ʿAbd al-Raḥmān I (r. AH 138–72/756–88 CE) had a large garden estate, called the Munya al-Ruṣāfa, planted outside Córdoba. He had this estate filled with rare and exotic plants.[66] Similar gardens were planted in Abbasid Baghdad and Samarra. In Byzantium, the continuation of the chronicle by Theophanes notes that Basil I (r. 867–86) had a garden planted, called the Mesokēpion,

that abounded "in every kind of plant."[67] By the tenth century we also encounter agricultural compilations in both the Arabic and Greek, including, Ibn Waḥshiyya's early tenth-century *Kitāb al-filāḥa al-nabaṭiyya* in Arabic and the mid-tenth-century *Geoponica* in Greek.[68]

It is in this context of interest in the ancient botanical tradition that we find an illustrated Greek manuscript of Dioscorides serving as a diplomatic gift between Byzantium and Umayyad Iberia. In his *'Uyūn al-anbā' fī ṭabaqāt al-aṭibbā'*, Ibn Abī Uṣaybi'a quotes Ibn Juljul's report that a Byzantine emperor named Armāniyūs (that is, Romanos), sent an illustrated Dioscorides to the Umayyad ruler 'Abd al-Raḥman III (r. AH 316–50/929–61 CE).[69] No one in Córdoba could make sense of the old Greek (*ighrīqī*) language of the book, so Romanos sent a monk named Nicholas (Niqūlā) to help with the identification of the plants.[70] In Córdoba, Nicholas joined a team that included the Jewish physician Ḥasdāy ibn Shaprūṭ, Muḥammad al-Shajjār ("the botanist"), someone known as al-Shabānisī, Abū 'Uthmān al-Jazzār al-Yābisa (perhaps a misspelling for al-Yābisī, "from Ibiza"), Muḥammad ibn Sa'īd al-Ṭabīb, 'Abd al-Raḥmān ibn Isḥāq ibn Haytham, and the Sicilian pharmacologist Abū 'Abd Allāh al-Ṣiqillī. The group worked together to translate the text. At that time, Isṭifan's translation of Dioscorides would have been the main

FIGURE 5.2 Floor mosaic of grapevine inhabited by animals and humans. From the Church of the Holy Martyrs Lot and Procopius, Khirbat al-Mukhayyat, Jordan. Photo: Sean Leatherbury/Manar al-Athar.

Arabic translation available.[71] Yvette Hunt has suggested that ʿAbd al-Raḥmān III's request may have been intended to show the insufficiency of this earlier Abbasid translation.[72] As the identification of ancient plant names provided a substantial obstacle to the use of the text even in Greek, Maria Mavroudi has suggested that Nicholas may have been called upon for his pharmacological expertise.[73] The ʿUyūn al-anbāʾ adds that the Sicilian Abū ʿAbd Allāh al-Ṣiqillī "spoke Greek and recognized the characteristics of the drugs."[74] Besides the Sicilian's knowledge of drug properties, the passage might indicate his familiarity with plant morphology—and perhaps even the illustrations that typically accompanied Dioscorides.[75]

The close relationship between the Arabic and Greek botanical traditions is evident, too, in illustrated versions of Dioscorides in Arabic. Scholars have long suggested that the illustrations from the Greek original Dioscorides were adapted directly for some illustrated Arabic Dioscorides manuscripts.[76] For example, a twelfth-century copy of an Arabic redaction (taḥrīr) of Dioscorides now in Paris, the only extant Arabic Dioscorides on parchment, has a number of illustrations that closely resemble those in the Old Paris Dioscorides.[77] Edmond Bonnet has shown that almost nine-tenths (89 percent) of the chapters have pictures similar to those in the Old Paris Dioscorides, while about two-fifths (39 percent) of those are nearly identical.[78] For example, the illustration of ṭrāghūs (fig. 5.3), an Arabic transliteration of the Greek tragos, closely resembles the illustration from the Old Paris Dioscorides (fig. 4.8).[79] In this case, the reason for the plant's appearance, the fact that the Greek word tragos also means "he-goat," was lost. Although the illustration follows the tradition of the Old Paris Dioscorides in showing tragos as a peculiar plant, Arabic botanists would later identify it as a barley product.[80] And while the illustration of a lupine (fig. 5.4, Arabic turmus, Greek thermos) belongs to a part of the text that is no longer extant in the Old Paris Dioscorides, it nevertheless closely resembles the illustration from the Alphabetical Dioscorides (fig. 5.5).[81] While these similarities speak to connections between the traditions, the history of the illustrated Arabic Dioscorides, however, proves to be more complicated than a simple reception of earlier Greek pictures. Some scholars have supposed, for example, that the Old Paris Dioscorides may have been illustrated on the basis of an earlier illustrated Arabic Dioscorides.[82] As we have already seen here, the Old Paris Dioscorides carries earlier Greek minuscule transliterations of Arabic plant names from around the same time as its creation that could have served as instructions to the illustrator—a possible indication that some illustrations were derived from an Arabic source. As Marie Cronier has noted, this Arabic source would have had different plant names than those found in later Arabic translations.[83]

The legacy of the illustrated Dioscorides on the medieval Latinate Mediterranean is less evident than it is in the Byzantine and Islamicate worlds. Before the thirteenth century, if Latin scholars and physicians looked for an illustrated herbal, they would have turned to the *Herbarius* of Ps.-Apuleius Platonicus (see fig. 2.1), as well as the *Liber medicinae ex herbis femininis* and the *Curae herbarum*.[84] The illustrations

FIGURE 5.3 Illustration of *ṭrāghūs* and oat (*brūmūs*). Paris, Bibliothèque nationale de France, MS ar. 4947 (the "Parchment Arabic Dioscorides"), twelfth century, fol. 23v. Photo: Bibliothèque nationale de France.

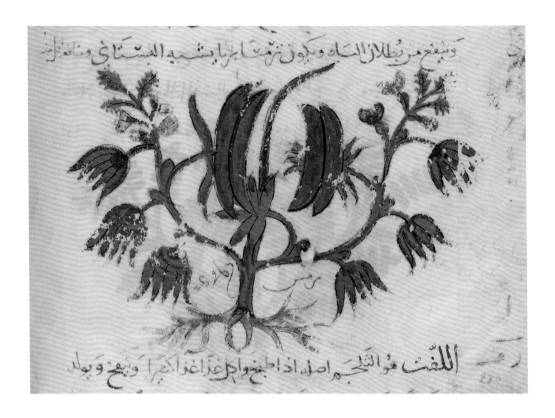

in these works descend from ancient herbals, though much changed and adapted for their new contexts. By the late thirteenth century, we encounter new illustrated herbals in Latin, most notably the illustrated *Tractatus* manuscripts. At that time we also find evidence for Latin reception of the Alphabetical Herbarium in an unusual codex made around 1300, now part of the Thott Collection in the Royal Danish Library in Copenhagen and published by Alain Touwaide (fig. 5.6).[85] I will return to this manuscript in chapter 8. It is worth noting here, however, that the Thott Codex only has pictures from the Alphabetical Herbarium and none of the text.

Illustrated Latin translations of Dioscorides were apparently rare in the early and high Middle Ages. Most early medieval Latin copies of Dioscorides are unillustrated, though the fact that Cassiodorus recommends Dioscorides partly on account of its illustrations could suggest that once there were more illustrated early medieval copies of Dioscorides in Latin than have survived. Only one illustrated copy of an early Latin translation of Dioscorides is known to have come down to us today.[86] Now kept in the Bavarian State Library in Munich, this manuscript was likely produced in Southern Italy (Naples?) in the late tenth century, and intriguingly has Greek numerals running beside each chapter (fig. 5.7). Scholars have speculated that the illus-

FIGURE 5.4 Illustration of a lupine. Parchment Arabic Dioscorides, fol. 28v. Photo: Bibliothèque nationale de France.

FIGURE 5.5 Illustration
of a lupine. Vienna
Dioscorides, fol. 135r.
Courtesy of Österreichische
Nationalbibliothek.

trations were influenced by illustrated Byzantine herbals, such as the
Morgan Dioscorides. The Munich Dioscorides includes, for example, illustrations of trees, branch-like plants, and oils, all of which are
present in the Morgan Dioscorides but absent in earlier extant copies
of Dioscorides. With some squinting, we can see similarities between
some of the illustrations in the Munich Dioscorides and the Greek
Alphabetical Herbarium, such as the depiction of the lupine (fig. 5.8,
compare to fig. 5.5).

Besides their striking stylistic and aspectual modifications, already
evident in the sixth-century Leiden *Herbarius*, early and high medieval Latin herbals often include additional figures that tend to act as
mnemonics related to each substance's medicinal effects, as seen in the
vomiting marginal figure from the Munich Dioscorides (fig. 5.7).[87]
Such inclusions hint that the illustrated Latin herbals were perhaps
used more for their accounts of the therapeutic uses of *materia medica*
than for the identification and close study of plants.

malû terre.

FIGURE 5.6 Latinate copy of Byzantine illustrations of cyclamen, left, and eryngium, right, with marginal illustrations of a swallow and a peacock. Copenhagen, Kongelige Bibliotek, MS Thott 190 4° (the "Thott Codex"), c. 1300, fols. 46v–47r. Courtesy of Det Kongelige Bibliotek.

pringue.

quon si est · t · sant ou
scra qui lise coneio ha
la cous si il de grant
bonte · e qui il eſpar
che il aie · e quant il
ha sait il ſera pris
de celuy bien · e cau
me uere chauche les
ens de coſtuy oiſeau
uaut a ſaiſir eſtendre
mp · e si le paon ch nuer
poi de la char de poitrine ne puer ch
com si fuſſe pleſne de bons ſp
ces · la careile eſt · t · breuage de au
le cuer qui le puet auoir bien uola
de · e lengorte · t · boit le ſane ſa
le ſiable · sc buato · e le mener se
eauencences chaſes · t · mils mauuai
ſe · a quat lom le cuer ge le paro s
rifſe de diſſenteria · t · ſoucto ſes en
rifſe de mauuaiſe mel

FIGURE 5.7 Illustration of white hellebore accompanied by a figure illustrating its properties. Munich, Bayerische Staatsbibliothek, Cod. Clm. 337 (the "Munich Dioscorides"), late tenth century, fol. 123v. Courtesy of Bayerische Staatsbibliothek München.

FIGURE 5.8 Illustration of a lupine and a snake. Munich Dioscorides, fol. 54r. Courtesy of Bayerische Staatsbibliothek München.

Illustrated copies of Dioscorides' *De materia medica* in Arabic also often feature additional figures that interact with the illustrations of plants. A warrior and a sage or physician flank betony, for example, in an illustrated leaf today in the Harvard Art Museums (fig. 5.9).[88] This folio is one of thirty-one folios that were removed from a manuscript in the Aya Sofya Library in Istanbul before 1910.[89] This manuscript was copied in 1224 (621 AH) by the scribe ʿAbd Allāh ibn al-Faḍl, perhaps in Baghdad. Unlike the vomiting figure in the Munich Dioscorides,

أو من ذراع أو أكثر مربع وورق طوال لينة شبيهة في شكلها

مة ؟ للموامشة فه طبه الراحة ومالي الأرض من

لورق هو عظم من ساير الورق دعاجرن النبات بزر

صورة نظرا

قرب من أح النبات شبه بالدي للشعر الذي

قاله ثمرا وورق هذا النبات ينبغي أن يجمع وأن يجفف

FIGURE 5.9 Warrior and physician with betony, folio from a manuscript of Dioscorides' *De materia medica* in Arabic, copied in 1224. Folio is now in the Harvard Art Museums (Object no. 1960.193). Harvard Art Museums/Arthur M. Sackler Museum, Bequest of Abby Aldrich Rockefeller; photo © President and Fellows of Harvard College.

the two figures here do not indicate the medicinal properties of the betony, but rather draw attention to the plant and enhance the scene's human interest.[90] With their detailed, vibrant garments, they prompt us to wonder who they are and why they are interested in this rather plain-looking plant. These two figures further echo the incorporation of bystanders and scenes of daily life in other, roughly contemporary Baghdadi manuscripts.[91]

Institutions

New institutions in the medieval Mediterranean also ensured the transmission of medical and botanical knowledge. Already in late antiquity, monasteries and Christian charitable foundations had undertaken to provide care for the sick. These charitable institutions included guesthouses, poorhouses, and sickhouses, variously called in Greek *xenōnes*, *xenodocheia*, and *nosokomeia*. The degree to which these charitable foundations were medicalized, especially during the early Byzantine period, has been the subject of considerable debate.[92] In most cases, such institutions only offered what we would call palliative and hospice care. Christian monasteries in the Near East, however, became particularly associated with the Greek medical tradition in the first centuries of Islamic rule.[93] Umayyad rulers initially maintained or emulated preexisting Byzantine foundations.[94] Hospitals in the Islamicate world, called *bīmāristānāt* (sing. *bīmāristān*) from the Persian for "places for the sick," begin to appear over the course of the ninth and tenth centuries.[95] By the mid-tenth century, a network of *bīmāristānāt* spread out over the city of Baghdad.[96] Itinerant physicians extended the reach of this network throughout the region of lower Iraq.[97] Baghdadi *bīmāristānāt* inspired others across the Islamicate world.[98]

While Baghdadi *bīmāristānāt* no longer survive, the Nūrī Bīmāristān in Damascus does. It provides us with a concrete example of a *bīmāristān*, and an architectural context in which we can imagine medical teaching and practice.[99] Established in 1154 (548 AH) by the Zangid sultan Nūr al-Dīn Maḥmūd ibn al-Zankī (r. AH 541–69/1146–74 CE), this lavish *bīmāristān* centers on a spacious cross-axial courtyard with a central fountain and four adjoining open "rooms" called iwans. The complex was once equipped with running water, a bathhouse, and latrines. Ibn Abī Uṣaybiʿa gives us a sense of the daily routines of physicians there.[100] In the morning, a physician would examine patients. He would then retire to his home or to the main iwan of the *bīmāristān* to read or copy medical texts, lecture on them, and discuss them with students. The cross-axial, four-iwan plan was especially well suited to teaching and was widely adapted in madrasa architecture. Ibn Abī Uṣaybiʿa also notes that Nūr al-Dīn donated valuable medical texts to his *bīmāristān* that were stored in cupboards in the main iwan. These books would have served not only for reference but also for teaching. Ibn Abī Uṣaybiʿa notes, for example, that his teacher Muhadhdhab al-Dīn al-Dakhwār (d. AH 628/1230 CE) "never taught anyone unless there was a copy of that book at his disposal for the student to read."[101] Qurʾanic passages referring to sickness and healing that embellished the main iwan's interior also linked medical practice there to Qurʾanic tradition.[102] The decoration of the main entrance of the complex, however, alluded to the pre-Islamic origins of medicine. A spoliated antique lintel over the building's main entrance seems to acknowledge the ancient origins of the medical tradition, while the muqarnas above the lintel reflect Islamicate architectural precedents that speak to the medical traditions of Iraq (fig. 5.10).[103]

An idealized image of botanical instruction appears in a famous illustrated Arabic redaction of Dioscorides now in the Topkapi Library in Istanbul (fig. 5.11).[104] The manuscript was copied in 1228 (626 AH) for Shams al-Dīn Abuʾ-l Fadāʾil Muḥammad, an otherwise unidentified ruler from the Jazira region of northern Iraq and Syria.[105] The scribe responsible for the manuscript, one Abū Yūsuf Bihnām ibn Mūsā al-Mawṣilī (that is, from Mosul), claims to have been educated in medicine.[106] In a note inserted into the Parchment Arabic Dioscorides in Paris, he further identifies himself as a Christian (*al-Masīḥī*).[107] The Topkapi manuscript's frontispieces famously recall Byzantine author portraits, including the frontispieces of the Vienna Dioscorides (fols. 1v–2v, figs. 1.3 and 1.4).[108] As a result, scholars have supposed that the frontispieces were made either by Byzantine artists or in emulation of Byzantine traditions of illustration. At the same time, however, the Topkapi frontispieces emphasize the transmission of knowledge from master to pupil in line with Arabic traditions of author portraiture.[109] Unlike the illustrations from the frontispieces of the Vienna Dioscorides that focused exclusively on Dioscorides' discovery of the mandrake, we see here that, while the student holds a mandrake, Dioscorides presents another (now badly flaked) plant. The scene invites us to weigh the merits of one prototypically powerful plant against those of another. The image recalls another anecdote from Ibn Waḥshiyya's *Kitāb al-filaḥā al-nabāṭiyya*, in which the marshmallow plant (*khiṭmī*) requests Shabāhā al-Jarmaqānī to ask the magicians of Babylon (*Bābil*) whether she or mandrake (*al-yabrūḥ*) is deserving of higher station. Although the Babylonian magicians side

FIGURE 5.10 Exterior façade and entrance of the Nūrī Bīmāristān, Damascus. © Creswell Archive, Ashmolean Museum, neg. EA.CA.5496.

FIGURE 5.11 Frontispiece showing Dioscorides with a student. Istanbul, Topkapi Library, MS Sultanahmet III 2127 (the "Topkapi Dioscorides"), made 1228, fol. 2v. Photo: akg-images (Werner Forman, Roland & Sabrina Michaud Collection).

with the mandrake, marshmallow actually finds this to be a ruling in her favor. They only honor the mandrake out of fear, she reasons. Their fear of the mandrake proves its evil and confirms her better nature.[110] It is tempting to think that Dioscorides here once held a plant similar to marshmallow that may be more useful in everyday medicine than mandrake. The broader lesson is that physicians should select reliable remedies appropriate to the needs of their patients, rather than famous and powerful ones.

We also begin to have greater evidence for the medicalization of hospital facilities at Byzantine monastic institutions by the early twelfth century.[111] A foundation document or *typikon*, dated to 1136, for the Pantokrator monastery in Constantinople, founded by Empress Eirene and her husband John II Komnenos, gives instructions for the operation of a public medical hospital, elderly home, off-site leper sanatorium, and private monastic infirmary.[112] Notably, among the hospital's personnel, the *typikon* provides for a teacher of medicine to "teach the student doctors of the hospital the knowledge of medicine in a consistent and zealous manner."[113] Alexandre Philipsborn has speculated that the creation of this position may have followed the model of medical schooling at *bīmāristānāt*.[114] Later Byzantine monasteries continued to offer medical care and instruction. By the fourteenth and fifteenth centuries, the monastery of St. John the Forerunner in the Petra district of Constantinople, particularly its Xenōn of the Kral, founded by the Serbian king (or "kral") Stefan Uroš II Milutin (r. 1282–1321), appears to have become a hotspot for the study of medical botany.[115] Several deluxe illustrated Dioscorides manuscripts wound up in this monastery, where they acquired marginalia and were rebound and copied.[116] As a result, the monastery fostered research in medicine and botany in the last days of the Byzantine state.[117] It was in this context that illustrations from the Morgan and Vienna Dioscorides manuscripts were copied into a paper codex in the mid-fourteenth century, now in Padua (fig. 5.12). In the fifteenth century, Isidore of Kiev directed the copying of illustrations from both manuscripts into a deluxe parchment codex that is now in the Chigi collection in the Vatican Library (see chapter 6).[118]

In the early fifteenth century, while the Vienna Dioscorides was in the library of the Petra monastery, a monastic nurse (*nosokomos*) from the Xenōn of the Kral requested the scholar John Chortasmenos to repair and rebind the codex.[119] Chortasmenos and his team made additional repairs, rematched pictures and text, noting instances where no picture was available, and transliterated its uncial text into minuscule.[120] These modifications and repairs speak to the ongoing use of

FIGURE 5.12 Folio from a fourteenth-century copy of Dioscorides. Padua, Biblioteca del Seminario, Cod. 194 (the "Padua Dioscorides"), fourteenth century, fol. 190r. Courtesy of Biblioteca Antica del Seminario di Padova.

the codex during this period and interest in its pictures. At this time, we also find that the personifications of *heuresis* and *epinoia*, and the depiction of the Byzantine princess Anicia Juliana, in the frontispiece illustrations were all relabeled, perhaps unintentionally, as personifications of Wisdom (*sophia*). In this reworking, it is Wisdom personified and not Discovery that reveals the mandrake to Dioscorides (fig. 1.3). Wisdom helps him and his painter record their findings (fig. 1.4). Finally, Wisdom, and not Anicia Juliana, appears enthroned, flanked by the virtues of Magnanimity (*Megalopsychia*) and Prudence (*Phronesis*), while an Eros now labeled "Desire for Wisdom" (*pothos tēs sophias*) holds open a book to receive Wisdom's largesse. This reworking of the frontispieces as a narrative about Wisdom hints at the expansive and elevated role that medicine assumed among the sciences in the late Middle Ages. To Chortasmenos' contemporaries, the scene of Wisdom enthroned may have recalled figurations of Divine Wisdom from Proverbs (9:1–5), as celebrated in the proliferation of Holy Wisdom imagery in late Byzantine churches.[121] We find a similar exaltation of medicine in a set of frontispieces, completed sometime between 1341 and 1345, for a collection of Hippocratic texts now in Paris.[122] The frontispieces feature the manuscript's recipient, Alexios Apokaukos, in dialogue with Hippocrates. In verses surrounding the frontispieces, Apokaukos describes medicine as "the most powerful of the sciences."[123] He explains that he undertook the study of medicine in order to "learn the plans of God."[124] As Joseph Munitiz notes, medicine is here presented "almost as if it was a synonym for philosophy—embracing all creation and theology."[125]

The exaltation of medicine as a divine science also appears in roughly contemporary medical writings outside of Byzantium.[126] We can cite, for example, a frontispiece from a mid-fourteenth-century adaptation of the *Tractatus de herbis*, called the *Liber de herbis et plantis*, by Manfredus of Monte Imperiale, which is now in Paris (fig. 5.13).[127] Manfredus notes that he both copied and illustrated the text, with many of the illustrations based on earlier *Tractatus* manuscripts. In the frontispiece, we find a seated man, perhaps Manfredus himself, with a bevy of students clamoring toward him, waving betony and feverfew in the air.[128] These two plants were frequently used in the medieval Latin herbal tradition. Betony had the honor of having the longest text devoted to it in the manuscript.[129] The seated master admonishes the crowd: "abstinence is the first and last medicine for body and soul."[130] The Hand of God emerges above, adding: "but test all things; hold fast that which is good...."[131] This passage from Thessalonians I (5:21–22) elevates Empiricist medicine as holy writ.[132]

FIGURE 5.13 Frontispiece to Manfredus de Monte Imperiali, *Liber de herbis et plantis*. Paris, Bibliothèque nationale de France, MS lat. 6823, c. 1330–40, fol. 1r. Photo: Bibliothèque nationale de France.

Manfredus de monte imperialj
de herbis &c

334

4996

6823.

Omnia probatu: Quoniu est venece.
fugiet ergo: Quoniu est incipibiletemalis.

Prima vultima medicina apter corpus
animu est abstinentia

Making Illustrated Herbals in the Middle Ages

The production of illustrated books was also transformed in the Middle Ages. Over the course of late antiquity, monasteries and monastics increasingly controlled significant aspects of the book trade. In late antique Egypt, monastics were active in all stages of book production.[133] The large libraries, philosophical schools, temples, and book dealers of the ancient world fade from the material record.[134] In surviving documentation of book production from late antique Egypt—including personal letters, lists of books, church inventories, and accounts of monastic collections—Chrysi Kotsifou found "no reference to pagan or even secular works whatsoever after the fourth century."[135] Clearly, such works were still produced and read, but they apparently took a back seat to the copying of religious texts. These patterns may echo what happened elsewhere in the Mediterranean, when monastics controlled book production.[136] Still, large urban centers, such as Constantinople, may have maintained libraries and scriptoria like those of the classical world longer than other parts of the Mediterranean. Some monasteries and monastics may have specialized in the copying of technical treatises.

Over this time, the process whereby herbals were illustrated shifted. While I noted in earlier chapters that ancient herbals were typically illustrated prior to the copying of text, medieval herbals were typically illustrated after the copying of text.[137] The text-first mode of illustration may have replaced other systems of illustration because the faithful transmission of texts remained the main priority in the production of most books, especially religious texts. As monastic scribes streamlined book production to privilege textual transmission, they may have been less likely to make special accommodations for pictorial transmission.[138] But monasticism had different impacts on book production in the medieval Latin West and in Byzantium. In Latin monasteries, book production largely took place in scriptoria. These enabled a degree of specialization and collaboration between monastic scribes and illustrators.[139] In Byzantium, however, copyists tended to remain as independently contracted labor.[140] Although Byzantine monasteries sometimes did support scriptoria, such as the Stoudios Monastery in Constantinople, there were also many lay copyists.[141] Throughout the Middle Ages, Latin and Greek scholars and physicians also commissioned scribes or acted as copyists and illustrators for their own copies.[142] Scholars also hired and collaborated with painters to illustrate their own manuscripts.[143] These modes of production may have accelerated in the later Middle Ages, due to the availability of paper and developments in medical professionalization. In the Latin West, the rise of universities also resulted in the expansion of the book market and book production by lay copyists, perhaps even those with some medical specialization.

Book production in the Islamicate world developed along different lines. By the ninth century, we find independent, professional stationers—*warrāqūn*, the makers and sellers of "leaves," and *nāskhūn*, or copyists.[144] These trades proliferated with the rise of paper production in the eighth and ninth centuries.[145] Copyists maintained shops or were itinerant. Larger libraries and madrasas also employed or provided space for copyists.[146] An illustrated Arabic Dioscorides now in Oxford, for example, was completed by its scribe in 1240 (637 AH) at one of the Niẓāmiyya madrasas, named after the Seljuq vizier Niẓām al-Mulk (d. AH 485/1092 CE).[147] In other cases, commissioners provided room and board to copyists for long-term publishing assignments.[148] We also encounter specialist copyists knowledgeable in medicine, such as Bihnām, the copyist of the Topkapi Dioscorides.

Despite notable differences in the production of Latin, Greek, and Arabic herbals, some practices were shared. Typically scribes obtained the writing supports, either parchment or paper, determined the page layout, and copied the text first. They might also be responsible for ornamentation and even the execution of underdrawings. In deluxe manuscripts, scribes would have directed and subcontracted illustration by professional painters.[149] In other cases, text and illustrations were executed by the same people, as seen in Manfredus' herbal and al-Nātilī's rectification of Dioscorides.[150] The widespread adoption of paper—beginning in the eighth century in the eastern Islamicate world, and spreading westward between the eleventh and fourteenth centuries—meant that herbals could be copied more cheaply. Yet some illustrated herbals such as the Chigi Dioscorides and the Parchment Arabic Dioscorides are on parchment even though paper was more common when they were made. This choice suggests that their makers considered the illustrated herbal a special kind of book that was meant to last. In a world dominated by paper, parchment would have given these volumes an antique air.

Evidence for picture-first modes of herbal production and illustration reappears by the end of the thirteenth century. An early example of this picture-first mode of illustration in a Dioscoridean herbal after late antiquity can be found in the Thott Codex, dated to c. 1300, which has illustrations from Byzantine manuscripts of Dioscorides (fig. 5.6), and again in the mid-fourteenth-century Padua Dioscorides (fig. 5.12).[151] In both codices, texts were added only after the copying of illustrations, if at all. As in the ancient herbal, this mode of production emphasizes the transmission of pictures over text, with the result that text is often lost or truncated.

Early evidence for the separation of text and pictures into distinct volumes can be found in a note at the beginning of an Arabic Dioscorides manuscript dated to 1219 (616 AH).[152] The note states that a separate volume with pictures of "plants, trees, animals, and minerals" was prepared "in order to facilitate access to them for . . . whoever needs to know something of that sort."[153] The note adds, however, that the pictures were still not entirely without text, as they included the "name of the plant, its strength, and some of its effects." The volume of illustrations does not survive. George Saliba and Linda Komaroff suggest that it may have had abridged texts, as seen in some other illustrated Arabic Dioscorides manuscripts. This parallels developments already evident in ancient illustrated herbals, such as the Antinoopolis Codex. While it is tempting to view this situation as the continuation of a practice established in those earlier herbals, it could have arisen simply from the privileging of pictures over text. The distinction between an illustrated and an unillustrated volume nevertheless suggests that contemporaries recognized two kinds of botanical literature. A logical development of this distinction comes in the form of late medieval "botanical atlases," volumes of botanical illustrations that lack all explanatory text. A single manuscript leaf from an Arabic Dioscorides dated to the thirteenth century and currently in Toronto may represent an early example of this phenomenon. [154] A number of botanical atlases survive from Northern Italian and Byzantine contexts from the fourteenth and fifteenth centuries.[155] Such volumes may have functioned as visual reference works, perhaps used in conjunction with other pharmacological texts, especially botanical lexica.[156] The existence of these volumes testifies to late medieval interest in botanical illustrations as a way of conveying specifically visual knowledge.

Disentangling patronage and readership in the case of medieval illustrated herbals in Greek, Arabic, and Latin remains a difficult task. Evidence is often missing or incomplete. We have seen that physicians and scholars, such as Manfredus or Isidore of Kiev, sometimes copied or commissioned manuscripts for personal use. Elites were often the recipients (and commissioners) of especially deluxe books, as we have seen in al-Nātilī's rectification of Dioscorides for a Samanid prince and Bihnam's redacted copy for Shams al-Dīn. We have also seen how Alexios Apokaukos, the wealthy recipient of a Hippocratic manuscript now in Paris, had himself depicted opposite Hippocrates

FIGURE 5.14 Frontispiece to the *Kitab al-diryāq* illustrating courtly life. Vienna, Österreichische Nationalbibliothek, Cod. A.F. 10 (the "Vienna *Kitab al-diryāq*"), mid-thirteenth century, fol. 3r. Courtesy of Österreichische Nationalbibliothek.

FIGURE 5.15 Frontispiece to the *Kitab al-diryāq* with portraits of famous physicians.
Vienna *Kitab al-diryāq*, fol. 3v. Courtesy of Österreichische Nationalbibliothek.

in the frontispieces to that manuscript. To these examples we can add two deluxe illustrated copies of an Arabic toxicological treatise called the *Kitāb al-diryāq* (*Book of Antidotes*). The earliest of the two, now in Paris, was initially copied in 1199 (595 AH) for the library of a learned *imām* about whom little is known.[157] Oya Pancaroğlu has pointed out that toxicology was part of "general learned discourse" with some practical utility for prominent people at risk of being poisoned.[158] Although the original recipient of the other manuscript, dated to the mid-thirteenth century and now in Vienna, remains unknown, one of its frontispieces depicts a lively court scene: in the center, an attendant grills meat before his patron, while gardeners toil out back (fig. 5.14).[159] A hunt occupies the upper register, while a procession fills the lower. The other side of the folio, however, bears the staid frontal portraits of nine physicians (fig. 5.15). Medical practice and palace life are here literally two sides of the same folio. While medicine supported and enriched courtly life, courtly patronage was also vital for sustaining and encouraging the medical tradition. As the twelfth-century Egyptian physician Ibn Jumayʿ (d. AH 594/1198 CE) wrote to the Ayyubid sultan, Ṣalāḥ al-Dīn (r. AH 570–89/1174–93 CE), "The first and most important ground [for the renewal of medicine] is the prince's concern for it."[160] By supporting the art of medicine through its teachers and its students, sovereigns demonstrated their wisdom, liberality, and concern for the wellbeing of their subjects.

Regardless of their origins and original recipients, scientific works often end up in the hands of medical scholars and institutions. That we know so little about the production and patronage of illustrated herbals speaks to the ways they circulated over the centuries. An illustrated herbal is preeminently useful, or at least promises utility. Over the long history of its use, it is subject to damages, especially to its front and back, the usual location of colophons and ownership marks. As it passes from owner to owner, information related to previous owners also becomes less relevant, and may even be an irritation to subsequent owners. The recasting of Anicia Juliana's munificence as an allegory about Wisdom is emblematic of this process of erasure and adaptation to new contexts.

Conclusion

Over the course of the Middle Ages, the ancient botanical tradition changed along with a changing world that saw the rise of Galenic medicine, empirical practice, and new institutions with new patterns of book patronage, production, and circulation. Despite the ruptures and transformations that the Mediterranean world underwent during the Middle Ages, the ancient botanical tradition, including botanical illustration, continued and even flourished. While medieval physicians and scholars held the ancient medical tradition in high esteem, they made their own corrections and innovations. This chapter has only skimmed the surface of what were much deeper and more complicated realities. The next two chapters take a closer look at the dynamic, critical practices evident in the botanical illustration of the Middle Ages.

6

THE CRITICAL
COPY

How useful could copies of ancient botanical illustrations have been to medieval scholars and physicians? Just as the intervening centuries had seen the transformation of texts and languages, so, too, the pictures had changed as they were transmitted from manuscript to manuscript. As Pliny had recognized centuries earlier, the process of copying pictures tends to introduce errors.[1] Misunderstandings of plant morphology, limitations in skills or available materials, and adjustments made to pictures so that they might conform to different stylistic norms could all have caused unintended transformations of visual content.[2] Such transformations would be expected whenever supplies and source manuscripts were scarce. In Northern and Western Europe difficulties identifying ancient plant remedies were compounded by the fact many Mediterranean plants struggled to live there due to differences in climate and terrain.[3] This was less the case, however, for the principal urban centers of the Mediterranean world—Constantinople, in particular.

This chapter considers what happened when copyists had access to multiple source manuscripts as well as many plants from the ancient botanical tradition.[4] In these cases copying provided opportunities for illustrators to improve, elaborate, and even expand the visual corpus. These findings contrast with the usual portrayal of medieval botanical illustration as a moribund tradition ineluctably degenerated by centuries of copyists' errors. The chapter focuses on critical copying practices evident in several illustrated manuscripts of Dioscorides that were all produced in Constantinople between the late ninth and early fifteenth centuries. I restrict this chapter to a handful of manuscripts from Constantinople because they all descend from the ancient botanical tradition and because we can establish copying relationships between them confidently.

Recognition of Variants

Textual traditions, whether in print or handwritten, generate variants. By "variants," scholars typically mean transformations in content, including (intentional and unintentional) rearrangements of text, shifts in wording, and transcriptional alterations. Analogous transformations can also occur in the transmission of pictures. One of the best witnesses to medieval awareness of pictorial variance can be found in the Morgan Dioscorides, dated to the late ninth or early tenth century on the basis of its script.[5] The commissioner and original recipient of this manuscript remain unknown.[6] It is the earliest surviving illustrated Dioscorides to combine text from both the original and the alphabetical versions of the *De materia medica*. Marie Cronier has argued that the text can be sourced according to three separate philological units: an unillustrated original Dioscorides; an illustrated Alphabetical Herbarium similar to the Vienna Dioscorides; and another illustrated version, composed of the chapters excised from the Alphabetical Herbarium. While scholars usually assume that pictures and texts in manuscripts were copied together, the situation in the Morgan Dioscorides appears to be more complicated.[7] Pictures based on the Alphabetical Herbarium sometimes appear with text based on the original Dioscorides. Moreover, some books in the Morgan Dioscorides seem to have many more pictorial sources than would be expected based on the sourcing of the text. In short, the Morgan Dioscorides handpicks texts and pictures from several different sources.

A particularly puzzling aspect of the codex is that individual chapters occasionally include two or more pictures illustrating what seems to be the same plant (see figs. 6.1, 6.2, 6.3, 6.6, and 6.7).[8] In both the Alphabetical Herbarium and the illustrated original Dioscorides, the norm is usually a single picture for each plant mentioned in the text, with the result that each chapter typically has only one picture. The Morgan Dioscorides confounds this pattern by occasionally including more pictures than are warranted by the text alone.

The largest set of these illustrations is ultimately derived from the Alphabetical Herbarium and usually involves one picture that is larger, darker, of a more bluish hue, and closer to the tradition represented by the Vienna and Naples Dioscorides manuscripts. The other picture is smaller, paler, and involves a yellowish green palette. The two pictures appear to have been derived from different branches of the same manuscript tradition of the Alphabetical Herbarium. For example, the chapter on eryngo (*ēryngion*) includes two pictures with similar features, notably a Medusa head at the end of the plant's long taproot (fig. 6.1).[9] The two pictures are similar enough to suggest that they ultimately derive from the same manuscript tradition. The smaller, lighter green plant appears to have been added after the larger, darker plant.[10] It is hard to date these additions. Given their similar appearance and palette, it seems likely that all of the smaller pictures were added at roughly the same time, perhaps a short time after the codex was completed. They were certainly already in the codex by the fourteenth century, when they were copied into another manuscript now in Padua.[11]

We can see how these variant pictures might have been generated, by looking at the chapter on hedgemustard (*erysimon*, fig. 6.2).[12] There we find one picture oddly cramped on its left side, as though it had previously been forced into a small space. We can surmise that it was probably copied from a manuscript such as the Naples Dioscorides, with limited lateral space. Though rooted in the same tradition, the pictures are visibly different from each other. As a result, we can suppose that there were at least two different copies of the Alphabetical Herbarium—or a version of Dioscorides derived from it—available in medieval Constantinople.

What did these different pictures represent to the users of the codex? Did they represent different plants, or different kinds of the same plant, or were they simply different pictures attesting to the same plant, as variants from different manuscripts or perhaps as views of different stages of plant life? Conveniently, one user of the codex wrote above the second picture of hedgemustard, "another hedgemustard."[13] In a botanical context this inscription has the sense of distinguishing between subtypes. This inscription consequently suggests that the user regarded the second picture as a separate variety of hedgemustard rather than as a different picture of the same plant. Looking again at the pictures, we can see why. The righthand picture shows a hedgemustard with shorter, broader leaves that contain fewer sinuses. As a result, it looks like a different plant. Contemporaries might, for example, have recognized the righthand picture as a representation of the plant *Sisymbrium orientale* L., and the lefthand picture as *Sisymbrium irio* L. The unintended generation of variants

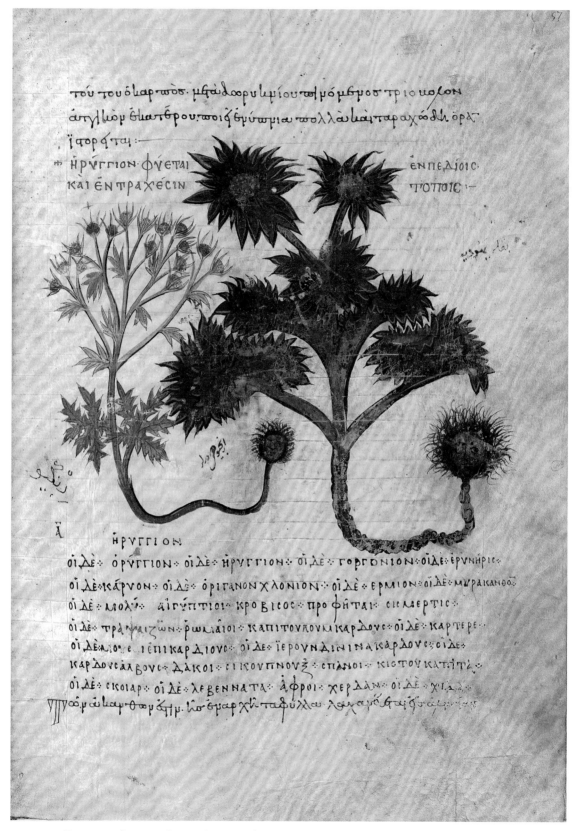

τοῦ τοῦ ὁ ϗαρπῶϲ· μᾶ δαωρυ ϊμίου ϗ μόμϗϗροϲ τρίο μοϲον
ἀγϗϊμϗν ϗ διαϗϗρου ποιϲ ϛ ϛμύωμια πολλα ϊϗϊ παραχῶ ϗϗ ὁρϗ
ϊϲορ ϛ ται :—

ᴴ ΗΡΥΓΓΙΟΝ ΦΥΕΤΑΙ   ΕΝ ΠΕΔΙΟΙϹ
ΚΑΙ ΕΝ ΤΡΑΧΕΣΙΝ   ΤΟΠΟΙΣ :—

ᾱ

ΗΡΥΓΓΙΟΝ

οἱ Δὲ· ΟΡΥΓΓΙΟΝ · οἱ Δὲ· ΗΡΥΓΓΙΟΝ · οἱ Δὲ· ΓΟΡΓΟΝΙΟΝ · οἱ Δὲ· ΕΡΥΝΗΡΙϹ
οἱ Δὲ· ΚΑΡΝΟΝ · οἱ Δὲ· ΟΡΙΓΑΝΟΝ ΧΛΟΝΙΟΝ · οἱ Δὲ· ΕΡΜΙΟΝ · οἱ Δὲ· ΜΥΡΑΚΑΝΘΟ
οἱ Δὲ· ΜΟΛΥ· ΑΙΓΥΠΤΙΟΙ· ΚΡΟΒΙΣΟϹ· ΠΡΟΦΗΤΑΙ· ϹΙϹΜΕΡΤΙϹ
οἱ Δὲ· ΤΡΑ ϗ μαιζῶν· ΡΩΜΑΙΟΙ· ΚΑΠΙΤΟΥΛΟΥΜ ΚΑΡΔΟΥϹ· οἱ Δὲ· ΚΑΡΤΕΡΕ
οἱ Δὲ ΜΟ ϛ ε ϊ ΕΠΙ ΚΑΡΔΙΟϹ· οἱ Δὲ· ΙΕΡΟΥΝ ΔΙ ΝΙ ΝΑ ΚΑΡΔΟΥϹ· οἱ Δὲ
ΚΑΡΔΟΥϹ ΑΛΒΟϹ· ΔΑΚΟΙ ·ϲ ΙΚΟΥΠΝΟΥ ϛ · ϹΠΑΝΟΙ · ΚΙϹ ΤΟΥ ΚΑ ΤΙ ΤϞ
οἱ Δὲ· ϹΚΟΙΑΡ· οἱ Δὲ· ΛΕΒΕΝΝΑΤΑ ἄφροι· ΧΕΡΔΑΝ· οἱ Δὲ· ΧΙΛ ϗ
ϒΥ ω ϛ μ ὦ ϗαρ ϛ ωρ ϛ ῆρ ϛ· ϊ ο ϛ ραρ χ ϗ· ϊ ταφύλλα λα ϗ α ϛ ϗα ϛ ϛ α ϛ μ ϛ

FIGURE 6.1 Depiction of eryngoes (*ēryngion*). New York, Morgan Library, MS M 652 (the "Morgan Dioscorides"),
late ninth or early tenth century, fol. 57r. Photo: The Morgan Library & Museum, New York.

FIGURE 6.2 Depiction of hedgemustards (*erysimon*). Morgan Dioscorides, fol. 46v. Photo: The Morgan Library & Museum, New York.

through copying thereby could have enabled contemporaries to recognize and then formally distinguish between two different plants.

We can draw similar conclusions from the illustrations of eryngo. The eryngo on the left is tall and pale green with persistent basal leaves and long, slender spine-like bracts subtending the floral capitula, recalling features of field eryngo.[14] The eryngo on the right is of a broader, bluish, tufted appearance with pointed, leafy bracts that are more in line with sea holly.[15] The picture of sea holly can be directly traced to the earlier Alphabetical Herbarium, but the depiction of field eryngo is actually closer to the plant described in Dioscorides' text.[16] Perhaps the earlier Alphabetical Herbarium had confused the two eryngoes, or at least failed to distinguish between them. The addition of the picture of field eryngo, therefore, signaled contemporary recognition of something amiss in the earlier Alphabetical Herbarium. Unlike the two hedgemustards, where difference seems to have arisen through the process of copying, the clarity of the distinction between the eryngoes could indicate a more critical and intentional process. It may be that the recognition of differences between the variant eryngoes reinforced the transmission of one picture as field eryngo and the other as sea holly, even if the original archetype for both was a sea holly. Or vice versa: an earlier picture of sea holly could have been modified so that it matched field eryngo, either because of differences recognized in the field, or because the makers realized the text better described field eryngo than sea holly. That "new" illustrations referring to distinct subtypes might still be based on "old" illustrations suggests that we should take the differences between pictorial variants seriously, rather than simply regarding them as mistakes. Even if a change occurs by chance, the decision to copy it signals contemporaries' recognition of its possible value as an indication of another kind or subtype of plant.

That these different pictures of eryngo were eventually understood to represent two different kinds of eryngo is confirmed by labels accompanying copies of them in a deluxe botanical atlas from the first half of the fifteenth century, now part of the Chigi collection in the Vatican Library.[17] This Chigi Dioscorides lacks Dioscorides' text, and contains only pictures of plants and animals copied from both the Morgan and Vienna Dioscorides manuscripts.[18] Isidore of Kiev, the metropolitan of Kiev and all Russia from 1436 to 1439 and later the Latin patriarch of Constantinople from 1459 until his death in 1463, apparently directed the copying of the Chigi Dioscorides, perhaps while resident in Constantinople in the 1430s or early 1450s.[19] When Isidore had the illustrations of the two eryngoes from the Morgan Dioscorides copied into the Chigi manuscript, he had the smaller eryngo labeled "the narrowleaf eryngo" and the larger eryngo labeled "the big eryngo."[20] These labels confirm that by the fifteenth century the two illustrations were understood to represent two different kinds of eryngo. The recognition that eryngoes constitute a broader "generic" category including at least two distinct subtypes is absent in the ancient botanical literature.

More visual differences can be seen in the two pictures in the chapter on woad (*isatis*, fig. 6.3).[21] The picture on the left is taken directly from the Alphabetical Herbarium, while the righthand picture is from another, now unknown source, but again perhaps from the same manuscript tradition. The two pictures clearly represent plants with different characteristics. The lefthand picture shows a plant with a basal rosette of large, pointed elliptical leaves with apparently sessile reddish fruits, some of which are borne on peculiar cross-like structures. The lefthand picture is perhaps closer to Dioscorides' written description of wild woad (*isatis agria*), although the Morgan Dioscorides includes a separate chapter on wild woad. Neither the lefthand picture of woad nor that of the wild woad bears much resemblance to woad as it is recognized today.[22] Instead, it is more like a dock plant with its large basal leaves and reddish seeds. (That Pliny compared the leaves of *isatis* to those of dock could hint at a reason for the confusion.[23] Like woad, several kinds of dock can also be used as dyes.[24] Docks typically yield yellow, red, and green dyed fabrics, and not the blue color of woad.) The smaller picture on the right, however, shows a single shoot with clasping pointed leaves along its lower stem while its branches above have clusters of pedicellate (that is, raceme) flowers. In other words, the righthand picture depicts a plant more in line with woad as we now know it. The inclusion of a second, more "accurate" picture of woad here may echo one Byzantine commentator's bold assertion that Dioscorides' description of woad was faulty.[25] The commentator provides his own description as a correction.[26] In general, the inclusion of a more accurate depiction of woad could indicate contemporary recognition of disagreements among the image, the text, and actual woad.[27] Recognition of such disagreements may underlie the decision to include multiple pictures in this case. The field eryngo could be visibly related to the earlier Alphabetical Herbarium, but the same cannot be said of the second depiction of woad. Perhaps it goes back

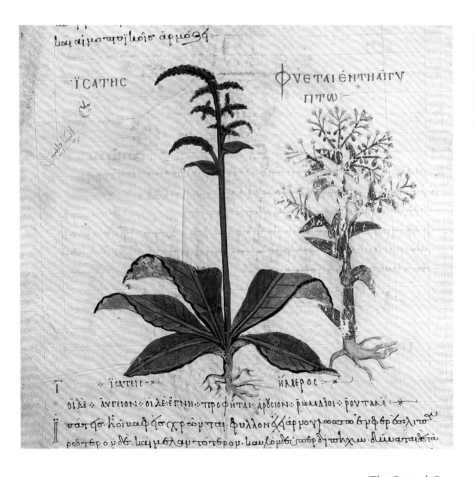

FIGURE 6.3 Depiction of plants labeled as "woad" (*isatis*). Morgan Dioscorides, fol. 68v. Photo: The Morgan Library & Museum, New York.

to an ancient manuscript tradition now lost. Perhaps, as we will see in chapter 7, it could have been created *ex novo*, "from scratch," because the makers of the manuscript recognized that the original picture did not resemble woad.

What complicates our understanding of the variants in the Morgan Dioscorides is the simple fact that most of them lack labels. On the one hand, that absence could suggest that the visual differences between variants were regarded as obvious in themselves. After all, Isidore had apparently had no trouble labeling the pictures of the different eryngoes in the Chigi Dioscorides. And we could argue that a user labeled the different hedgemustard—the one case where we do encounter an inscription—because the visual distinction between variants there is so easily explained as the result of a distortion from the copying process. On the other hand, the absence of labels may also indicate uncertainty about the classification of subtypes with respect to the botanical tradition. The absence of labels for pictorial variants in the Morgan Dioscorides effectively hedges on the identification and classification of similar plants. As a result, while contemporaries clearly recognized variance and ambiguity within the ancient botanical tradition, they may have been uncomfortable with complicating that tradition further by introducing novelties, at least on the level of text. This reticence can be seen in Photios' letter to Zacharias, quoted in chapter 5, in which Photios complained about the strange plants and names used by contemporary physicians. The Chigi Dioscorides, by contrast, demonstrates less anxiety about the introduction of new names, perhaps due to shifts in attitude about the ancient botanical tradition as well as developments in late Byzantine botanical lexicography.[28]

In addition to this group of pictures, there are instances where multiple illustrations of plants were added to earlier chapters in the Morgan Dioscorides. For example, the codex includes a separate folio, now at the front of the codex, that features a large illustration of balm (*melissophyllon*, fig. 6.4), even though there is a separate illustration of balm deeper in the volume (fig. 6.5).[29] The similarity between the red title accompanying the balm on fol. iv and the red titles elsewhere in the volume suggest that it could have been added to the codex at the time of its initial creation, or not long after. As with the eryngo and hedgemustard, these pictures of balm could both have been derived from the tradition of the Alphabetical Herbarium, particularly due to the similarities of their lower shoots. And as we saw with the other illustrations, later copyists also recognized these distinct depictions of balms as different types of balm.[30]

FIGURE 6.4 Depiction of balm (*melissophyllon*). Morgan Dioscorides, fol. iv. Photo: The Morgan Library & Museum, New York.

ΜΕΛΙϹ

ϹΟΦΥΛΛΟΝ

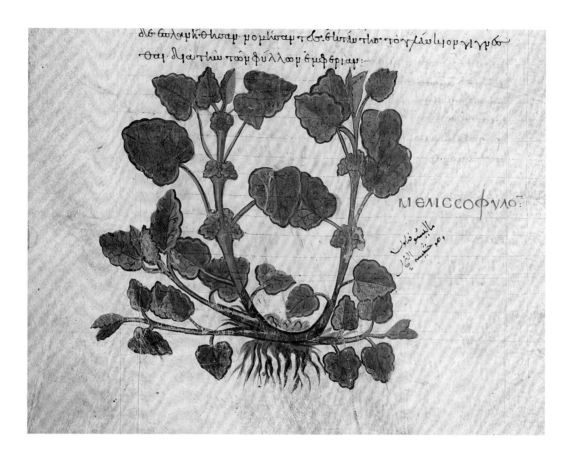

Expanding Views

Besides the illustrations derived from the Alphabetical Herbarium, the Morgan Dioscorides includes many illustrations from other illustrated versions of Dioscorides, perhaps from an illustrated original version of the text similar to the Yerevan Fragment (see chapter 4). As with the illustrations from the Alphabetical Herbarium, this group occasionally includes multiple pictures of plants in single chapters. The chapter on lotus (*kyamoi heteroi*, "other beans," that is, "the Egyptian bean," fig. 6.6) shows three different illustrations: the leftmost picture shows the flower in profile, with the stamens, receptacle or carpels, and petals; the center picture shows a bud, fruit or seed pod, and leaves both head-on and in profile; the righthand picture shows an open flower, head-on.[31] Each picture conveys different information about the plant. In this way, the illustrations echo approaches to representation seen in the Alphabetical Herbarium that emphasized as many different distinguishing features of the plant as possible (see chapter 3).

Some doublings, however, remain difficult to explain. The chapter on fava bean (*kyamos*, fig. 6.7) is accompanied by two pictures that

FIGURE 6.5 Depiction of balm (*melissophyllon*). Morgan Dioscorides, fol. 102v. Photo: The Morgan Library & Museum, New York.

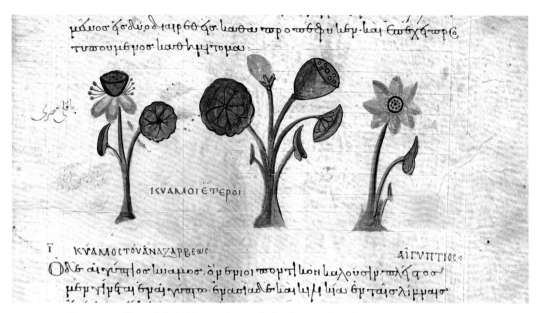

FIGURE 6.6 Depiction of lotus (labeled *kyamoi heteroi*, "other beans," also "the Egyptian bean").
Morgan Dioscorides, fol. 75r. Photo: The Morgan Library & Museum, New York.

FIGURE 6.7 Depiction of fava bean (*kyamos*). Morgan Dioscorides, fol. 74v. Photo: The Morgan
Library & Museum, New York.

are virtually identical.[32] It seems possible that they derived from an earlier manuscript where they originally showed different kinds of information, but that they came to resemble each other as they were copied together from manuscript to manuscript over time. This loss of visual distinction through copying accords with the familiar scholarly understanding of the results of successive uncritical copying. At the same time, if this is the case for these illustrations, then we can see that the copying of pictorial variants into the Morgan Dioscorides represents the continuation of earlier practices of pictorial compilation. Interestingly, a marginal note, reading "do not paint [this]," is perhaps a signal that this may have been the end of the line for this duo in at least one later copy.[33]

The Morgan Dioscorides challenges common views of medieval botanical illustration as a stagnant tradition based on the uncritical copying of older manuscripts. It shows that medieval people looked closely and critically at earlier botanical illustrations. Differences within the same manuscript traditions allowed them to recognize morphological distinctions between similar but different kinds of plants. While variants might arise randomly, the recognition of them as reflecting different kinds subsequently reinforced the elaboration of the tradition.

Descent with Modification

While the makers of the Morgan Dioscorides drew upon multiple pictorial sources when they initially made (and possibly refurbished) the codex, later users also modified some of the pictures after they were copied.[34] For example, in the depiction of the geranium (*geranion*, likely *Geranium tuberosum* L., fig. 6.8), someone added an entirely new leaf and an inflorescence, evidently so as to define and further clarify the plant's bracts, pods, pedicels, and persisting sepals.[35] The added leaf visibly contrasts with the other leaves, too. The distinctly stalked pinnate leaf with deep sinuses and obtuse lobes could refer to another geranium entirely, such as *G. robertianum* L., a widespread species with medicinal uses that differ from Dioscorides' recommendations.[36] Extant copies of these illustrations from the Morgan Dioscorides do not retain these modifications. Later copyists may have found them too incongruous, or perhaps redundant. Still, the act of modifying earlier pictures indicates critical engagement with the visual content of the manuscript.

This attention to visual content may go hand in hand with the poor quality of the main text in the Morgan Dioscorides, as well as the

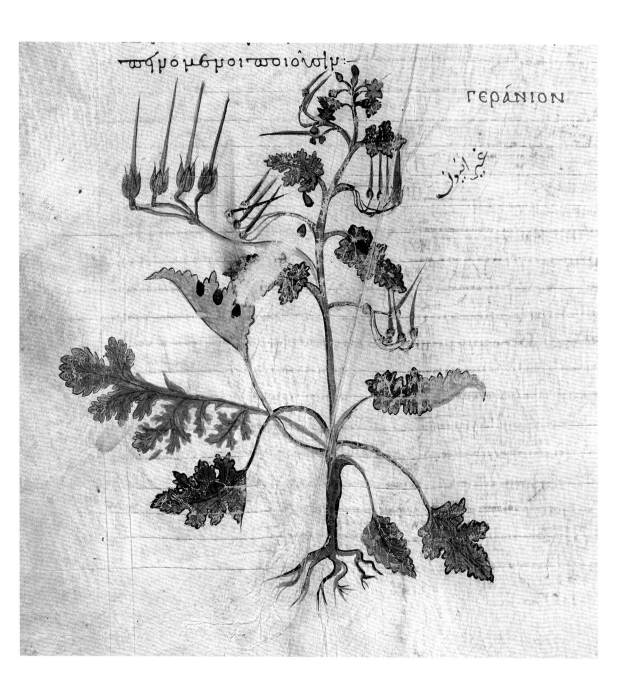

ϲφρόμϐροιϖοιο͡υϲιν:−

ΓΕΡΑΝΙΟΝ

عجم اِبوز

absence of labels accompanying the added illustrations. The emphasis on pictures over text in the Morgan Dioscorides echoes trends that we noted in the early illustrated herbal. It also anticipates the development of botanical atlases devoid of text in the later Middle Ages, as exemplified by the Chigi Dioscorides. It is worth noting that the Morgan Dioscorides contains strikingly little in the way of annotations, with the notable exception of many later notes in Arabic script, which may date to the early modern period. As a result, while scholars

FIGURE 6.8 Depiction of geranium (*geranion*). Morgan Dioscorides, fol. 30v. Photo: The Morgan Library & Museum, New York.

often gauge the usage of a medieval codex by its written annotations, such an approach does not work well for the Morgan Dioscorides, or any manuscript that was primarily used for its pictures. In fact, other indications of use—including the accumulation of grime, drips of candle wax, and subsequent additions and modifications of pictures—although difficult to date, all suggest extensive historical use and engagement with the manuscript.

That users modified earlier illustrations raises the possibility that illustrations could have been shaped and incrementally improved as they were copied over time. For the most part, such gradual change is difficult to demonstrate, as so few manuscripts survive. We can, however, point to direct evidence for such processes in the modifications and copies of a depiction of Spanish broom (*spartos*) originally from the sixth-century Vienna Dioscorides.[37]

Although useful as a generic portrayal of the Spanish broom's upright, rush-like habit, the original sixth-century depiction of the plant in the Vienna Dioscorides (fig. 6.9) can but vex when it comes to more specific plant morphology.[38] The abstracted flowers resemble butterflies more than the actual flowers of a Spanish broom, as though a kaleidoscope of clouded yellows had just settled on the jaunty, whisk-like sprays. One medieval user of the codex added peculiar ink loops along the lower shoots—perhaps an attempt to capture the plant's sparse, narrow leaves.[39] While these loops make the picture more informative, they seem out of place here and, lacking detail and color, are ambiguous. Knowing less about the plant, we might wonder if these loops indicate leaves, fruits, or something else entirely (such as insects).

In the fourteenth century, another user of the codex addressed these problems head on, by adding an entirely new sketch of the plant along with an explanatory note below the main text on the facing folio (fig. 6.9).[40] Although the sketch appears rough, it delineates the leguminous seedpods and pea-like flowers of the Spanish broom more precisely and accurately than the sixth-century illustration does. The note beside the sketch confirms as much, as it zeroes in on the flowers and the fruits: "Spanish broom has this shape [*schēma*]. Its flower is similar to the calavance bean or the fava bean except yellow. Its seed [resembles] vetch or lentil."[41] The note and sketch are suggestive of the prominent role that comparison played in the observational practices of late Byzantine botany. The introduction of this "new" illustration into the Vienna Dioscorides also shows that contemporaries added new imagery into other early codices, besides the Morgan Dioscorides.

The sketch of the Spanish broom in the Vienna Dioscorides was later copied into an illustrated Dioscorides now in Padua (fig. 6.10).[42] That the sketch of Spanish broom was copied into the Padua codex suggests that contemporaries considered it worthy of copying, perhaps because it supplied additional information, absent in the sixth-century painting in the Vienna Dioscorides.[43] It was, moreover, copied without the accompanying note, possibly indicating that its contributions were recognized as obvious in themselves. This new iteration of the sketch of the Spanish broom is larger than the original and straddles the line between simple line drawing and a more elaborated painting.

When it came to copying the illustration of the Spanish broom from the Vienna Dioscorides into the Chigi Dioscorides (fig. 6.11), the painter chose to make only one picture by combining details from both the sketch and the original sixth-century picture.[44] In particular, they added the leguminous seedpods conspicuously absent in the original, which were supplied by the sketch. As a result, we encounter an exact copy of neither the original illustration nor the sketch, but rather a combination of the two. That the Chigi illustrator included the idiosyncratic loops that were added to the stem of the plant further indicates that the Vienna Dioscorides was the main source for this illustration, and not another, now lost manuscript of Dioscorides.

The fruiting of the Spanish broom between the Vienna Dioscorides and the Chigi codex provides a useful model for how we might envision the gradual transformation of botanical illustrations in the premodern world. Researchers tend to think such illustrations were simply made "from life" or copied from a model. The reality could be much more complicated. There were always a number of possible sources of information and levels of mediation that resulted from various ways of gathering, retaining, and conveying information, including, but not limited to, sketches, modifications to earlier illustrations, and marginal annotations. It may be that it was only through the circulation of information in these myriad forms, as well as direct observation and personal experience, that artists and scholars arrived at emergent early modern forms of botanical illustration putatively, but preeminently, defined as "from life."

The fruiting Spanish broom is not the only novelty in the Chigi codex. The illustration of creeping cinquefoil (*pentaphyllon*, fig. 6.12), for example, includes fruit that do not appear in the Vienna Dioscorides (fig. 6.13).[45] There are also a number of illustrations that appear in neither the Morgan nor the Vienna Dioscorides.[46] But unlike the fruiting Spanish broom, it is more difficult to find the sources

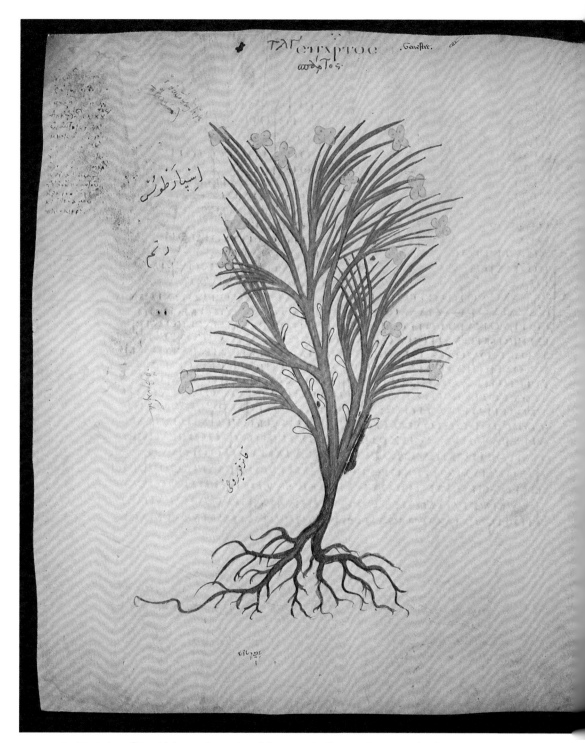

FIGURE 6.9 Depiction of Spanish broom (*spartos*) on fol. 327v and a fourteenth-century sketch of Spanish broom on fol. 328r. Vienna Dioscorides, fols. 327v–28r. Courtesy of Österreichische Nationalbibliothek.

ΟΙΛϹ   ΛΟΒΟΝ
ΟΙΛϹ   ΛΥΓΟΝΚΛΛΟΥϹΙΝ
ΟΜΙΝΟϹΕϹΤΙΝϕΕΡϕΗΝ ΡΑΒΔΟΥϹΜΛΚΡΑϹΕΝ ϕΥΛΛω
ΛΥϹΟΡΛΥϹΤΟΥϹ ΜΕΤΛϹΛΙΤΕΛΟΥϹΛϹϹΜΕϹΟΥϹΙ
ϕΕΡΕΙΛΟΛΟΒΟΥϹϕΕϹΤΕϕΛϹΤΟΛΟΥϹ ΕΝΟΙϹΕϹΤΙ
ΜΛΤΛϕΛΚΟϹΕΛΗΛΝΟϹϹΜΗϹ ΓΟΝΟϹΤΕΡΛΕΥΚΟΓⲞ
ΤΟΥΤΟΥΟΚΛΡΠΟϹΚΛΡΓΛϹϹ ΓΠΟΟϹΗΓΛϹΥΗΜΟ
ΛΙΚΡΛΤϕΟΛΚΝΟΒΟϕΙΤΓΩΤΟ ΚΛΟΛΤΡΩΛΜΩΜΕ
ΤΛϹΗΓΛϹϹϕϹΤϹΛΙΙϹϕΕΤϹΕΡΕΛΛΕΒΟΡΟϹΛΚΗΝ
ΛΥΝϕϹ ΟΛϹΚΛΡΠΟϹΚΙΝΩϹΚΛΡΓΗΝΚΛΤϕΚΛΟΛΡⲅⲅ
ΚΛΙΛΥΤΛΙΛΟΜΡΛΒΔΟΙϹΠΥΛΛΤΚΡΛΧΕΙϹΛΙ ϹΓⲅⲕⲞ
ΤΓϹΙϹΜΚΛΧΥⲓⲥⲟⲟⲥⲉⲗⲧⲉⲭⲓⲗⲗⲕⲟⲛⲉⲥⲧⲓⲛⲃⲟⲛ
ⲟⲛⲏⲙⲗⲟⲥⲟⲏⲕⲩⲗⲟⲟⲥⲉⲧⲓⲗⲟⲙⲥⲏⲟⲥⲏⲥⲧⲉⲥⲏ · ⲉⲏⲓ
ⲟⲓⲗⲥⲉⲏⲭⲓⲅⲟⲃⲣⲥⲉⲗⲭⲧⲉⲥⲗⲙⲏⲟⲗⲗⲧⲅⲧⲓⲗⲥⲅⲕⲁⲩⲍⲟⲩ
ⲥⲏⲓⲧⲟⲩⲥⲓⲥⲭⲓⲭⲗⲓⲕⲟⲩⲥ ⲙⲅⲉⲗⲁⲥⲕⲙⲁⲙⲁⲧⲟⲇⲁ ⲉⲥⲕⲁⲓ
ⲍⲩⲥⲙⲁⲧⲫⲇⲁⲥⲥ ⲕⲵ

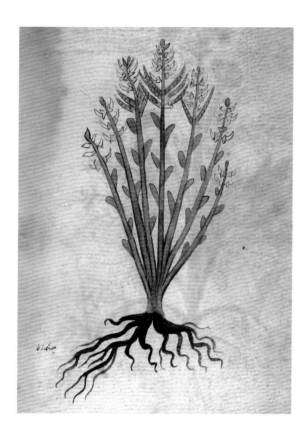

FIGURE 6.10 Copy of the sketch of Spanish broom from the Vienna Dioscorides. Padua Dioscorides, fol. 155r. Courtesy of Biblioteca Antica del Seminario di Padova.

for the creeping cinquefoil's additions. Some of them may reflect other, ancient copies of Dioscorides, now lost, but some may also go back to individual firsthand observations of nature, or, as in the case of the Spanish broom, to the accumulation of observations carried out by a number of different individuals.

The most extreme form of improvement, however, would be the complete replacement of an illustration from one manuscript with another illustration. This, too, may have happened in the making of the Chigi codex. In the Morgan Dioscorides, the evergreen rose (*kynosbaton*) appears not as a climbing rose but as a thorny tree with a wide trunk (fig. 6.14).[47] The minuscule white flowers and thorns were probably supplied by a reading of the text.[48] In the Chigi Dioscorides, however, we find an entirely different picture, more in line with our expectations for what the plant looks like on the basis of its name, even though most of the illustrations on the folio were copied directly from the Morgan Dioscorides (fig. 6.15). As before, it is hard to know where this illustration came from—it could have been created *ex novo* or perhaps copied from a source that is no longer extant.

The fact that the painters added these details or changed out entire pictures suggests that we should rethink our general understanding of

FIGURE 6.11 Depiction of (clockwise from top right) Spanish broom (*spartos*), pondweed or water lettuce (*stratiotis*), and garlic germander (*scordion*). Vatican, Biblioteca Apostolica Vaticana, MS Chigi F.VII.159 (the "Chigi Dioscorides"), early fifteenth century, fol. 157r. © 2022 Biblioteca Apostolica Vaticana.

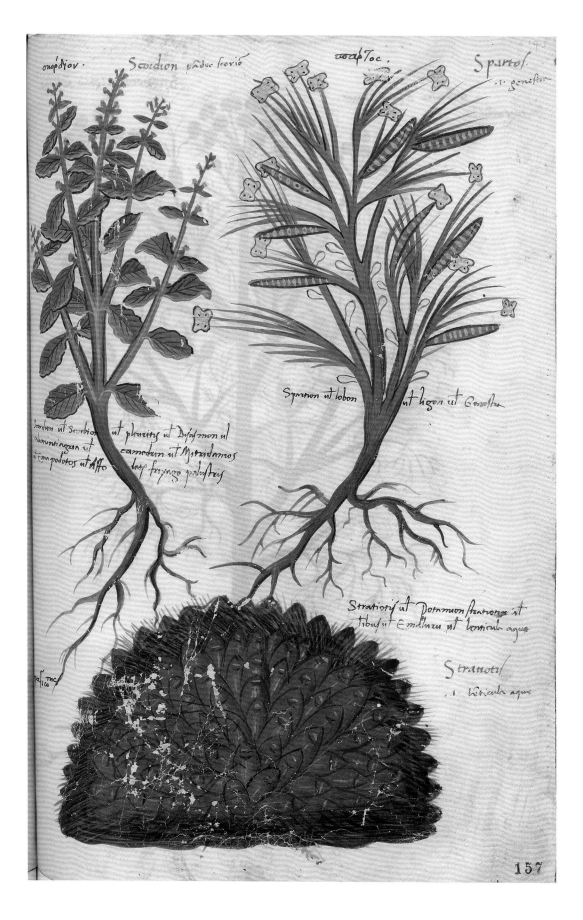

ουερδιον.　　Scordion φ̄ dec scorió　　ασαρτος.　　Spartos.
.i. genesta

Spartion ut lobon　　ut ligon ut Genesta

Scordion ut Scorbion　　ut pleuritis ut Disafmon ul
ϸ̄ανuntingan ut　　camedun ut Mytridamios
ut em podotos ut Affo　　lioφ frigago palustris

Stratiotis ut Potamion stratiota ut
tibus ut Emathin ut homiouli agua

Stratiotis
.i. lisricula agua

157

how details are affected by the process of copying.[49] Copying does not necessarily lead to the ineluctable and deleterious loss of accurate detail: sometimes it can lead to the addition of new details. Recognition that such processes occurred in the fourteenth and fifteenth centuries also suggests that we should rethink what may have happened in earlier manuscript transmission. If late medieval copyists could improve pictures and build upon those improvements over time, the same could have occurred in antiquity and the early Middle Ages. In fact,

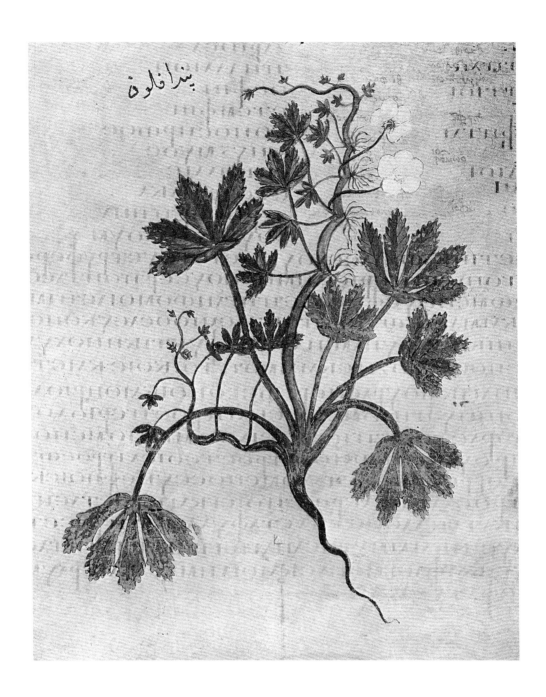

such activity would explain how some of the variants in the Morgan Dioscorides, such as the eryngo and woad, were originally generated. The dearth of surviving material from earlier periods is perhaps the only reason we do not see direct evidence of these practices earlier. It may be the case that the profusion of details in the illustrations of blackberry, anemone, and geranium in the Alphabetical Herbarium emerged over a long period of time through the continual addition of new features and properties by different artists.

FIGURE 6.13 Depiction of creeping cinquefoil (*pentaphyllon*). Vienna Dioscorides, fol. 273r. Courtesy of Österreichische Nationalbibliothek.

The Critical Copy

FIGURE 6.14 Illustration of an evergreen rose (*kynosbaton*). Morgan Dioscorides, fol. 252v. Photo: The Morgan Library & Museum, New York.

FIGURE 6.15 Illustration of an evergreen rose (*kynosbaton*). Chigi Dioscorides, fol. 214v. © 2022 Biblioteca Apostolica Vaticana.

Observation and Experience

Throughout this chapter, we have seen that Byzantine readers and image-makers repeatedly checked ancient botanical illustrations against other sources of information, including the accompanying text and pictures in other manuscripts, as well as actual plants. In many ways, the compilation of pictures in the Morgan Dioscorides echoes a larger movement in ninth- and tenth-century Byzantine intellectual culture, which Paul Lemerle has called "encyclopédisme," and Paolo Odorico has more recently and less anachronistically called the "cultura della *syllogē* [culture of gathering]."[50] These terms refer to Byzantine scholars' reassembly of ancient learning, first through the copying and transliterating of earlier texts, and second by collecting and organizing textual excerpts into manuals. The compilation of texts and pictures in the Morgan Dioscorides certainly fits within this broader movement.

At the same time, we should not confuse textual with pictorial compilation. In the Morgan Dioscorides pictures were used to expand and elaborate the botanical tradition. Pictures operated as autonomous sources of visual knowledge. They are distinct from the text, as much as they illustrate it. The critical copying and compiling of illustrations in the Morgan Dioscorides was then a selective process carried out independently of, if concurrently with, textual compilation. Scholar and painter had to come to terms with each variant illustration according to its visual properties and resemblance to actual plants. The modification of pictures among the Vienna, Morgan, and Chigi Dioscorides manuscripts in particular speaks to the continuity of these practices within the Byzantine botanical tradition.

The gradual, critical modification of illustrations over time was thus a central means by which the premodern botanical tradition was elaborated and corrected. This scale and pace of change stands apart from conceptions of scientific progress popular today. As George Kubler noted in *The Shape of Time* (1962):

> All things and acts and symbols—or the whole of human experience—consist of replicas, gradually changing by minute alterations more than by abrupt leaps of invention. People have long assumed that only the large changes were significant, like those represented by great discoveries, as of gravitation or the circulation of blood. The little changes separated by infinitesimal alterations like the changes appearing in copies of the same document prepared by different scribes, are dismissed as trivial. . . . large-interval changes are similar to small-interval changes. Furthermore, many changes assumed to be large are really small when seen in full context.[51]

The accumulation of minor changes over time is a decisive factor in the botanical traditions of the premodern Mediterranean. In light of Kubler's larger project on the nature of change among successions of artworks, we can view a picture in a Byzantine botanical manuscript as an instance within a longer sequence of copies. An illustration series has its own temporal qualities—its own rate and predisposition to change according to factors such as its familiarity to contemporaries or its medical utility.[52]

The historian of science Hans-Jörg Rheinberger later adapted Kubler's conception of internal time in his analysis of scientific research according to material processes. Rheinberger's work thus helps us to reconsider how a picture series might also act as a substrate for exploratory scientific research. According to Rheinberger, a research device has to fulfill two basic requirements. First, it has to be stable enough so that the knowledge which is implemented in its functioning does not simply deteriorate in the course of continuing cycles of realization. This is a necessary but not sufficient condition for its becoming a device that is endowed with internal time and is thus able to act as a historial arrangement. Second, it has to be sufficiently loosely woven so that in principle something unpredictable can happen and over many rounds of performance must happen. In everyday life and in most of our social contexts this is a situation that one tries to avoid as an inconvenience. Within the research context it is a situation that has to be actively promoted.[53]

By "historial," Rheinberger refers to a Derridean conception of history as ungrounded and unbounded by chronology and teleology, in which recurrence is inherent in the very act of looking back. By juxtaposing pictures from different sources, the users of the manuscript compared different moments in a historial assemblage. The manuscript tradition thus constituted a vast set of individual picture series, each with its own time, each with its own rate and predisposition to change. It is an obvious anachronism to suggest that a Byzantine illustrated manuscript constituted part of an experimental system. A medieval copying apparatus was hardly intended to result in nonidentical reproduction, even if that was the usual result. Despite this, copying could still have operated in a similar way: it is preeminently aimed at stability, while allowing for the selective and gradual generation and preservation of variants. Copying in and of itself is not enough to make this a research system: there must be unpredictable outcomes that are apparent in hindsight, which is to say, there must be later recurrences and retrospection. Indeed, copyists have no idea how their copies will be understood and copied. Small changes are recognized through later comparison to other pictures. Together with the accumulation of divergences and even errors that demanded later correction, this system of knowledge production moves toward ever greater complexity. This series of copies, of minute changes and variances, is thus the substance of an ever evolving botanical tradition.

A sustained tradition of critical copying amounts to a kind of collective "empiricism" drawn out over a long period of time. While varieties of empiricism are manifold, I refer here to the establishment of knowledge through the accretion of observations based in sensory experience (*empeiria*). Consistent with what we know about the ancient Empiricist "school" of medicine, experience here derives both from personal observation (*tērēsis* or *autopsia*), as well as the accounts of others (*historia*, whether written or painted).[54] *Empeiria* is conceptually related to the repeated performance of trials (*peirai*), a practice suggested by the juxtaposition of variant pictures in the Morgan Dioscorides. The critical action of comparing and judging is analogous to the position that Photios rehearses in his letter to Zacharias. There Photios argued that the manifest benefits of bloodletting would refute (*elenchon*) with proof (*peira*) Zacharias' physicians' recommendations against it. In other words, Zacharias can judge the truth of the matter for himself by putting it to a test. Critical practices in Byzantine botanical illustration display a similar kind of "testing" of pictures against personal observation as well as the broader botanical tradition.

However empirical these practices may seem, they remained rather limited. Medieval critical observations tended to be narrow in scope and often limited to the identification of errors or anomalies within the existing botanical tradition. Similar to the empirical practices evident in some Western medieval sciences, such as astronomy, critical work in the Byzantine botanical tradition was "scattershot and individualistic."[55] A principal factor limiting the impact of empirical inquiry within the Byzantine context may have been what Gianna Pomata has called the longstanding "stigma of inferiority . . . attached to observational knowledge in the Greek philosophical tradition."[56] Such attitudes doubtless prevailed not only in textual, but also in pictorial approaches to the creation of knowledge. As we saw in chapter 2, knowledge gained through direct, firsthand experience was to be tempered and developed through *epinoia*, an abstracted, conceptual definition of a subject. There is no reason to suppose that such views disappeared in the Middle Ages. In general, too, Byzantine scholarly deference to antique authorities hampered the systematic introduction of novelty into Byzantine sciences; we see this in Photios' hostility to the "strange" botanical substitutions of his contemporaries.

Specific institutional contexts may have played a large role in defining the critical practices and traditions in medieval botany. Most of the manuscripts discussed here can be connected to the monastery of St. John the Forerunner (Prodromos) in the Petra district of Constantinople from the fourteenth until the mid-fifteenth century.[57] That monastery had an impressive library and was the site of the important royal Serbian hospital, the Xenon of the Kral.[58] It is tempting to link many of the practices noted here to that specific context. The Morgan Dioscorides, however, suggests that critical copying practices likely existed prior to the fourteenth century. While we do not know where these manuscripts were prior to their incorporation into the library of the Petra monastery, they may have been in similar settings.[59] Such institutional contexts were not limited to Byzantium. Charitable foundations in the medieval Islamic world and Western Europe could have allowed for similar concentrations of manuscripts in proximity to specialist practice and training that may have enabled the formation of critical practices of botanical illustration.

Conclusion

Medieval copyists employed a variety of critical practices when copying ancient botanical illustrations. This complicates the usual view of the medieval tradition of botanical illustration as at best stagnate and limited to uncritical copying, and, at worst, degenerated by the ineluctable accumulation of careless errors. Instead, medieval copyists recognized variants within the manuscript tradition and harnessed them to make sense of Dioscorides and to better reflect the diversity of the natural world. They modified earlier pictures by adding to them and by clarifying and correcting specific details. All of these practices suggest that the Byzantine botanical tradition was gradually elaborated through piecemeal processes of selective revision, a limited form of empiricism based on the comparison of pictures, texts, and actual plants. The attention lavished on the pictures also demonstrates the depth of medieval interest and faith in the visual as a way of knowing, and in the pictorial as a way to create visual knowledge. There is no reason to suppose that these critical practices of copying illustrations were unique to Byzantium. They certainly may have occurred at other times elsewhere in the premodern world. In the next chapter we will find that diverse practices of nature observation and novel botanical illustration appear in the medieval Greek, Arabic, and Latin botanical traditions.

7

EX NOVO

Scholars tracing the rediscovery of direct nature observation in the history of botanical illustration typically look back to the *Tractatus de herbis* herbals of thirteenth- and fourteenth-century Italy. The earliest extant example is a manuscript now in the British Library known by its shelfmark, Egerton 747. The illustrations in this herbal are often cited as the earliest examples since antiquity of pictures of plants created through the observation of nature. This usual account posits that medieval people neglected to look and draw from nature for at least seven hundred years, if not longer. This chapter, however, shows that Greek- and Arabic-speaking illustrators created entirely new, *ex novo* observational studies of plants prior to the creation of Egerton 747. These illustrators experimented with a variety of ways to observe and depict plants. At the same time, similarities among observational plant depictions from the Greek, Arabic, and Latin traditions point to their common inheritance of the ancient botanical tradition and their ongoing dialogue with each other.

The chapter begins with Egerton 747 and why scholars have assigned it such a privileged position in history. It then identifies novel botanical illustrations in earlier manuscripts, particularly observational nature studies in the late ninth- or early tenth-century Morgan Dioscorides and an early thirteenth-century Arabic Dioscorides now in the Topkapi library in Istanbul. These manuscripts reveal experimentation with different ways of creating accurate *ex novo* depictions of plants. They consequently provide a fuller view into the emergence of novel modes of plant depiction in the Middle Ages that were nonetheless rooted in earlier traditions going back to Dioscorides' Empiricist approach to botanical inquiry. The juxtaposition of these manuscripts demonstrates that their makers shared similar concerns over the legibility of botanical form with an attendant emphasis on distinguishing features and visual clarity. By identifying earlier modes of botanical observation and depiction, the chapter further addresses the thorny questions of how we can know if an illustration involved nature observation and how we can suppose it is new rather than merely a copy of an ancient picture.

The Illustrated *Tractatus de herbis*

Egerton 747 was produced in Central or Southern Italy, perhaps in Naples or Salerno, between 1280 and 1315. With its 406 pen-and-wash illustrations of plants, it represents a landmark in the history of botanical illustration.[1] Otto Pächt identified the significance of the illustrations for the history of the emergence of nature observation in Italian art.[2] Subsequent scholarship has confirmed Pächt's findings.[3] Minta Collins has referred to the illustrations as the "first nature studies of plants since Classical times."[4] The illustrations accompany a text called the *Tractatus de herbis*.[5] The *Tractatus* was itself based on an earlier, unillustrated herbal known by its opening line as the *Circa instans*, written by Matthaeus Platearius of Salerno in the third quarter of the twelfth century.[6] The *Circa instans* is a compilation of excerpted texts from earlier herbals.[7] The *Tractatus* included more medicinal substances than the *Circa instans*, as well as additional authors.[8] Despite their dependence on earlier texts, the *Circa instans* and the *Tractatus* were novel compilations. Egerton 747 may be the earliest surviving copy of the *Tractatus*.[9] Its production appears to have been somewhat *ad hoc* and to have involved close collaboration between scribes and artists. The scribes even changed their ruling of the folios partway through production in order to better accommodate the illustrations.[10]

The composition, treatment, size, and coloring of the botanical illustrations throughout the codex vary tremendously. Minta Collins has divided the illustrations into three basic groups: "some invented for unfamiliar plants, some made up of characteristic features but not in proportion, and others drawn from life."[11] In her first category, we can look to the invented illustrations of the exotic nutmeg (*nux muscata*) and coconut (*nux indica*, fig. 7.1). At the time these pictures were executed, coconut palms were only cultivated in the Indo-Pacific, while nutmeg grew in what is now Eastern Indonesia. For the second category we can look to the more accurate but somewhat vague shepherd's purse (*bursa pastoris*). For the third category, we can turn to the detailed and accurate depictions of sorrel (*brictanica*) and white bryony (*brionia*) on the same folio (fig. 7.2).[12]

In general, the terms "observation" and "nature study" should be understood here to indicate only the relative accuracy and novelty of the *Tractatus* cycle as a whole.[13] Neither term here suggests that the pictures in Egerton 747 are themselves copied directly from nature.

FIGURE 7.1 Illustrations of nutmeg and coconut trees (*nux muscata* and *nux indica*, respectively). London, British Library, MS Egerton 747, 1280–1315, fol. 67v. © The British Library Board.

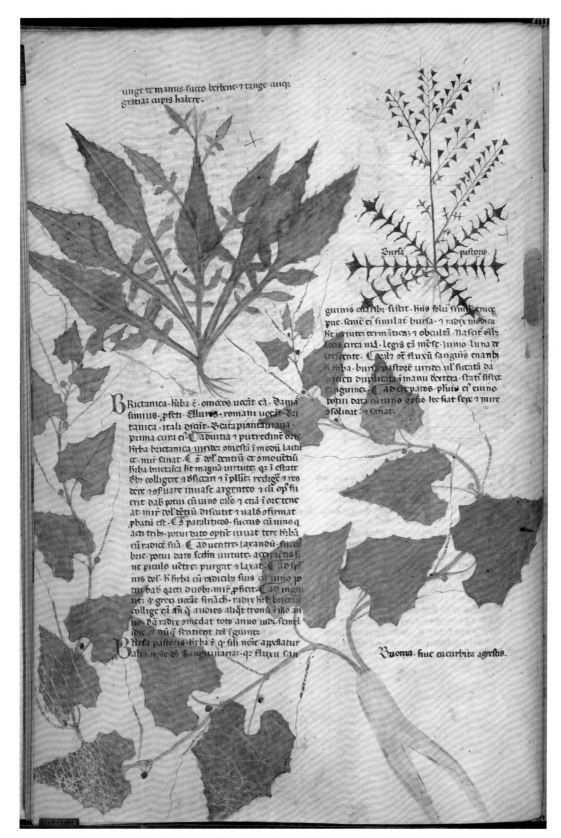

FIGURE 7.2 Illustrations of shepherd's purse (*bursa pastoris*), sorrel (*brictanica*), and white bryony (*brionia*). MS Egerton 747, fol. 16v. © The British Library Board.

The *Tractatus* illustrations are confident, often simplified pen-and-ink drawings with washes. They look like clean copies of earlier pictures or sketches. But they are observational nature studies in the sense that they are relatively accurate and relatively new. They have little in common with earlier herbal illustrations and thus reflect relatively recent acts of direct observation. This novelty is further confirmed by the date of the *Tractatus* itself: it is not antique but is rather based on the twelfth-century *Circa instans*.

Despite the accuracy of many illustrations, they tend to appear diagrammatic and flat, as though pinned and pressed. There are few indications of volumetric form through modeling and shading. This way of illustrating plants emphasizes visual clarity, particularly the visibility of leaf shape, margin, and general habit. As Otto Pächt has noted, the pictures of Egerton MS 747

> cannot yet be counted as portraits in the full sense of the word. . . .
> They contain a great mass of intimate observation of details, but
> for the plant as a whole they still fail to supply the organic con-
> nection of the parts. They either adhere to decorative pattern . . .
> or else press naturalistic details into conventional plant schemes as
> for instance in the case of the pine-tree. The needles are correctly
> rendered and the pine-cones are correct, but the whole has the
> appearance of a small plant, not of a tree. Moreover, even where
> the single parts and their mutual relationship are satisfactorily
> depicted, the botanic specimens never resemble living plants; they
> are pressed flat—into profile or full frontal views—artificially ar-
> ranged, prepared for the Herbarium; half picture, half diagram.[14]

Pächt places Egerton 747 as a missing link between the illustrations in previous Latin herbals (for instance, the illustrated copies
of Ps.-Apuleius Platonicus, as in fig. 2.1) and later, more naturalistic
and illusionistic illustrations represented by the likes of the late four-
teenth- or early fifteenth-century Carrara herbal (see chapter 8). But
Pächt's observations about the rendering of the stone pine (*pinea*, fig.
7.9) also relate to difficulties specific to the rendering of trees, most
notably the challenge of simultaneously capturing a tree's characteristic features and its general habit.[15]

While Pächt compared the illustrations to pressed specimens in
order to underscore their didactic form and function, subsequent researchers have suspected that pressed plants may actually have been
used as models for some of the illustrations in Egerton 747.[16] Many
medicinal herbs were obtained in their dried, though not necessarily

flattened, forms from herb collectors.[17] The choice to emulate dried, flattened forms could have arisen from the use of such plants in teaching. This would go hand in hand with the codex's purported scholarly origin, perhaps at a medical "school" in Salerno.[18] We might also recall here the connection between Hellenistic botanical illustrations and collections of *exemplaria* (see chapter 2). At the same time, however, the flattened aspect of the illustrations may have simply maximized the legibility of leaf shape.

Egerton 747 enjoys a prominent place among historians of art and science because so many of its illustrations are demonstrably accurate and novel. This does not mean that *ex novo* plant depiction based on the observation of nature originated with this codex. The poor rate of survival of manuscripts generally makes it difficult to determine a start date for the direct observation of nature in medieval botany. In chapter 6, we saw examples of copying informed by varying degrees of nature observation, perhaps as early as the ninth century. As we will see in this chapter, there is indeed evidence of nature observation in botanical illustrations preceding Egerton 747.

Addressing Absence in the Morgan Dioscorides

The botanical illustrations from the *Tractatus de herbis* seem especially novel because they accompany a relatively new text.[19] But a text does not need to be new in order to receive novel illustrations.[20] Even within illustrated Greek manuscripts of Dioscorides, where hundreds of illustrations were typically copied from earlier, even ancient, manuscripts, we encounter *ex novo* illustrations. This is due to gaps and absences within the manuscript tradition. While Dioscorides' *De materia medica* had been illustrated in antiquity, there is little evidence to suggest that it was ever illustrated in its entirety. The illustrated version of Dioscorides known as the Alphabetical Herbarium originally excluded illustrations of larger plants, such as trees, not to mention illustrations of animals, metals, wines, and oils. The Old Paris Dioscorides would have been more comprehensive, but it is today fragmented.[21] And while the now-fragmented Old Paris Dioscorides would have once contained illustrations of trees, there is no evidence to suggest that tradition was available to the makers of the Morgan Dioscorides.

The earliest extant illustrations of trees in a Byzantine herbal come from the late ninth- or early tenth-century Morgan Dioscorides. As noted in chapter 6, the Morgan Dioscorides represents a version of the text that combines the Alphabetical Herbarium and original versions of Dioscorides. The book on trees in the Morgan codex contains many illustrations that had been excluded from the Alphabetical Herbarium. The only illustration in it that can be traced back to the Alphabetical Herbarium is that of the wild vine (*ampelos agria*).[22] Most of the illustrations in the fourth book derive from other sources.[23] We can divide most of the illustrations of trees into three broad groups according to their details and proportions: the first group is relatively small and includes few specific details; the second group is more schematic, but fairly accurate, with many observed details such as leaf shape and fruit; and the third is more modeled and proportionately tree-like, but ultimately less detailed and sometimes more inaccurate.[24]

Examples from the first and second groups coincide above a short chapter on juniper "berries" (actually fleshy cones, *kedrides*, fig. 7.3), which was isolated from the main chapter on juniper and cedar (*kedros*, fig. 7.4). The codex has two completely different pictures: a laurel and a juniper. While juniper represents the content of the chapter, the picture of the laurel illustrates the erroneous title of this chapter, "a different laurel" (*daphnē hetera*).[25] This error likely arose during the compilation of the text: in the original *De materia medica*, laurel and *kedros* juniper appear right after each other.[26] The mistake could have occurred when Dioscorides' order by drug action was adapted to the alphabetical order of the Morgan Dioscorides. A scribe evidently first excerpted the juniper berries from the main chapter on *kedros* but then neglected to move the text on juniper berries to be with the other plants starting with the letter kappa. This is essentially a copy-and-paste error. The scribe writing the titles then confused this chapter on juniper berries for a chapter on another kind of laurel, as it follows the chapter on laurel. Such errors suggest that the scribe may have worked from an index that had been made to aid in the compilation of the text.[27]

The laurel is vague on details but conforms to a general understanding of what laurels look like. The fact that the laurel appears on the left side closer to the red title, which spills over into the next line, suggests that it preceded the juniper. Similar observations pertain to the tree heath (*ereikē*, here spelled *eirēkē*) depicted below on the same folio. Most of these generic depictions of trees tend to illustrate chapters apparently taken from an original version of Dioscorides.[28] These account, however, for only a few of the chapters and illustrations in the fourth book of the Morgan Dioscorides.[29]

The juniper, however, includes all the main details found in the more elaborated pictures from the Alphabetical Herbarium—not

FIGURE 7.3 Illustration of another laurel (*daphnē hetera*) and a prickly juniper with "berries" (*kedrides*), above, and tree heath (*eirēkē*). Morgan Dioscorides, fol. 246r. Photo: The Morgan Library & Museum, New York.

ΚΕΔΡΟΣ ⁖

ὒ Ϗ ὄυὓἦ Ϗ ρ ὀμ Ϗρ ϕ ρ Ϗ κϗδρ ί ᾳ συ ρϕ ϥ ϥᾳι. Ϯαρ ᾰὼν

ρϖαρ ί σ σου· μαυϗρό τϵρ ὀμϗϐϗτ οι ϖ∂ρ ατσοϕϝυ·

only its shrubby habit, roots, and shoots, but also its prickly needles, berries at multiple stages of maturation, and even a dead branch
on the lower right side. In a Constantinopolitan context, the picture
would have specifically identified an Eastern prickly juniper.[30] The
main illustration of laurel (*daphnē*) on the preceding folio (fig. 7.5),
has details more in common with prickly juniper than the other laurel.[31] It accurately portrays the laurel's alternating, lanceolate leaves
with subtly undulating margins and its fruit at multiple stages of maturation. The two illustrations of laurel invite different ways of seeing
and knowing: the detailed laurel directs us to the specifics of morphological detail, while the abstracted laurel shows us the essentials of
"laurelness," namely, a shrubby, upright habit, lanceolate leaves, and
dark berries.

   The details visible in the picture of the main laurel and the prickly
juniper echo those seen in other pictures of the second group. There
pictures account for about one-fifth of the pictures in the fourth

FIGURE 7.4 Illustration
of a cedar and Phoenician
juniper (both different
kinds of *kedros*). Morgan
Dioscorides, fol. 251v.
Photo: The Morgan Library
& Museum, New York.

ΔΑΦΝΗ

FIGURE 7.5 Illustration of a laurel (*daphnē*). Morgan Dioscorides, fol. 245v. Photo: The Morgan Library & Museum, New York.

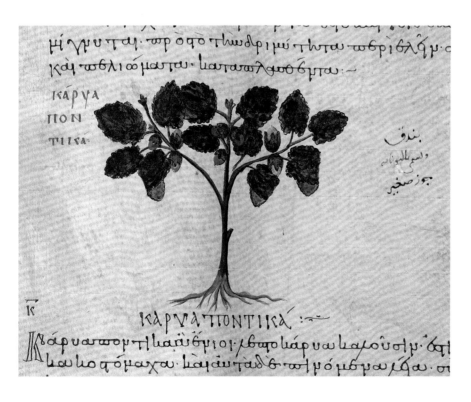

KΑΡΥΑ ΠΟΝ ΤΙΚΑ

FIGURE 7.6 Illustration of a hazel plant (*karya ponti-ka*). Morgan Dioscorides, fol. 255r. Photo: The Morgan Library & Museum, New York.

ΜΟ ΡΕ
α:

book.[32] This group tends to emphasize leaves and fruits, at various stages of development, as in the depictions of the hazelnut (*karya pontika*, fig. 7.6), mulberry (*morea*, fig. 7.7), and cherry (*kerasea*, fig. 7.8).[33] In general, the leaves and fruit appear as though they were applied to a frame and adjusted as necessary to match the plant depicted. This emphasis on the roots, fruits, and leaves reflects the contemporary understandings of how to identify plants. As we have seen in earlier chapters, ancient botanists did not understand the function of flowers in plant life cycles. Flowers were also typically less available for identification.

The emphasis on fruit, however, responds to an abiding philosophical interest in the aim or purpose of plant life, that is, its *telos*. In *De generatione animalium* (*On the Generation of Animals*), Aristotle claimed that plants existed only to produce seeds.[34] His successor, Theophrastus, although less teleological in his thinking, still devoted considerable attention to the maturation and development of fruit.[35]

FIGURE 7.7 Illustration of a mulberry tree (*morea*). Morgan Dioscorides, fol. 259v. Photo: The Morgan Library & Museum, New York.

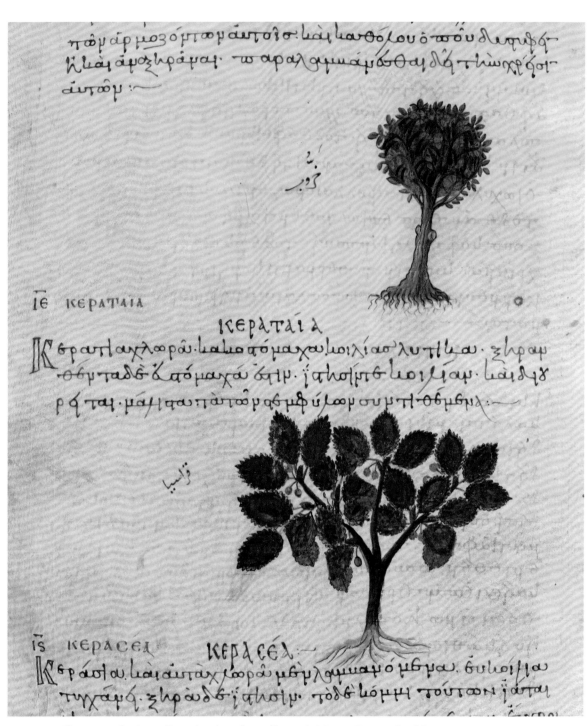

τῶν ἁρμοζόντων αὐτοῖσ· καὶ καθόλου ὁ ττῶν διαφέρῃ
δ' καὶ ἀποτράψαι· ττ αραλαμμαψέσθαι διὰ τὴν χρέσι
αὐτῶν :⁓

ΙΕ ΚΕΡΑΤΑΙΑ

ΚΕΡΑΤΑΙΑ

Κεραττ αχλωρὰ· καὶ αιο τόμαχου κοιλίασ λυτικα· ζηραν
δὲ ττ αδξ ἃ τόμαχά ὅτιν· ἡ αττ οιρτὸ κοιλίασ· καὶ δῆξ
ρά ται· μάῃ ιται τὰ τὸῷ ραφ μῆ ὕ λωρ συ μττι τόμβηλ:⁓

ΙΣ ΚΕΡΑCΕΛ ΚΕΡΑCΕΛ :⁓

Κεράσι αι· καὶ αὐτ αχλωρὰ μῆρλ αμμαψ ό μῆραι· ὑ λωῃ ια
τυγχά ρᾳ· ζηρὰ δὲ ἡ αττ οιρ· τὸδῆ κόμῃ τούτων ἡ ατ αι

FIGURE 7.8 Illustration of a cherry (*kerasea*) and carob tree (*kerataia*). Morgan Dioscorides, fol. 253v. Photo: The Morgan Library & Museum, New York.

According to ancient botanical thinking, the illustrations specifically portray the maturation or "concoction" of fruit. This process was said to involve the sun's heat stewing the fruit's interior and causing its juices to evaporate, which were in turn replaced by moisture and nutrients from the soil.[36] In the case of a fruit such as the mulberry, this process caused the fruit to change color, sweeten, and become juicier. The accompanying text says none of this. The illustrations thus enabled visual reflection on the *telos* and principle (*logos*) of fruits. The pictures here allowed the manuscript to serve for "speculations in philosophy and natural science," as we saw Photios recommend (see chapter 5).[37] Such reflections doubtless entertained some readers. The Byzantine intellectual Michael Psellos (d. after 1081), for example, delighted in speculating on the "machinery" of nature for the very reason that it provided joy.[38] In one letter Psellos playfully muses on the armored defenses and brain-like form of the walnut; in another, the curiously hard shell and cosmic form of the chestnut.[39]

While the illustrations in this group contain the main elements associated with the most elaborated illustrations from the Alphabetical Dioscorides, they cannot have come from there. The Alphabetical Herbarium had excluded trees. These pictures do not resemble ancient botanical illustrations, and they were not likely adapted from other ancient depictions in other media, on account of their peculiar proportions, large leaves, and thin trunks. Their proportions are in fact ideally suited to the layout of the text in the Morgan Dioscorides. They also could not have come from a reading of Dioscorides, as none of the texts associated with these pictures includes substantial descriptions of their leaves and fruits. The text on the mulberry, for example, excludes description and merely states that the tree is "known to all."[40] At the same time, the illustrations' accurate details suggest that they were not far removed from pictures based on the direct observations of nature. The trees depicted in this group were cultivated in and around Constantinople for culinary, economic, and ornamental uses.[41] Plants suited to hotter climates, such as date palms, pistachio, and carob trees, though common in the ancient Mediterranean, do not appear in this group. The error in the collation of the text in the chapter on the juniper berries further supports the idea that these illustrations were relatively recent, Byzantine creations. They may have even been created for the Morgan Dioscorides itself, in emulation of the more detailed pictures from the Alphabetical Dioscorides.

These illustrations in the Morgan Dioscorides bear some superficial resemblance to those in the *Tractatus de herbis*.[42] For example, we can compare the stone pine (*pinea*) in Egerton 747 (fig. 7.9), which

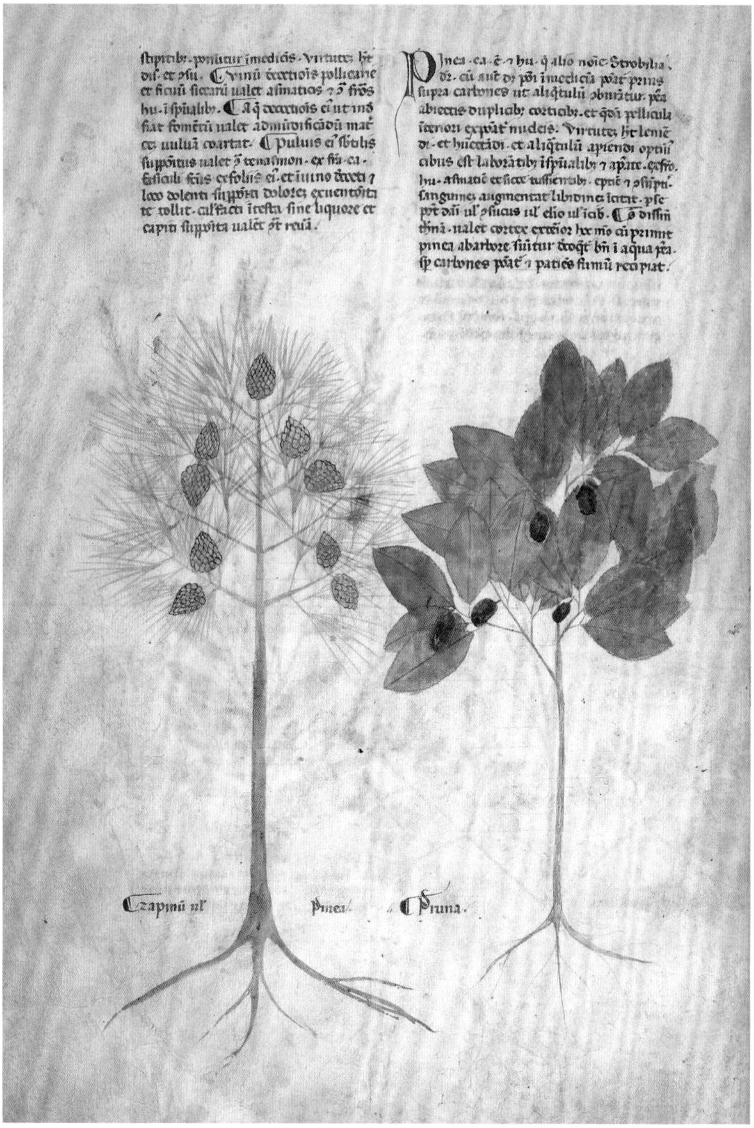

FIGURE 7.9 Illustration of a stone pine (*pinea*) and plum (*pruna*). MS Egerton 747, fol. 74v. © The British Library Board.

Otto Pächt had singled out for its accurately observed detail and schematism, directly to the one (*strobylea*) depicted in the Morgan Dioscorides (fig. 7.10).[43] (Egerton 747 in fact lists *strobilia* as a synonym for *pinea*.)[44] They both show trees more as saplings than as adults, evidently to emphasize distinguishing features, such as needles and cones. This does not mean the manuscripts are connected; rather, they reflect similar ways of thinking about and visualizing distinguishing features. The Morgan illustration actually includes more details than does Egerton 747, such as the needles growing in sheathed pairs, and the three different cones: the small strobili at the ends of the branches; the small, immature cones; and the large, woody

mature cones. These differences, particularly the greater attention to diachronicity across plant parts, again suggest that the makers of the images in the Morgan Dioscorides emulated the Alphabetical Herbarium or were inspired by philosophical interest in botanical *telos* and Dioscorides' empirical concerns.

Turning to the main chapter on the *kedros*, we find two illustrations that represent a third approach to tree depiction (fig. 7.4).[45] Here we find trees rendered more proportionally with a greater emphasis on their trunks. The roots, though not always represented, are typically tangled, short, and wavy. The two pictures of *kedros* correspond to two varieties mentioned in the text: a big *kedros* and a small, thorny one.[46] That the tree depicted on the left has large, upright cones suggests that it is a cedar.[47] (The Greek *kedros* can designate cedars as well as junipers.) Wellmann's edition describes the cones (literally, the "fruit") of the big *kedros* as smaller than those of a cypress.[48] In contrast, the

FIGURE 7.10 Illustration of a stone pine (*strobylea*). Morgan Dioscorides, fol. 269v. Photo: The Morgan Library & Museum, New York.

Morgan Dioscordes says the cones of the big *kedros* are *bigger* than those of the cypress.[49] The manuscript's makers clearly understood the big *kedros* to be cedar. The small, thorny *kedros* on the right is likely a juniper, perhaps Phoenician juniper due to its yellow fruit.[50] The pictures in this group often share compositional similarities that hint at the painter's basic approach. The right *kedros* and the *kynosbaton* (see fig. 6.14) are similarly composed, for example.[51] In them, the painter first sketched a trunk with three divisions leading up to a dense ball of vegetation, and then blocked in the main areas and added details. This third group of illustrations is the largest in the fourth book of the Morgan Dioscorides.[52]

This third group of pictures can be fairly accurate, as seen in the chapter on *kedros*. Still they are not as detailed as the second group of pictures.[53] As before, the most accurate depictions are of trees that would have been more familiar to contemporaries in and around ninth- and tenth-century Constantinople.[54] A notable exception, however, is the carob tree (*kerataia*, fig. 7.8), which, although commonly found in southern Anatolia and in the Near East, is unsuited to the cooler climate of Constantinople.[55] The depiction of carob is accurate in capturing the shape of its pods, but the rest of the plant is

FIGURE 7.11 Illustration of a sumac tree (*rhous dendron*). Morgan Dioscorides, fol. 264v. Photo: The Morgan Library & Museum, New York.

fairly generic. Carob pods may have been available in contemporary Constantinopolitan markets. It is possible then that part of the illustration is based on familiarity with only part of the carob tree. Other illustrations are based on the description of the plant given by the text. For example, the depiction of the sumac tree (*rhous dendron*, fig. 7.11) shows its fruit hanging down at the bottom like grapes, whereas the actual tree has its fruit at the top of the plant.[56] The text describes the fruit as growing in "grape-like clusters, close-packed, corresponding in size to terebinth fruit and somewhat flat," but does not specify where on the tree the fruit is found.[57] The painter apparently based the picture on the description of the plant in Dioscorides.

In short, the depictions of trees in Morgan Dioscorides were based on earlier manuscript illustrations, readings of the text, or observations of actual plants and plant parts. Many of these illustrations could have been made *ex novo* for the Morgan Dioscorides, especially as their accuracy and approach seem to vary according to contemporaries' familiarity with the plant. In general, while the more generic illustrations might be connected to one of the hypothetical textual sources for the Morgan Dioscorides, the variety of approaches employed in the second and third groups tends to suggest that they were not derived from a single source. If they had been copied from one manuscript source, we would expect them to be more homogeneous. But this is not the case. The hypothetical source could already have been rather heterogeneous. Alternatively, we can imagine that the other hypothetical textual source was unillustrated, or only partly illustrated. *Ex novo* depiction is especially likely in the case of the prickly juniper, given that it corrects an error that arose in the compilation of the manuscript. The accuracy and apparent novelty of many of the illustrations suggest that Byzantine painters occasionally depicted nature from observation.

Accepting that some of these illustrations were based (at least in part) on the observation of actual plants raises the question of the artist' working methods. They generally worked by adding features to an underlying frame. In some cases, the painters needed only to have seen parts of the plant. Consequently, they may have worked from plant parts obtained from a market or collected from another site. In other cases, however, the fact that multiple stages of fruit maturation are depicted suggests that the painters had experience with living plants, as encountered in an orchard or garden. Many of these trees could have been found in gardens and orchards within the capital at that time, such as Basil I's Mesokēpion, which purportedly abounded "in every kind of plant."[58] There were likely many other smaller, private gardens in the city. The writer John Geometres (d. 1000), for example, provides us with a vivid description of his own garden.[59] That the more accurate and detailed depictions of plants were cultivated and familiar in Constantinople suggests that the originators of these illustrations worked in that urban (and suburban) context and may not have ventured far beyond it.[60]

The illustrations of trees in the Morgan Dioscorides thus provide a limited view into the plants of the Byzantine capital in the late ninth or early tenth century. This view contrasts with the realities that we encounter outside of the capital. Alexander Olson has shown that the wider Aegean region was at this time dominated by forests of evergreen and deciduous oak.[61] While planting limited subsistence crops, Byzantine peasants managed these forests for wood, fodder, and pasturage. The Morgan Dioscorides hints at a disconnect between the lived experience of the natural environment in the gardens of the capital and the forests and fields of the countryside in the rest of the empire. At the same time, the Morgan Dioscorides adds to a broader picture of a medieval botanical world. The diverse plantings in Constantinople likely existed as an ideal for many people living in the Balkans, the Aegean, and Anatolia.

Invention in Illustrated Arabic Herbals

Novel observational depictions of nature also appear in illustrated Arabic herbals. While many of the illustrations in Arabic translations of Dioscorides go back to earlier Greek antecedents in the Alphabetical Herbarium and the illustrated original Dioscorides, we invariably encounter illustrations that do not reflect those earlier traditions. For example, we find an illustration of aloe in the twelfth-century Parchment Arabic Dioscorides in Paris (fig. 7.12) that is strikingly different from the illustration of aloe in the Old Paris Dioscorides (fig. 4.9). The Parchment Arabic Dioscorides shares many illustrations from the earlier tradition of the Old Paris Dioscorides. The picture of aloe in the Old Paris Dioscorides shows few morphological details and gives an inaccurate impression of aloe leaves. By contrast, the picture of aloe in the Parchment Arabic Dioscorides is more detailed and accurate. It conveys the correct shape of the aloe's leaves, its spiny leaf margins, a spike inflorescence, and dead, reddish leaves at the base of the plant. These different illustrations suggest either the existence of a no-longer-extant version of an illustrated Dioscorides from antiquity, or a critical development within the medieval Greek or Arabic tradition based on observation of or familiarity with aloe.[62]

Arabic-speaking scholars also added new illustrations when they ran into gaps in the existing tradition. Mahmoud Sadek has observed, for example, that one of the earliest surviving illustrated Arabic herbals, the 1083 CE (AH 475) Leiden copy of al-Natilī's rectification (*iṣlāḥ*) of Iṣṭifan's Arabic translation of Dioscorides (fig. 5.1), included many illustrations of plants and animals that had been absent from earlier illustrated copies of Dioscorides in Greek.[63] As Arabic authors updated their translations of Dioscorides' *De materia medica* and wrote new herbals, they also needed to include medicinal substances that had been unavailable in the ancient Mediterranean. Already in the late tenth century, the Andalusi author Ibn Juljul (d. after AH

FIGURE 7.12 Illustration of aloe (*ṣabir*). Parchment Arabic Dioscorides, fol. 50v. Photo: Bibliothèque nationale de France.

377/987 CE) compiled a list of substances absent in Dioscorides.[64] Some centuries later, al-Ghāfiqī (d. c. AH 560/1165 CE), another Andalusi author, authored a new herbal, his *Kitāb fī l-adwiya al-mufrada* (*Book of Simple Drugs*), which included many of these medicinal substances. Two later illustrated copies of al-Ghāfiqī's herbal have been found. The earliest of the two, now in the Osler Library in Montreal, can be dated by its colophon to 1256 (AH 654).[65] Many of the illustrations in the manuscript go back to the tradition of illustration in the Old Paris Dioscorides.[66] But some illustrations in the Osler codex cannot have come from that earlier tradition because al-Ghāfiqī included plants that had been absent from the original Dioscorides. For example, the Osler codex contains a recognizable illustration of an eggplant (also called an aubergine, *bādhinjān*), complete with light purple flowers and mature, dark purple fruit, along with simplified, unlobed leaves (fig. 7.13).[67] Eggplants are endemic to South Asia and were unknown to Dioscorides.[68] Ibn Juljul had earlier included the eggplant in his list of substances absent from Dioscorides.[69] The illustration of the eggplant in al-Ghāfiqī's herbal therefore has to have been a postclassical addition, an *ex novo* creation within the Arabic botanical tradition based on direct observation or habitual familiarity with this common vegetable. Overall, however, the Osler codex emulates the existing established visual traditions of the Arabic Dioscorides, despite its novelties.

The Arabic botanical tradition, however, also provides the earliest evidence for botanical scholars and artists going out into the field to document plant life directly from nature. Ibn Abī Uṣaybiʿa (d. AH 668/1270 CE) records in his ʿUyūn al-anbāʾ that the physician Rashīd al-Dīn ibn al-Ṣūrī (d. AH 639/1242 CE) employed a painter to depict plants from life:

> [Al-Ṣūrī] began to compose [his *Kitāb al-adwiya al-mufrada* (*Book of Simple Drugs*)] during the reign of al-Malik al-Muʿaẓẓam [r. 1218–27], to whom he dedicated it. The book gives a full account of simple drugs, and also provides insight into simples of which the author had acquired knowledge, and which had not been mentioned by his predecessors. Rashīd al-Dīn ibn al-Ṣūrī would go to places in which plants grew, such as Mount Lebanon and other spots in which particular plants were found, taking along with him a painter who had at his disposal all kinds of colors and brushes. Rashīd al-Dīn would observe and examine the plants, and then he would show them to the painter, who would look at their color, measure their leaves, branches and roots, and then paint them, and strive to

جزء وىتولد عنه اذما به الامراض السوداويه وبفتح سدد الكبد والطحال والخل
والدهن بطحانه وانما ينقى الحده والجراه سنه المشوى منه بلا دهن وشربه ما يوكل
منه المشوى الى غيره اذا اشوى على الملح حتى يآوه وتذهب مرارته لمتين
له ضرر فان اكل على هذه الصفه الخل اطفى الصفرا ونفع من الخفقان والمرض
بالبين ولا يبال ان اكله واذا الرعمل به ذلك كان نفعا لسدد الكبد والطحال
وخاصته اذا اكل الا انه
ردى الغذا متولد للسدد
مفسد للبشره ويبغر اللون
وبفسد ويورث اذما انه
الكلف والسرطانات
والبواسير والصلابات
والجذام ويصدع الراس ويبير
العراب سيما الفى منه
ردى والجلد شاسلم وعذاب
ما تخرجه انه ازد والصحم
انه حار ز بانتي الثابته
وهو متولد لفساد الكبد والطحال الا انه اذا طبخ بالخل فمن اكله فتحا ولينه اطفل البطن
ولا بطلته ويجى اقماعه المجففه فى الطلا طلاه نافعا للبواسير واذا اخذ
من جوف الباذنجان المسلوق اوقيه ومزج بنصف الشراب مزتا وسقى اذر اللبن
غيره اذا غلى الزيت واذيب فيه شمع كان منه قبروطى نافع من شقاق
الاطراف من البرد واذا اجرق وعجن رماده بالخل قلع الثآليل
بصل هذا الرابعه من درجات الاشياء التى يسخن
وجوهر جوهر غليظ فهو متولد للنب اذا ادخل المعده فتح افواه العروق

FIGURE 7.13 Illustration of eggplant (*bādhinjān*). Montreal, McGill University, Osler Library, MS 7508, fol. 73r. Courtesy of Osler Library of the History of Medicine, McGill University.

represent them. Rashīd al-Dīn had an instructive method for these illustrations: first he would show them to the painter at the time of sprouting and tenderness, and would have him paint them at that stage. Then, he would show them to him when they were fully grown and in full bloom, and the painter would depict them at that specific stage. Finally, he would show him the plants when they were withered and dried up, and the painter would sketch them at that stage. In this way, the reader of the book could see the plants as he would encounter them in the field, and this would enable him to obtain more perfect information and clearer notions.[70]

In al-Ṣūrī we find an acute interest in what Dioscorides had called *autopsia*—seeing with one's own eyes. In chapter 2, we saw that pictures simulated the experience of *autopsia*. Ibn Abī Uṣaybiʿa explains that the pictures in al-Ṣūrī's herbal had a similar impact: "the reader of the book could see the plants as he would encounter them in the field." Al-Ṣūrī's working methods also clearly respond to Dioscorides' insistence that plants be seen at multiple stages of growth. Dioscorides' preface may have served as an important inspiration for empirical practice in the Arabic botanical tradition. In referring to Mount Lebanon, Ibn Abī Uṣaybiʿa also makes Al-Ṣūrī follow the example of Dioscorides. Ibn Abī Uṣaybiʿa elsewhere notes, citing Ḥunayn ibn Isḥāq, that Dioscorides "lived isolated from his community in the mountains and in regions rich in vegetation."[71] This description echoes the situation of Mount Lebanon.

Al-Ṣūrī's interest in direct observation was also shared by his mentors and contemporaries. His mentor, the philosopher and physician ʿAbd al-Laṭīf al-Baghdādī (d. AH 629/1231 CE), repeatedly emphasized his own observations in his extant writings.[72] The Andalusi herbalist Abū l-ʿAbbās al-Nabātī (d. 1238, also called Abū l-ʿAbbās al-Jayyānī), another one of al-Ṣūrī's mentors, traveled to the Eastern Mediterranean as part of the Hajj, but no doubt took it as an opportunity to study plants unknown in his homeland.[73] Ibn Abī Uṣaybiʿa similarly reports that Ibn al-Bayṭār (d. AH 646/1248 CE), another Andalusi scholar and a contemporary of al-Ṣūrī, went to the land of the Greeks (*al-Aghāriqa*), into Asia Minor (*bilād al-Rūm*), "where he not only met experts . . . from whom he obtained a great deal of knowledge about plants, but also observed the plants in their natural environment."[74] Ibn al-Bayṭār would go on to author a vast herbal, his *Kitāb al-jāmiʿ li-mufradāt al-adwiya wa-l-aghdhiya*.

When al-Baghdādī and Ibn al-Bayṭār traveled out to see plants in the world with their own eyes, they echoed the Empiricist concerns that Dioscorides had long before voiced in his preface. Through their medical training, they would also have encountered similar concerns in other texts. Galen, for example, extolled a combination of observation and theory as an ideal in his synthesis of Empiricist and Dogmatist medicine. The famous physician Al-Rāzī (d. c. AH 313/925 CE) followed this example by maintaining detailed records of his observations and experiences while working as a physician in a hospital, though he also warned his students of the dangers of relying upon "unqualified experience"—experience without theory.[75] The late twelfth and early thirteenth centuries, however, saw a greater emphasis on experience-based medicine, particularly among a circle of largely Damascene physicians associated with Ibn Abī Uṣaybiʿa's mentor, Muhadhdhab al-Dīn al-Dakhwār (d. AH 628/1230 CE).[76] Many of the physicians in al-Dakhwār's circle would go on to author their own works based on their practical experience. This group and their writings would have contributed to a broader culture of medical inquiry in the Levant that emphasized direct, practical experience.

Travel plays an important part in the observational studies of al-Baghdādī and Ibn al-Bayṭār. In doing so, they, like al-Ṣūrī, followed the example of Dioscorides. Ibn Abī Uṣaybiʿa had noted that Dioscorides "wore himself out traveling the world."[77] At the same time, however, their extensive travels also respond to the concerns of the Arabic geographical tradition. Starting in the ninth century, Arabic geographers had given direct observation (ʿiyān) a central place in their writings.[78] This legacy is apparent in the title and preface to the *Kitāb al-ifāda wa-l-iʿtibār fī l-umūr al mushāhada wa-l-ḥawādith al muʿāyana bi-arḍ Miṣr* (*The Book of Edification and Admonition: Things Eye-Witnessed and Events Personally Observed in the Land of Egypt*) by al-Ṣūrī's mentor al-Baghdādī.[79] The fact that Ibn al-Bayṭār traveled in order to carry out botanical inquiries indicates a keen awareness that botanical life is specific to place and could thus be understood according to geography, the science of place.

It is difficult to gauge the impact of al-Ṣūrī's illustrated herbal today. Except for some textual fragments, his herbal has not survived.[80] Ibn Abī Uṣaybiʿa's account of al-Ṣūrī's life and work was, moreover, informed by his own acquaintance with (and high regard for) the man. Even if al-Ṣūrī's herbal did not achieve a wider distribution, perhaps on account of the difficulty of copying it, his ideas and methods could easily have circulated. Ibn Abī Uṣaybiʿa tells us that al-Ṣūrī traveled between the Levant and Egypt multiple times over the course of his distinguished career as a prominent physician to several Ayyubid emirs. Later in life, al-Ṣūrī settled in Damascus, "where he established a scholarly salon that was frequented by many persons wishing to study the medical art."[81] We can suppose that al-Ṣūrī's ideas and methods for botanical inquiry may have spread to his contemporaries initially through his prominent positions at the courts of several emirs and later during his residency in Damascus.

Although al-Ṣūrī's herbal does not survive, we can look to a contemporary manuscript: a redaction of Dioscorides in Arabic now in the Topkapi Library in Istanbul.[82] We considered one of the frontispiece illustrations from this manuscript in chapter 5. The text in this manuscript is related to that in the twelfth-century Parchment Arabic Dioscorides now in Paris.[83] While many of the illustrations in the manuscript tend to follow earlier Arabic illustrated manuscripts of Dioscorides, many of which go all the way back to the tradition of the Old Paris Dioscorides, there are also a variety of botanical illustrations clearly drawn from other sources or created *ex novo*. These include a naturalistic depiction of a vine (fig. 7.14). Richard Ettinghausen has gone so far as to say: "It is a faithful copy of a classical type of illustration, so faithful indeed that if it were not painted on paper one would be inclined to regard it as a Greek 'original' inserted into the Arabic volume."[84] This particular illustration, however, need not have been copied from a classical or Byzantine manuscript source, as the vine was a common decorative motif in the medieval Mediterranean. Whatever the source and resonance of the image, the illustrator likely intended to evoke the imagery of an ancient illustrated herbal of the kind that Dioscorides himself may have authored and illustrated, at least according to common medieval understandings of what Dioscorides' original work looked like.[85]

Sergio Toresella has further identified several illustrations of garlic (*thawm*) in the codex as the earliest depictions of plants executed from life (fig. 7.15).[86] We have already seen in this chapter that there may be earlier examples of the observation of nature. But Toresella is right to emphasize the artist's peculiar attention to the play of light across the lined surfaces of individual cloves. This attention to three-dimensionality is striking and without many contemporary parallels. Despite this care, it remains unclear how the illustration might be based on direct observation. The garlic heads here seem to lack their papery outer sheathing. Moreover, an illustration of three garlic plants once occurred in the Parchment Arabic Dioscorides.[87] The garlic in the Topkapi manuscript may thus represent an instance of a painter improving upon an illustration in the process of copying it. In doing so, the copyist may have drawn upon an actual garlic plant, perhaps shucked to show individual cloves.

بـ نوعا كرم نرى نوعان الاول لاجل له عنبا لكنه يطحل من زا ويستى نفاح

الكرم والنوع الثاني يكل له عنبا وعنبه صغير اسود والنوع الثالث

والاول قوته ومنفعته كالكرم البستاني

الطبع

FIGURE 7.14 Illustration of a vine. Istanbul, Topkapi Library, MS Sultanahmet III 2127 (the "Topkapi Dioscorides"), made 1228/9, fol. 252v. Image: © Presidency of the Republic of Türkiye and Directorate of National Palaces Administration.

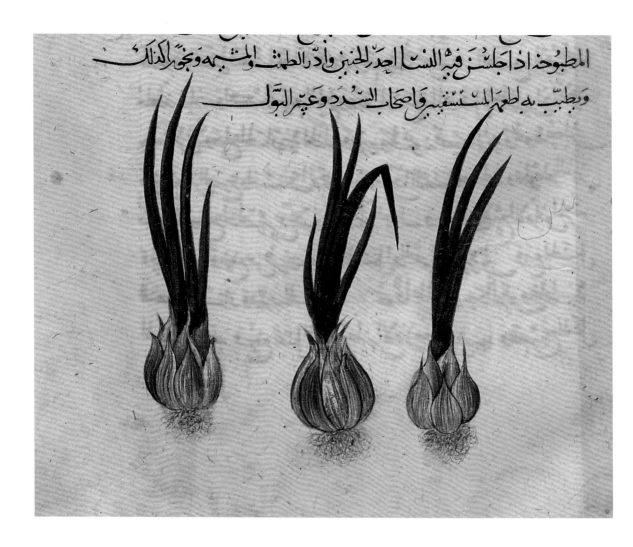

المطبوخة اذا جلس فيه النساء احد الجبن واذرا الطمث المتشبه ونحو ذلك

ويطيب به لطعم المستفير واصحاب السدد وعسر البول

The Topkapi Dioscorides may also contain the earliest surviving examples of nature printing in a scientific treatise (figs. 7.16 and 7.17).[88] The illustrator here inked the leaves from two different kinds of *karafs*, a plant name that can designate celery, parsley, and similar plants, and then stamped them on the page. Nature printing here may have been employed to more fully distinguish between the two *karafs* depicted. They represent cultivated celery (*karafs bustānī*, fig. 7.16) and parsley (*karafs ṣakhrī*, fig. 7.17).[89] (Pictures usually follow rather than precede the relevant text in illustrated Arabic herbals.) There were, however, many kinds of *karafs*, which could go by different names.[90] All *karafs*, moreover, belong to the carrot family (Apiaceae) which includes visually similar yet lethal species such as poison hemlock, deadly carrot, and fool's parsley. The nature prints may have thus helped to secure the reference of the text to these specific *karafs*. Nature prints have a different relationship to their referents than do

FIGURE 7.15 Illustration of garlic (*thawm*). Topkapi Dioscorides, fol. 97r. Image: © Presidency of the Republic of Türkiye and Directorate of National Palaces Administration.

regular botanical illustrations. They are a trace of direct, physical contact between the leaf and the manuscript page. The leaf is then itself causally implicated in the production of the image. The nature print could thus be said to have an indexical relationship to its referent; the role that direct contact plays in generating the image guarantees the accuracy of the image and thus its capacity to refer.[91]

The ultimate inspiration for nature printing here remains unclear. Sergio Toresella has suggested that it goes back to traditions regarding holy images, literally not made by human hands [*acheiropoiēta*], that miraculously reproduce through imprinting.[92] Block printing, however, had already been used in the Islamicate world for making talismans and Hajj certificates.[93] At the same time, nature printing represents a convenient solution to some of the core problem of authoritatively and accurately reproducing the form of a leaf, though it risks over-representing an individual's variant features. It could have also simply arisen through the use of leaf prints for decoration.[94]

The two *karafs* in the Topkapi manuscript are not only made up of nature prints, but also include depictions of roots and stems. The taproots and new growth at the plant's basal rosette resemble what we would expect for parsley and celery and clearly demonstrate that the artist has observed nature beyond merely using the leaf prints. Leaf prints along with these distinguishing features were carefully arranged into an overall composition that echoes what we find in earlier Arabic and Greek Dioscorides manuscripts. However novel or uncommon this technique was, the basic approach here involved applying an inked trace of a distinguishing feature to an underlying frame based on conventions established by earlier botanical illustrations. This basic approach conforms to what we saw in the Morgan Dioscorides, where an artist applied specific distinguishing features, such as leaves and fruits, to a frame. That the leaf imprints are arrayed together in the Topkapi Dioscorides with purely iconic elements such as the roots means that the pictures are both indexical and iconic in their function. Different parts of the picture operate in different ways. This clear differentiation of referential functionality extends the hybrid visual form seen in both the Morgan Dioscorides and the Egerton 747 codex, where carefully observed details are juxtaposed with more conventional structuring schematic devices.

All of the extant examples that I have cited here—the eggplant in al-Ghāfiqī's herbal; the aloe in the Parchment Arabic Dioscorides; and the vine, garlic, parsley, and celery in the Topkapi Dioscorides—are common, cultivated plants.[95] They speak to working methods similar to those observable in the Morgan Dioscorides. Botanical artists of the

144

تكو
الطّبع

كرفس صغرى سنك شيراى دياادوقنا وماقذونياى الاشافين
بزوه يشبه النانخواه لكنه اطيب واكثر حـ وونة وهومدرالبول والطمش
مسكن لغز المعده والقولن والمعص وليجاع الحواصر والكلي والمثاند شربا
وخلط ي الاد وبه المدرللبول وى المعاجين

FIGURE 7.16 Nature print of celery (*karafs al-bustānī*) and painting of "mountain celery" (*karafs jabalī*).
Topkapi Dioscorides, fols. 143v–44r. Image: © Presidency of the Republic of Türkiye and Directorate of
National Palaces Administration.

نكه كرفس جبلي له قضبان تعلو شبر واحد من اصل غليظ مستدير زنيه

قضبان صغار وله راس كراس الشوكران وله ثمره طويله حريفه طيبه تشبه

الكمون منبته في الصخور والمواضع الجبليه لثمرته قوه اذا اشربها الجمرادرا البول

واحدر الطمث وخلطان في المعاجين والابازير الحارّه

الطبع

كذ كفرنـ كبير... نسه قوم مورينون وهو كبر واسين من الكرفس الستاني
وقضيبه محوف عالٍ غبر وفيه وفيبه كالخط واورا فيه ماعرض كثير مع سبل
الحمره فيه ماڙهر واله ثمر راسود طويل اصمن جرتيب واصله ابيض طيب الرائح
عنہ غليظ ونبت ونفيه مواضع كثيره الأذى وفي الاجام ونوكل الكرفس
نيا ومطبوخا ويوكل... وحده اوم النبك ... لبزه قوه يدر
بها الطمث... شربا الحلفين وسفع المبرود درشا واذا اخلط بهر ومسح به
بدن من بہ نطا رالبول نفعه واصله كذلك

الطبّع

FIGURE 7.17 Nature print of parsley (*karafs ṣakhrī*). Topkapi Dioscorides, fol. 144v. Image: © Presidency of the Republic of Türkiye and Directorate of National Palaces Administration.

Islamicate world thus worked closely with the plants that they grew around them. They wanted their illustrations to accurately reflect the botanical world as they knew it. That al-Ṣūrī's herbal was illustrated by an artist working out in the field thus marks a notable departure from a broader norm.

Emulation and Inspiration

It is difficult to know the degree to which botanical scholars and artists were aware of each other in the thirteenth century. Completed in January 1228 (626 AH), the Topkapi Dioscorides is roughly contemporary to al-Ṣūrī's now lost herbal. While al-Ṣūrī dedicated his work to al-Malik al-Muʿaẓẓam, the Ayyubid emir of Damascus, Abū Yūsuf Bihnām ibn Mūsā al-Mawṣilī, the scribe responsible for the Topkapi manuscript, dedicated it to the otherwise unknown ruler Shams al-Dīn Abu'-l Fadāʾil Muḥammad.[96] The timing is close enough that we could view some illustrations in the Topkapi manuscript as a reaction to or imitation of al-Ṣūrī's work, or part of the same intellectual movement that found inspiration in the example of Dioscorides.

Observational study and *autopsia* in the thirteenth century went hand in hand with interest in antiquity. In al-Baghdādī's *Kitāb al-ifādah wa-l-iʿtibār*, for example, we find lengthy, sensitive appraisals of Egyptian antiquities alongside descriptions of local flora, fauna, and customs.[97] Contemporary rulers across the Mediterranean, such as the Holy Roman Emperor and king of Sicily Frederick II (r. 1220–50) and Kayqubād (r. 1219–37), the Seljuq sultan of Rūm, emulated classical forms in their self-representation, especially in coinage, and embedded the remnants of classical statuary and other *spolia* into their public architecture[98] It is against this background of contemporary awareness of antiquity that we should view the naturalism of the Topkapi manuscript. The depiction of the grapevine is not just a foray in naturalism, but also an emulation of ancient botanical illustrations. It is the pictorial equivalent of Ibn al-Bayṭār's journeys to study plants in Dioscorides' homeland. In Asia Minor (*bilād al-Rūm*), the land of the ancient Greeks (*al-Aghāriqa*), Ibn al-Bayṭār similarly sought to go back to the roots—both antique and natural—of the botanical tradition. Emulation of Dioscorides was itself an antiquarian impulse.

The thirteenth century was a fruitful time for intellectual exchange among the botanical traditions of the medieval Mediterranean. The novel illustrations for the *Tractatus de herbis* may have been inspired by the pioneering work of Arabic herbalists. Minta Collins has already noted that news of Ibn al-Bayṭār's innovative research could have reached Southern Italy by way of contacts between the Levant and the court of Frederick II.[99] Ibn al-Bayṭār served the Ayyubid sultan al-Kāmil (r. AH 615–35/1218–38 CE), with whom Frederick II corresponded leading up to and during the sixth crusade (1228–29). Any word of Al-Ṣūrī's and Ibn al-Bayṭār's work, and perhaps the broader interest in Dioscorides' empirical practice, would have fallen on fertile ground in Southern Italy at that time. Frederick II is said to have undertaken quasi-scientific experiments, some of which he reports in his book on falconry, *De arte venandi cum avibus* (*On the art of hunting with birds*), written in 1240s.[100] And it is by the end of the thirteenth century that we encounter evidence for renewed Latin reception of the illustrated Greek Alphabetical Dioscorides as represented by the Thott Codex (see chapter 5 and 8).

Conclusion

Depicting plants through direct observation was not a novel invention of the *Tractatus* manuscripts in the thirteenth century. The four main examples of *ex novo* botanical illustration that we have considered here—in the Morgan Dioscorides, the Osler al-Ghāfiqī, the Topkapi Dioscorides, and Egerton 747—share some approaches. All involve a degree of schematization, adjusting closely observed details and applying them to a general compositional framework that was repeated over and over again. While the Morgan and Topkapi Dioscorides manuscripts occasionally demonstrate an interest in three-dimensional botanical form, as is evident in their sporadic use of modeling, they typically eschew that interest in favor of a flattened visual aspect that gave priority to visual clarity. This flattened visual aspect may indicate the influence of dried or pressed plants—literally pressed against the page, in the case of the Topkapi nature prints. While the depictions of plants in all four manuscripts tend to focus on leaf shape and margins, the Morgan Dioscorides places a greater emphasis on the rendering of fruit at multiple stages of maturation. This interest may reflect greater familiarity with detailed diachronic illustrations from the Alphabetical Herbarium and perhaps an interest in distinguishing between immature and mature fruit for philosophical or medicinal purposes. Arabic, Byzantine, and Latin medieval botany developed similar solutions for similar needs and pressures. And yet they had similar interests and used similar tools and strategies, not only because of their common inheritance of the ancient botanical tradition, but also because they were frequently in dialogue with each other, especially by the end of the thirteenth century. What ultimately separates the novelty of the *Tractatus* herbals from other examples of novel illustration seen here is not the fact of their novelty or their observation of nature, but rather the sheer quantity of them—al-Ṣūrī's exceptional, no-longer-extant herbal notwithstanding. From the standpoint of the Greek and Arabic botanical traditions, the number of *ex novo* pictures in the *Tractatus* could be viewed as compensation for a striking deficiency—the Latin botanical tradition's distance from Dioscorides, and its relative isolation from the Greek and Arabic botanical traditions.

8

ECHOES AND REVERBERATIONS

From the thirteenth to the sixteenth century, scholars and physicians across the Mediterranean world explored new modes of botanical visualization that were nevertheless rooted in the botanical traditions of the ancient world. They used, reproduced, and disseminated ancient botanical illustrations while editing, transforming, and completely refashioning them. This chapter turns to the ways visual botanical knowledge circulated in the late Middle Ages and how ancient botanical illustrations met the advent of early modern botany and new humanist modes of botanical illustration. Ancient illustrations inspired new forms of illustration and promised to help early modern scholars and physicians make sense of the ancient botanical tradition—above all Dioscorides' *De materia medica*.

Into the Margins

For centuries scholars, physicians, artists, armies, and dignitaries traversed the Mediterranean, disseminating botanical images along the way. Botanical images traveled not only as paintings in books purchased, taken as booty, or given as gifts, but also as pen-and-ink sketches and as memories. While botanical images likely circulated in these ways beforehand, we begin to encounter far more evidence for it in the thirteenth century.

For example, pictures from the late ninth- or early tenth-century Morgan Dioscorides were copied into the margins of a late tenth-century Constantinopolitan manuscript now in the Vatican library containing a reworking of Galen's *De simplicium medicamentum facultatibus* (fig. 8.1).[1] Scholars are divided on when the miniatures were added to the codex, though they were likely added in or by the fourteenth century.[2] Because the Morgan Dioscorides and Vatican Galen contain different texts, the illustrators had to carefully match pictures from the Morgan Dioscorides to the corresponding chapters in the Vatican Galen. By copying botanical icons into the margins, the illustrator thus concretized the scholarly act of seeking out a picture of

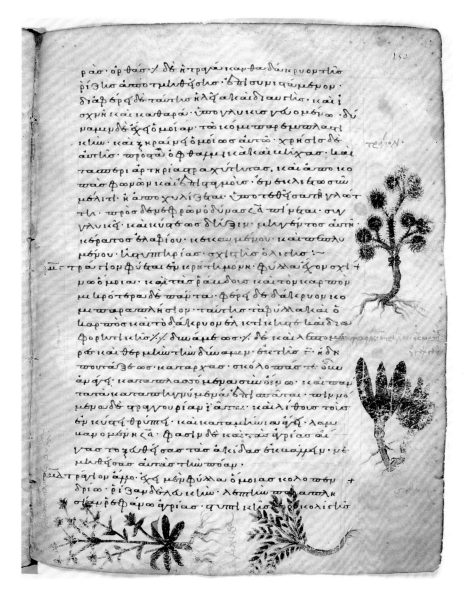

FIGURE 8.1 Marginal illustrations of plants added to Galen's *De simplicium facultatibus*. Vatican, Biblioteca Apostolica Vaticana, MS gr. 284 (the "Vatican Galen"), fol. 150r. © 2022 Biblioteca Apostolica Vaticana.

a plant from another source. Pictures were thus used in conjunction with other medical and pharmacological texts, which were unillustrated. Pictures from Dioscorides clearly had a life independent of the text within which they could be found.

Humbler marginal illustrations appear in two related fourteenth-century Byzantine manuscripts: one now in the Ambrosiana library in Milan (fig. 8.2), the other in the Marciana library in Venice.[3] The codex in the Ambrosiana is a composite notebook that developed in stages, while the Marciana codex is a cleaned-up miscellany handbook based on it.[4] Both contain texts that would have been useful to a physician of the time, including lengthy excerpts of Dioscorides, accompanied by marginal ink drawings of plants and animals likely copied from the Morgan Dioscorides (or a copy of it).[5] The marginal drawings give only a general impression of botanical form. Clear, discernible features, such as leaves, branches, buds, fruits, and flowers, dissolve in the fluid strokes of the copyist's pen.[6] They broadly convey only the most general view of growth habit. The sketches instead give the overall shape or *Gestalt* of a given plant as it appears in the Morgan Dioscorides. As a result, the sketches presuppose familiarity with pictures from the Morgan Dioscorides. By trimming out or abstracting details, the marginal sketches convey composition and configuration to readily cue memories of the pictures on which they are based. Like thumbnail images, they primarily orient viewers to their source images.

Both codices contain excerpts of *De opificio hominis* (*On the Making of Man*), in which the author states that a mere outline or shape is sufficient for unskilled observers to know the subject matter of a sculpture.[7] They do this through recognition of both a form and an intention. The sketches of plants similarly work as incomplete pictures that seek completion in the viewer's mind's eye. A few of the sketches even include small notes on the colors of the original paintings. For example, one note in the Ambrosiana codex accompanying a picture of wormwood reads, "green-blue and a purple root."[8] A few notes employ similes; for example, the note beside the aloe's roots reads, "a wine-colored root like lees of wine."[9] The sketches and color labels work together for the user to recall visual memories and thus reimagine the source picture of the plant. In cuing memories of the source picture, the sketch may also have called to mind a broader series of associated facts and knowledge related to the plant in question, including its habitat and medicinal properties.

These marginal sketches hint at how botanical images circulated in the late medieval Mediterranean. Given the humbler and more

FIGURE 8.2 Marginal sketches of plants. Milan, Biblioteca Ambrosiana, MS A 95 sup. (the "Ambrosiana Notebook"), fourteenth century, fol. 19v. © Veneranda Biblioteca Ambrosiana.

portable nature of the Ambrosiana and Marciana codices, their cursory sketches may have been *more* likely to have traveled widely and been used in the field than those in a large codex, such as the Morgan Dioscorides, which is perhaps more suited to indoor consultation on account of its ungainly size. Marie Cronier and Patrick Gautier Dalché have reconstructed some of the movements of the Ambrosiana Notebook.[10] From Constantinople, in the first half of the fourteenth century, its owner and copyist, likely a physician, took the codex to Cyprus, probably via Miletos, the main seaport for the emirate of Menteşe in the 1330s, where he was asked to cure Selman Pasha's son.[11] In 1345 or 1346, the owner noted a debt owed by the physician John (Iōannēs) indicated in *grossa*, a Venetian currency.[12] He also mentioned a "*maistro* Gianni," a name that could, according to Cronier and Dalché, reflect a Cypriot Italian dialect.[13] Another note recorded the birth of *maistro* Gianni's son, Nicola, on 22 December.[14] The last folio of the notebook bears a map of the island of Cyprus.[15] The Marciana codex also appears to have a connection with Cyprus: it was copied on unwatermarked papers, which led Cronier and Dalché to suggest that it was copied on Cyprus, where such papers were commonly used.[16] Thus, the Ambrosiana Notebook was perhaps first produced in Constantinople, possibly at the Petra monastery, and then taken to Cyprus, where it may later have been copied into a smaller, cleaner miscellany, the Marciana Handbook.

The Ambrosiana and Marciana codices demonstrate that botanical images did not need to circulate as finely finished paintings. They could travel as sketches and memories. A single encounter between a scholar and an illustrated manuscript may have been all that was needed to inspire different ways of making and seeing scientific images. While the Ambrosiana Notebook and Marciana Handbook demonstrate that some experts continued to use their notes and sketches, we can also imagine that wealthy individuals might have commissioned new manuscripts with pictures inspired by what they had seen or heard about elsewhere. Recognizing that such sketches could also operate as memory aids, we find, at least from the point of view of the material record, that rather simple, sketchy drawings could have served as important vectors for the transmission of ideas about what botanical illustrations could or should look like. As a result, a significant mode for the transmission of ideas about pictorial form may now be largely invisible in the material record.

The Ambrosiana and Marciana codices also illustrate the circulation of medical and botanical knowledge beyond the institutions and libraries of large urban centers. Cyprus, a diverse, independent kingdom ruled by the French house of Lusignan, may have played a pivotal role in the development and dissemination of various medical traditions in the late Middle Ages. A few decades prior to the arrival of the Ambrosiana Notebook in Cyprus, a practical medical text called the *Therapeutics*, attributed to an otherwise unknown physician named John, was adapted into several vernacular versions there.[17] A century or so later, John Argyropoulos, who studied at the University of Padua and taught at the Petra monastery in Constantinople, then home to the Morgan and Vienna Dioscorides manuscripts, wrote a letter to a Cypriot physician answering twelve questions on natural and medical philosophy.[18]

Latin Reception of Dioscorides in the Late Middle Ages

Evidence for renewed reception of the illustrated Greek Dioscorides in the Latin West dates to around the end of the thirteenth century. Prior to the thirteenth century, there may have been some limited instances of reception, as indicated by Cassiodorus' reference to an illustrated Dioscorides, and the existence of a Latin translation of Dioscorides, as preserved in a tenth-century illustrated manuscript now in Munich.[19] The Naples Dioscorides had also been in Italy for centuries, though it does not seem to have excited much interest—late medieval Latin annotations within it notwithstanding.[20] In general, these occasions of reception do not seem to have resulted in sustained reception or awareness of the Greek Dioscorides in the Latinate world until the thirteenth century.[21] This general picture holds despite the translation of other Greek and Arabic medical texts into Latin in the eleventh and twelfth centuries.[22] A complete and widely available translation of Dioscorides into Latin would remain a desideratum until the late fifteenth century.[23]

Latin artists and scholars may first have had more sustained access to Greek manuscripts of the Alphabetical Dioscorides in the thirteenth century.[24] Marginal notes in a Southern Gothic *rotunda* script appear in the Vienna Dioscorides at this time (for example, fig. 6.9, top of fol. 327v).[25] This kind of Gothic script suggests that a Latin-speaking scholar had access to the Vienna Dioscorides in the thirteenth century. By the end of that century, we also find illustrations from several different versions of the Alphabetical Dioscorides in the Thott Codex now in Copenhagen (fig. 5.6).[26] It is hard to know exactly what manuscripts served as the main sources for the illustrations in the Copenhagen codex. It may have drawn upon a manuscript source similar to the Naples or Morgan Dioscorides and one similar to the

Vienna Dioscorides.[27] The codex was illustrated by someone working within a "Western" tradition of painting.[28] And, as in the ancient herbals, the makers of the Thott Codex copied the plant illustrations prior to the text.[29] The ample margins could suggest that the codex had always been intended to receive a text, although it is hard to say what kind of text was originally envisioned. The Thott Codex may have been copied from an earlier archetype, which was perhaps also a picture book.[30] The emphasis on the transmission of pictorial content over the texts may speak to the value that its makers placed specifically on visual knowledge.

Scholars tend to assume that the notes in the Vienna Dioscorides and the illustrations in the Thott Codex could only have occurred after the Fourth Crusade during the Latin occupation of Constantinople between 1204 and 1261.[31] The Fourth Crusade devastated the city. Countless medieval and ancient works of art and literature were lost. Aristocratic Byzantine families fled and established rival states—at Trebizond, Nicaea, and Epirus. Michael VIII Palaiologos (r. 1259–82) of Nicaea, however, eventually retook Constantinople in 1261. While the Byzantine reconquest may have limited some Latin access to Greek manuscripts, it is unlikely to have prevented it outright. After 1261, Byzantine monks and Latin mendicants spearheaded several translation initiatives from Greek into Latin and vice versa, many of which focused on philosophical and scientific texts.[32] The Flemish scholar William of Moerbeke (d. 1286), for example, translated works by Archimedes, Aristotle, Galen, Hero, Plato, and Proclus, as well as commentaries on Aristotle by Simplicius, Themistius, and Alexander of Aphrodisias.[33] He notes in a colophon that he translated Alexander of Aphrodisias' commentary on Aristotle's *Meteorology* on April 24, 1260, in Nicaea.[34] The Dominican Simon of Constantinople wrote to the Byzantine monk and scholar Sophonias that he had seen a letter of St. Basil in an ancient book in the Greek monastery of Kyr Meletios in Attica.[35] Clearly, Latin scholars could view ancient manuscripts in Greek cities and institutions. Even after 1283, when Andronikos II (r. 1282–1328) expelled the mendicant orders from Constantinople, we find the Italian scholar Pietro d'Abano (d. 1316) noting that he encountered alphabetically arranged copies of Dioscorides' *De materia medica* during his stay in Constantinople from 1293 to 1303.[36] It is worth noting that Pietro d'Abano had connections to Padua, knew the painter Giotto, and may have had a hand in planning the large fresco program at the city hall and lawcourts in Padua (Palazzo della Ragione).[37]

Moreover, while Western-trained painters likely illustrated the Thott Codex, we cannot say whether they also illustrated its presumed model. Even if we suppose that they did, it remains possible that they did so after 1261. For example, a bilingual Latin-Greek gospel book now in Paris, once assumed to have been copied during the Latin occupation, has been convincingly redated to later in the thirteenth century.[38] Rather than a token of Crusader rule, this codex now appears to have been made as a gift for the pope in support of the union of the Roman and Byzantine churches.[39]

Latin and Greek scholars may even have collaborated on occasion. Manuel Holobolos (d. 1310/14) notes in his preface to the *De plantis* that a Latin scholar "from someplace or another" had given him a manuscript copy of the Latin text, itself a translation from Arabic.[40] Several decades earlier, we find a Latinate scribe named Andreas Telountas from Nauplion apparently collaborating with a Thessalonican painter of the Astrapas family in order to create an illustrated scientific manuscript largely devoted to Ptolemy's *Geography* and now in the Marciana Library in Venice.[41]

The illustrated Dioscorides arrived in Western Europe at a time when artists and elites were actively interested in more detailed and accurate naturalistic imagery as well as the emulation of classical art.[42] These efforts—already under way in the thirteenth century, as seen in the new illustrated cycles for the *Tractatus de herbis*—led further to the appearance of new forms of botanical illustration, exemplified by the late fourteenth- or early fifteenth-century Carrara Herbal.[43] Copied by Jacopo Filippo of Padua between 1390 and 1404 for Francesco Carrara the Younger, the Carrara Herbal is an illustrated translation of an Arabic herbal attributed to Serapion the Younger. Overall, the illustrations in the Carrara Herbal demonstrate meticulous attention to detail and careful, complex modeling created through blending, hatching, and highlights. Most of the illustrations give close views of the plants that emphasize specific morphological details (fig. 8.3). The volume was never completed—an indication of the effort and time required to create so many detailed depictions of plants. A few decades later, between 1445 and 1448, the painter Andrea Amadio illustrated an herbal, now in the Marciana Library in Venice, for the physician Niccolò Roccabonella.[44] Amadio copied and modified illustrations from the Carrara Herbal, but also added new illustrations, resulting in a deluxe volume of 454 illustrations. At roughly the same time, we find a similar herbal, now in the British Library, with novel illustrations of plants, that was copied in the region of Belluno.[45] Many of these manuscripts were copied in Northern Italy, particularly in the vicinity of Padua, and perhaps by extension in association with the University of Padua. After 1405, Padua fell into Venetian hands. That

FIGURE 8.3 Illustration of grapevines (*vigna*). London, British Library, Egerton MS 2020 (the "Carrara Herbal"), c. 1390–1404, fol. 28r. © The British Library Board.

De lauigna desmestega. Capitolo.

LA uigna desmestega che non se lauora: dixe galieno che la uertu e simele a la uigna saluega: ben de soura noma de te un pucho pui molele in uerta. Andea dixe galieno che te foiere questo e lici anch quanto li se tria e fasene embiasbo de questa: fai o cum el sugo: pelorcho in tiga te atostemation calxe che e in los sormigo. El sugo de se fere de questa uigna coa aquili che ha ulceration a li burcti e coa aquila che spua sanguen in quili che lo metese dol brun uerza sormigo. e coa ile mature che ha la perto corona. cosi fin cum ola te questa ca non el se la infusion in laquitro questa a po beuere. El la lagrema de lau igna e simele a la gumm che se cosiqua soura ti partam la uertu de la go ma. E questa che lo mando questa cum el uino tira sini la pne e sau ne lunicon a la uioleza esta cura e cosi a la regna ulcerosa e no ulcerosa e couense a ma nao che se facen luncion a questa uioleza lauare el mem bro cum el sal nitro e cum la olio spesse sie chrauera la piu. La lagre ma che uen fura vole uencele de la uigna quanco la e fresthamenir re colte de la uigna e po metri inro fugo cum un cauo chiura de la scio cauo uen fura questa lagrema a miuo surbreiba ue ru demoiure le ueruge che uen chiama niin ince. La uertu de la cendere. de questa uencela e de la cendere dele grasse de la uua de questa quanto sen fa empiastro cum

maritime republic's immense wealth and control over Mediterranean trade in *materia medica* likely further contributed to local interest in botanical visualization and study.

Ancient Pictures for an Early Modern Botany

The fifteenth and sixteenth centuries saw an acceleration in Latin scholars' access to and interest in illustrated Dioscorides manuscripts in Greek. During the fifteenth century, Latin scholars traveled to Byzantine cities to buy manuscripts and to learn Greek.[46] In the early 1420s, the scholar Giovanni Aurispa (d. 1459) wrote to Ambrogio Traversari (d. 1439) that he had seen a codex "of remarkable antiquity, in which are painted herbs, roots, and certain animals," in Constantinople.[47] And again, in the late 1430s, the humanist Giovanni Tortelli (d. before 1466) also wrote about having seen the Morgan or Vienna Dioscorides:

> I saw in Constantinople a manuscript of this author [i.e., Dioscorides] written in very old Greek letters, and in it were pictures, not only of herbs but also of birds, beasts, and reptiles, drawn, in my opinion, with as much artifice and detail as Nature herself could have produced.[48]

Tortelli emphasizes the naturalism of the pictures—their proximity to nature; they looked like they had been executed by "Nature herself." Latin scholars were not alone in their excitement for ancient botanical illustrations.

Byzantine scholars, too, esteemed the pictures in these manuscripts. Isidore of Kiev, for example, directed the production of the deluxe Chigi Dioscorides, filled with copies of illustrations from the Morgan and Vienna Dioscorides manuscripts, some of which were improved and elaborated in the process of copying (see chapter 6). This manuscript was copied entirely on parchment—a rare extravagance at a time when most manuscripts were copied on paper.

In the following decades, increasing demand for copies of Dioscorides in Greek contributed to a flurry of copying in Constantinople and, after its fall in 1453, on the islands of Crete and Corfu.[49] Latin scholars had become more interested in Dioscorides' *De materia medica* as a way to verify and identify plants described by Pliny the Elder in his *Natural History*. Latin scholars were at that time coming to terms with the *Natural History*'s errors and corrupted manuscript tradition. In 1492, Niccolò Leoniceno (d. 1524) published all of the errors that he had found in Pliny.[50] This work launched the so-called Ferrara debates, a series of heated exchanges on the value of Pliny between Leoniceno and the humanist Angelo Poliziano (d. 1494) and his lawyer, Pandolfo Collenuccio (d. 1504).[51] At about the same time, Ermolao Barbaro (d. 1493) independently published thousands of corrections to Pliny.[52] Barbaro himself became engaged in translating and commenting on Dioscorides, though he did not publish that work.[53] These early efforts culminated in the sixteenth century with several other translation projects, as well as Pietro Andrea Mattioli's extensive commentary on Dioscorides (1544).[54] Critical attention to Pliny further coincided with attempts to purge the Latin medical tradition of the postclassical contributions to botany made by medieval and especially Arabic authors. Latin botanists increasingly saw themselves as the sole heirs of the classical botanical tradition of Theophrastus and Dioscorides.

Emblematic of Latin scholars' interest in Dioscorides and its illustrations is a lavishly illustrated botanical atlas on paper, once owned by the naturalist Joseph Banks (d. 1820) and now in the Natural History Museum, London.[55] Francesca Marchetti has dated this manuscript to just before 1453, though its watermarks could date the manuscript to anytime between the 1450s and the early 1480s.[56] Except for the Greek plant names, the manuscript was originally without text, though pre-Linnaean polynomial plant names were later added to the codex in the seventeenth or eighteenth century, by which time it was likely in Italy, perhaps Padua.[57] Unlike other late Byzantine codices focused on illustrations from Dioscorides, such as Isidore's Chigi Dioscorides or the fourteenth-century Padua Dioscorides, which reproduce illustrations from both the Morgan and Vienna Dioscorides manuscripts, the Banks Dioscorides reproduces illustrations from only the Vienna Dioscorides. It may be that the makers chose to focus on the older of the two—the age difference was readily visible in the Vienna Dioscorides' uncial script. In doing so, the makers copied a manuscript closer to Dioscorides' time, and therefore presumably free of any accumulated errors, but also missing many later corrections.

The Banks Dioscorides has a handful of bilingual notes in Latin and Greek, many of which refer to plant names in Pliny.[58] One note unequivocally identifies its writer as a Latin speaker.[59] The humanist hand of these bilingual notes is similar to that responsible for the Greek plant names. The scribe, commissioner, and principal recipient of the codex may all have been the same humanist scholar. But while the scribe was likely a Latin scholar, the painters responsible for the codex clearly worked within a Byzantine tradition of painting and perhaps came from Constantinople or Venetian Crete.[60] The manuscript was thus a product of close collaboration between a Latin-speaking humanist scholar and a Byzantine artist.

While most of the pictures in the Banks Dioscorides are direct copies of those in the Vienna Dioscorides, there are several notable differences. First, the eryngo's Gorgon-headed root has been omitted in the Banks codex (fig. 8.4).[61] This absence may testify to greater familiarity with the plant's root or a more naturalistic conception of what a botanical illustration can and should represent. Another major difference is the inclusion of a picture of pale stonecrop labeled *aeizōon to amaranton* on the first folio (fig. 8.5).[62] This plant is absent from the Vienna Dioscorides and does not appear in any of its other

FIGURE 8.4 Illustration of eryngium missing the Gorgon-headed root. London, Natural History Museum, MS Banks Coll. Dio 1 (the "Banks Dioscorides"), late fifteenth century, fol. 124r. © The Trustees of the Natural History Museum, London.

FIGURE 8.5 Illustration of pale stonecrop (*aeizōon to amaranton*). Banks Dioscorides, fol. 1r. © The Trustees of the Natural History Museum, London.

ἀείζωον τὸ ἄκρον

Jos: Banks

Sempervivum minus flore luteo

n6

copies. The picture is likely a novel invention of the fifteenth century.[63] Like other pictures in the codex, it also seems to be the work of a Byzantine painter.[64] As suggested by the detail, the absence of major anatomical misunderstandings, the fact that the plant appears at a specific stage in its yearly cycle, and that it is unattested in earlier manuscripts, the picture was not far removed from a "from life" nature study. That some late Byzantine botanical lexica list this plant at the beginning of the text could further suggest that this picture might have operated as a kind of frontispiece.[65] The association with botanical lexica further suggests that the volume was intended to act as a kind of pictorial reference work, further confirmed by the bilingual annotations comparing Latin and Greek plant names. As a frontispiece, *aeizōon to amaranton* foregrounds the Byzantine artist's skills of observation and ability to work between Byzantine and Italianate traditions of painting. Early modern Latin scholars thus accessed the ancient botanical tradition through living Byzantine arts and sciences.

The fall of Constantinople in 1453 accelerated the flight of many Byzantine scholars to Italy and the territories of the Venetian Republic. John Argyropoulos first fled to the Peloponnese before returning to Italy, where he took up teaching appointments in Florence and Rome. His student Michael Apostolēs spent some time in Italy before settling in Venetian Crete, where he made a living by copying manuscripts, including scientific treatises, between 1455 and 1470.[66] Another of Argyropoulos' students, Andronikos Eparchos, fled to Corfu, another Venetian territory, where he, his sons, and his grandsons supplemented their income by copying and selling manuscripts from the Petra monastery.[67] These Corfiot manuscripts would come to play a central role in the early modern reception of Dioscorides in the Latin West.[68] It is through this group of manuscripts that Aldo Manuzio (Aldus Manutius) may have prepared the first printed edition of Dioscorides in 1499.[69] Ermolao Barbaro consulted another Corfiot manuscript for his translation work.[70] In 1538, Andronikos' grandson Antonio Eparchos sold a cache of these manuscripts to the French king François I (r. 1515–47).

Isidore of Kiev, too, made Italy his home, taking the Chigi Dioscorides with him. The codex acquired numerous annotations in Latin, including a Latin index at its front. In the late fifteenth or early sixteenth century, some of the Chigi illustrations were copied into a Latin botanical atlas on parchment now in Vienna, perhaps via an intermediary.[71] The Vienna Atlas appears to have been an ambitious attempt to collate botanical pictures and texts from multiple sources, and perhaps also included some *ex novo* nature studies. The production of the atlas appears to have been sporadic and haphazard: in some areas the text was copied prior to illustration, while in others pictures preceded text. This atlas remains incomplete, as text accompanies less than half of the 574 illustrations, numerous sections of text remain unillustrated, and some paintings were never finished.[72]

The makers of the Vienna Atlas tried to make sense of the Dioscoridean illustrations. They attempted to identify unknown or unfamiliar ancient plants and to collate variant pictures of them so as to identify different subtypes with potential medicinal utility. For example, the Vienna Atlas includes two unusual Dioscoridean illustrations of hypocists (*Cytinus* spp.), which are parasitic on rockrose (*Cistus*), even though the central illustration alone accurately shows the parasite *Cytinus hypocistis* L. growing up from the roots of the rockrose (fig. 8.6).[73] The makers clearly knew what *Cytinus hypocistis* looked

FIGURE 8.6 Rockrose (*rosa canina*) and hypocist parasites (*hypocisis altera, hypocisi(s), hypochistidos*), left and right illustrations copied from the Vienna Dioscorides via the Chigi Dioscorides. Vienna, Österreichische National-bibliothek, cod. 2277 (the "Vienna Atlas"), fol. 23r. Courtesy of Österreichische Nationalbibliothek.

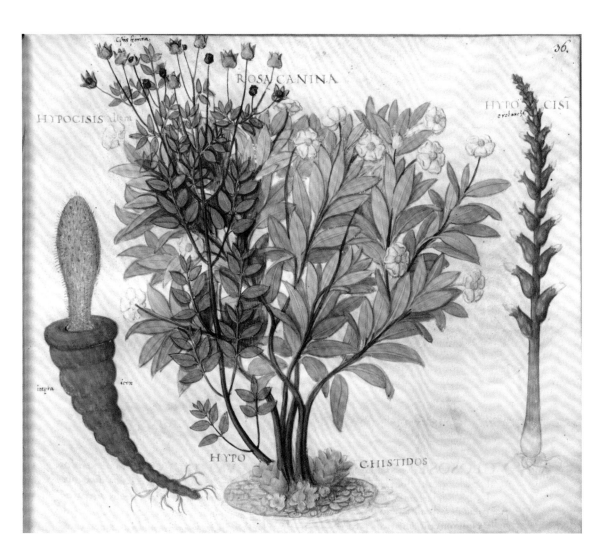

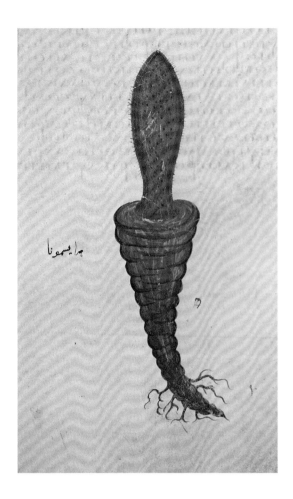

FIGURE 8.7 Another hypocist (*hypokisson heteron*). Vienna Dioscorides, fol. 356r. Courtesy of Österreichische Nationalbibliothek.

like, yet they included the other two illustrations that originally came from the Vienna Dioscorides (fig. 8.7) via the Chigi Dioscorides. The peculiarly tongue-like illustration of the left *hypocistus* may attempt to show a tubercle or an emergent shoot of *Cytinus*, while the right likely portrays another parasitic plant altogether—a broomrape, just prior to flowering. This approach recalls the handling of pictorial variants in the Morgan Dioscorides and the adaptation of pictures in the Chigi Dioscorides. In the case of these rockrose parasites, variant recognition has gone to an extreme: instead of the two rockrose parasites mentioned by Dioscorides, the makers of the Vienna Atlas entertain the possibility of three.

But inquiry did not stop with the initial illustration of these three different parasitic plants. At a later point, a painter added another branch to the central plant, showing the rockrose with matured fruit. This addition testifies to the fact that the artist(s) responsible made several observations of the rockrose over an extended period.[74] This direct observation of plant life thus went hand in hand with the critical reception of the ancient botanical tradition. Another later hand, perhaps that of János Zsámboky (Johannes Sambucus, d. 1548), humanist and court physician to the Holy Roman Emperor Maximilian II (r. 1564–76), further annotated plant depictions throughout the volume.[75] Zsámboky labeled the tongue-like depiction as "a nonsense image."[76] He also correctly reidentified the right plant as a broomrape (*orobanche*). Similar critical annotations appear throughout the codex.

During the late fifteenth century, the illustrations from the Banks Dioscorides were also copied into two manuscripts, one now in the Cambridge University Library (fig. 8.8), the other in the Biblioteca Ambrosiana in Milan.[77] The travels of these two codices are in fact broadly emblematic of the reception of the illustrated Dioscorides in the early modern world. The Cambridge Dioscorides is a close copy of the Banks Dioscorides. In its earliest form it was a botanical atlas with Greek plant names. Scholars have identified the copyist of the Ambrosiana Codex (not to be confused with the Ambrosiana Notebook) with Manuel Grēgoropoulos (d. 1532), a scribe and notary active on Crete in the early sixteenth century.[78] The Ambrosiana Codex was copied in different stages, with some illustrations coming from the Banks Dioscorides and others from another Byzantine botanical codex, one perhaps related to another codex now in Bologna.[79]

Despite their connection to the Banks Dioscorides, these two manuscripts eventually took different paths through the early modern Mediterranean. Like the Banks Dioscorides, the Ambrosiana Codex acquired a smattering of humanist annotations. Both the Ambrosiana Codex and the Ambrosiana Notebook eventually came into the possession of the noted Neapolitan humanist and botanist Gian Vincenzo Pinelli (d. 1601), from whom they passed into the Ambrosian library in Milan. Pinelli may have been responsible for the red underlines added to the Latin synonyms listed in the Ambrosiana Notebook.

While the Banks and Ambrosiana codices made their way to Italy sometime in the sixteenth century, if not sooner, the Cambridge Dioscorides was in Constantinople. Hebrew transliterations of the Greek names were added to the codex, along with translations of those names into Latin, but written with Hebrew characters and inserted into Hebrew sentences, sometimes with Hebrew plant names.[80] An ownership note in Hebrew at the back of the codex notes that the codex was acquired from an individual named Eliezar in Constantinople on 8 Elul 5397 (August 28, 1637).[81] Over the course of the manuscript's history, its front and back reversed: later Hebrew page numbers run right to left, that is, against the original left-to-right orientation of the book. These page numbers signal the transformation of the volume from a Greek reading format to a Hebrew one. After this change to the manuscript's orientation, Ottoman transliterations and notes were added throughout the codex, starting at its new beginning. Longer notes were also added at different stages. Many of these notes include detailed descriptions of medical properties. Red frames, characteristic of Ottoman books, were also added to a few folios toward the new beginning of the codex (fig. 8.8). In 1682, the manuscript was acquired in Smyrna and given to Cambridge University.

Ottoman reception and use was not limited to the Cambridge Dioscorides; both the Morgan and Vienna Dioscorides manuscripts acquired numerous annotations in Arabic, Hebrew, and Ottoman Turkish that testify to their extensive use by Ottoman physicians and scholars. It currently remains unclear what exactly happened to these two manuscripts in the aftermath of the fall of Constantinople in 1453. While the historian Doukas tells us that janissaries ransacked the Petra monastery, then home of the manuscripts, it may have continued

FIGURE 8.8 Illustration of a coral. Cambridge, University Library, MS Ee. 5.7 (the "Cambridge Dioscorides"), fols. 379v–80r. Reproduced by kind permission of the Syndics of Cambridge University Library.

to operate for some years.[82] Mehmed II (r. 1444–46, 1451–81) gifted the monastery to the Christian mother of Mahmud Pasha in 1462.[83] By the time Maximilian II (r. 1564–76) had the Vienna Dioscorides purchased in 1569 at the urging of Ogier Ghislain de Busbecq, it was in the possession of the son of Moses Hamon (d. 1554), one of the chief Jewish physicians to Suleiman I (r. 1520–66).[84] Moses Hamon or a member of his family may have been responsible for many of the Hebrew inscriptions in the Vienna Dioscorides.[85] A sixteenth-century Spanish travel book, the *Viaje de Turquía*, reports that Moses Hamon boasted of having spent eight thousand ducats on his personal library, which was estimated at his death to have been worth five thousand ducats.[86]

The Hebrew annotations in the Vienna and Cambridge Dioscorides manuscripts speak to the reception of the Greek botanical tradition among Jewish physicians and scholars in the Ottoman empire. Some Ottoman Jewish physicians, such as the Italian Jacopo di Gaeta (d. 1484), personal physician to Mehmed II, had received medical training in Italian universities.[87] Jewish physicians' wide connections and familiarity with the Latin medical tradition gave them a unique position in the Ottoman empire. Through them patients could access not only the Hebrew medical tradition, but also the Latin medical tradition, without having to consult foreign doctors, such as Venetians, with whom the Ottomans were often at war. There was at that time no translation of Dioscorides into Hebrew, so Jewish physicians would have had to rely on the Greek text or translations into Latin or Arabic. Volumes such as the Vienna and Cambridge Dioscorides manuscripts would have had an obvious appeal.

At the same time, the Ottomans established charitable foundations with libraries and medical facilities following earlier models for such establishments in the Islamicate world.[88] Medical positions at such foundations were not necessarily restricted to Muslim physicians. An endowment document for a charitable foundation from Mehmed II's reign makes provisions for two doctors "of whatever community" (that is, inclusive of non-Muslims).[89] The same document notes that among householders with adjacent properties, there were three Jewish physicians, one Christian, and no Muslims.[90] As the state expanded, the Ottomans acquired earlier Arabic manuscripts of Dioscorides and gifted them to individuals or donated them to charitable foundations. Illustrated manuscripts of Dioscorides in Greek and Arabic continued to be read, circulated, and produced in the Eastern Mediterranean into the modern era.[91]

Botanical Icons Remade

While Latin-speaking physicians and scholars sought out deluxe illustrated manuscripts of Dioscorides in the fifteenth century, the pictures themselves were not reproduced widely. During this time, the humanist botanical tradition offered new illustrations of plants that adopted an increasingly narrow conception of what constituted suitable botanical illustration. Pictorial references to plant names and properties through mnemonic devices, especially zoomorphic and anthropomorphic details, were also increasingly excluded from illustrated humanist botanical works, though they continued to appear in other kinds of herbals of the period, such as alchemical herbals.[92] The humanist synthesis favored naturalistic and plastic treatment of botanical forms, though it also remained indebted to pictorial forms from earlier manuscripts, including a shallow, blank ground with a didactic or pinned-out appearance.[93] While this synthesis coexisted with other modes of botanical illustration, it would eventually become the primary mode of botanical illustration within the modern botanical tradition. As we will see, the printing press likely played a major role in the lasting success of the humanist mode of botanical illustration.

The advent of the printing press, however, did not immediately herald the wide dissemination of new, accurate, and detailed depictions of plants. The earliest known illustrated printed herbal is an edition of the Latin herbal attributed to Ps.-Apuleius that was printed anonymously in the early 1480s by the Sicilian publisher Giovanni Filippo De Lignamine (Johannes Philippus de Lignamine) active at Montecassino, near Rome.[94] The edition does not include new illustrations; rather, the publisher had simple woodcuts created, based on pictures in an ancient manuscript in the library of the monastery at Montecassino. Lignamine thought the manuscript dated to Roman times and included original illustrations. Scholars now believe this manuscript was made in the ninth century. It can only be examined in facsimile, though, as the original was destroyed in the aerial bombing of Montecassino in 1944.[95] Despite Lignamine's faith in the authority of the ancient illustrations, his edition was largely unsuccessful.[96] Several factors likely contributed to this failure. First, the illustrations from the herbal of Ps.-Apuleius Platonicus may have been out of fashion compared to botanical illustrations in more recent Latin manuscripts as well as copies of Dioscorides' *De materia medica*. Second, practices of nature printing and of gluing actual plant specimens into herbals in Italy indicate contemporary interest in direct study of plants.[97] These practices complement a variety of methods of "ocular demonstration" that developed in the botanical pedagogy of the fifteenth and sixteenth centuries, including the development of dried herbaria, botanical gardens, and botanical field-trips (*herbationes*).[98] It is possible that manuscripts, rather than printed books, continued to play a central role in botanical inquiry in late fifteenth- and early sixteenth-century Italy due to an awareness of Pliny's critique that reproducing botanical illustrations introduced errors into them.[99] Such thoughts may have reinforced contemporary skepticism of printed illustrated herbals, while lending greater authority not only to the newest observational illustrations but also to the most ancient. This may explain why there are so many fifteenth- and sixteenth-century observational nature studies and so few illustrated printed herbals from the Italian peninsula.

More successful attempts to harness the medium of print for botanical illustrations occurred north of the Alps. With his publication of *Herbarum vivae eicones* (*Lively Icons of Herbs*, 1530–36), Otto Brunfels set out to resuscitate ancient botanical knowledge both textually, through his critical work with ancient texts, and pictorially, through accurate illustrations newly executed from nature.[100] That Brunfels's title pairs an insistence on liveliness (*vivae*) with the Greek word for images (*eicones*) speaks to the humanist aspirations of the project— an intention to revive the ancient botanical tradition of Dioscorides and Crateuas with pictures that were in line with that tradition, as opposed to the tradition of the medieval Latin herbal, above all associated with the herbal of Ps.-Apuleius Platonicus.[101] Otto Brunfels had Hans Weiditz draw and prepare the woodcuts for his *eicones*.[102] Weiditz attended closely to every detail of the particular specimens before him, including incidental and idiosyncratic features, such as drooping, shriveled leaves (fig. 8.9). *Contrafrayt Kreütterbuch*, the German title of Brunfels's *Herbarum vivae eicones*, may capture this intention to furnish the individual details of a specific specimen, that is, to "counterfeit" or make directly from the specimen.[103] Weiditz's inclusion of individual details amounts to a rhetorical claim to his pictures' truthfulness and therefore an attempt to persuade the viewer of their accuracy. This conceit was, however, alien to the ancient understanding of botanical illustration, where broken stems and wilted leaves, if present, primarily referred to morphological variation throughout a plant's life cycle. As we have seen in chapter 3, botanical illustrations were understood to abstract the defining characteristics of a plant from any single instance of perception. Brunfels's project had limited success, as he struggled to coordinate his textual production, Weiditz's illustration, and printing by the publisher, Johannes Schott.[104]

Borago.

Burtefcht.

FIGURE 8.9 Illustration of borage (*borago*). Otto Brunfels, *Herbarum vivae eicones* I (Strasbourg: Schott, 1531), p. 113. From the Collection of the Lloyd Library and Museum, Cincinnati.

Leonhart Fuchs would have greater success with his *De historia stirpium* (*On the History of Plants*), published in 1542. Like Brunfels, Fuchs aimed to revive the ancient botanical tradition as represented by Pliny, Dioscorides, and Galen. Mindful of the Plinian critique of herbal illustration, Fuchs defended his decision to illustrate his herbal on the grounds "that a picture expresses things more surely and fixes them more deeply in the mind than the bare words of the text."[105] Unlike Brunfels and his team, Fuchs and the illustrators Heinrich Füllmaurer and Albrecht Meyer aimed to present each plant through an ideal, generalizing picture, "as complete as possible [*pictura absolutissima*]," that is, showing all the characteristic features of the plant.[106] In doing so, Fuchs and his collaborators created *ex novo* pictures from nature that nevertheless echoed many of the concerns, compositions, and devices found in ancient botanical illustrations, notably the "didactic" arrangement of parts against a shallow blank ground, and the attention to botanical diachronicity (fig. 8.10). Fuchs, Meyer, and Füllmaurer also discarded some of the illusionistic innovations in earlier botanical illustrations from late medieval Latin herbals such as the Carrara Herbal; Fuchs was wary of complex modeling, favoring instead clear, linear compositions that could be reproduced easily in print.[107] Fuchs later published a version of his herbal in German and, echoing the circulation of plant pictures in textless atlases in the late Middle Ages, had an octavo version published that consisted simply of pictures of plants with their names in German, Latin, and Greek. This pictorial atlas was more portable than the earlier versions of his herbal.[108] Fuchs's attention to wide distribution and the readership of the general public helped to secure the lasting legacy of his herbal and, with it, a particular approach to botanical illustration. The woodcuts from Fuchs's herbal would also be widely reproduced in the following centuries in other authors' herbals.

It is hard to know how aware Brunfels, Fuchs, and their collaborators were of ancient botanical illustrations. While there were a variety of modes of botanical illustration available within the Latin botanical tradition by this time—as clearly seen in the illustrated *Tractatus* herbals and the Carrara Herbal—the forms of botanical illustration adopted by Fuchs and Braunfels cleaved more closely to ancient botanical illustrations than they did to those of the *Tractatus* or the Carrara Herbal. Both Brunfels and Fuchs eschewed the cropped, playful spontaneity and vivacity of botanical forms in the Carrara Herbal, as well as the flattened linear illustrations of the earlier *Tractatus* herbals. In their own ways, the illustrations in both Brunfels's and Fuchs's herbals present the viewer with a didactically "pinned" arrangement

FIGURE 8.10 Illustration of a sloe (*prunus sylvestris*). Leonhart Fuchs, *De historia stirpium* (Basel: Isingrin, 1542), p. 404. Cambridge, University Library, Se1.2.81. Reproduced by kind permission of the Syndics of Cambridge University Library.

of plant parts, and not what Pächt called "the subjective impression of the thing as spontaneously perceived."[109] Fuchs further avoided the complex, illusionistic modeling so evident in the Carrara Herbal. These botanical authors and their artists may simply have reinvented these approaches to botanical illustration by responding to ancient concerns such as those enumerated by Pliny the Elder. At the same time, however, the circulation of ancient botanical illustrations in multiple forms, particularly in Italy, does not exclude the possibility that such works helped to inspire "new" humanist modes of botanical illustration.

While print helped to ensure the lasting legacy of humanist herbals, it did not supplant earlier scientific manuscript cultures.[110] Early modern scholars often treated printed works similarly to how their medieval predecessors had used manuscripts—as sites for the accumulation of collective, empirical knowledge. They did so not only by writing annotations, but also by adding to the printed illustrations, as seen, for example, in later modifications to a copy of Fuchs's herbal, now in Minneapolis (fig. 8.11). Here a user has added drawings of flowers and leaves to a printed illustration of a violet. As we saw in the modifications and sketches that were added to the Morgan and Vienna Dioscorides manuscripts (figs. 6.8 and 6.9), these added details may have helped later users clarify aspects of the plant's morphology.

Early modern scholars and scientists also continued to consult ancient and medieval manuscripts. When the Belgian botanist Rembert Dodoens (d. 1585) reworked his Dutch herbal, his *Cruÿdeboeck*, first published in 1554, as his Latin *Stirpium historiae pemptades sex sive libri triginta* (*Six Pemptads or Thirty Books on the Histories of Plants*), published in 1583, he included some figures from the Vienna Dioscorides.[111] Labeled "from the Caesar's codex" (*ex codice Caesareo*), these additions generally represented foreign or unknown plants. For example, in justifying his inclusion of a picture of a birdsfoot trefoil (*Coronopus* or *Korōnopous* in Greek, literally "crowsfoot")[112] from the Vienna Dioscorides (figs. 8.12 and 8.13), Dodoens notes:

> But if we can indeed trust the image which is found in the Caesar's codex, *Herba Stella* is far different from *Coronopus*; for *Coronopus* is painted here with delicate and vine-like stems. It bears recurved pods like the toes of crows' feet, as the figure shows. Moreover, the most learned and diligent Bernard Paludanus [i.e., Berend ten Broeke] has defended the faithfulness of the old imperial manuscript, as he related that he personally found and observed this sort

VIOLA
PVRPVREA
Viola odorata Linn.

Mertzen veiel.

FIGURE 8.11 A violet (*viola purpurea*) with later additions. Leonhart Fuchs, *De historia stirpium* (Basel: Isingrin, 1542), p. 311. University of Minnesota Libraries, Owen H. Wangensteen Historical Library of Biology and Medicine, Folio 580.01 F95.

of *Coronopus*, which is portrayed in the image, not far from Tripoli and at the foot of Mount Lebanon.[113]

We can see similarities between Dodoens's treatment of visual evidence and the premodern compilation of pictures that was evident as early the late ninth or early tenth century in the Morgan Dioscorides. A principal difference, though, is Dodoens's attention to writing out his observations and justifying them through an appeal to the testimony of the observant traveler Berend ten Broeke (d. 1633). Elsewhere, Dodoens justifies the accuracy of another illustration from the Vienna Dioscorides by comparing it to a dried specimen that Berend ten Broeke had procured in his travels.[114] This additional proof further demonstrates the growing importance of scholarly access to dried plants in herbarium collections.

FIGURE 8.12 An illustration of birds-foot trefoil based on *Coronopus* from the Vienna Dioscorides, in Rembert Dodoens, *Stirpium historiae pemptades sex sive libri triginta* (Antwerp: Plantin, 1583), p. 109. Cambridge, University Library, Adams.3.58.3. Reproduced by kind permission of the Syndics of Cambridge University Library.

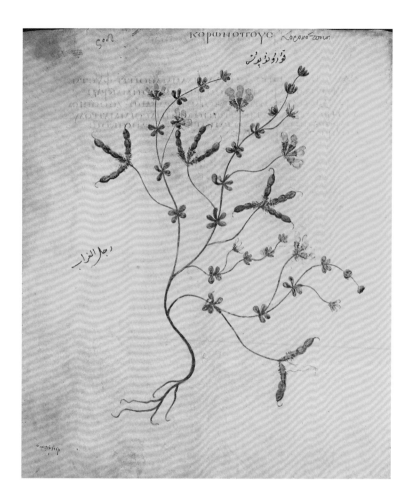

FIGURE 8.13 Illustration of birdsfoot trefoil (*Korōnopous*). Vienna Dioscorides, fol. 178v. Courtesy of Österreichische Nationalbibliothek.

In Dodoens's text accompanying this illustration of a birdsfoot trefoil in the *Pemptades*, he gives new inscriptions—a description based on the picture of *Coronopus* in the Vienna Dioscorides. He also verifies it by appealing to the testimony of the inquisitive, acquisitive, and especially mobile Berend ten Broeke. Novel inscriptions and mobilizations are two features that Bruno Latour has identified as necessary for explaining differences between premodern and modern sciences.[115] The modern sciences move people and things at large scales in order to make statements through inscriptions. By moving things and peoples through ever shifting and broadening networks, scientific discourse not only invents but also stabilizes its objects. As Latour notes, "You have to invent objects which have the properties of being *mobile* but also *immutable, presentable, readable* and *combinable* with one another."[116] The *Coronopus* from the Caesar's codex is out there near Tripoli and at the foot of Mount Lebanon, and Berend saw it. This attention to description and verification, reinforced by vast mobilizations, and above all recorded and circulated in print, illustrates some of the broader shifts between premodern and early modern science. The principal, emergent differences here, then, relate to scale and documentatio rather than to the realization of newfound attitudes and critical capacities for observation and empiricism. These differences would sharpen to define the contours of the botanical tradition as a modern science. The fact, moreover, that Dodoens had done this work with the Vienna Dioscorides also meant that others would not have to. Although Dodoens's *Pemptades* was widely read and circulated, fewer botanical scholars studied the Vienna Dioscorides for its pictures in the following centuries.

Ancient Pictures for a Modern Botany?

The last serious attempt in Western Europe to identify Dioscorides' plants according to the ancient illustrations from the illustrated Dioscorides occurred in the eighteenth century. The English botanist John Sibthorp (d. 1796) embarked on a botanical expedition to the Ottoman empire to characterize the flora of the Eastern Mediterranean. On his way, he stopped briefly in Vienna, where he hired the Austrian illustrator Ferdinand Bauer (d. 1826). He also acquired there a set of proofs (fig. 8.14) of engravings based on the Naples and Vienna Dioscorides manuscripts, both then in the Imperial Library. Writing to his friend John Hawkins (d. 1841), Sibthorp announced his intention to use the illustrations in his research: "The Grecian Flora has been little examined. I think I shall be able to throw some light on the Absurdity of Dioscorides[.] I have by the Friendship of Jacquin procured a Copy of the Drawings of the oldest Manuscript of which is extant which will facilitate my Enquiries."[117]

Nikolaus von Jacquin (d. 1817) was the director of the botanical gardens at the University of Vienna. This "Copy of the Drawings of the oldest Manuscript" was one of five from an ambitious but unfinished project initiated by Gerard van Swieten (d. 1772), the librarian to Empress Maria Theresa (r. 1745–65), to reproduce the ancient plant illustrations from the Naples and Vienna Dioscorides manuscripts.[118] Jacquin's "Copy" of the ancient illustrations from Dioscorides gave Sibthorp and Bauer a glimpse of ancient views of the botanical world. Jacquin also sent an annotated copy to Carl von Linné (d. 1778), the originator of modern binomial nomenclature. He, too, expressed interest in the ancient pictures, inquiring as to the age of the manuscripts.[119] The ancient illustrations still held the promise of giving botanists of the eighteenth century another way to make sense of Dioscorides' botanical world.

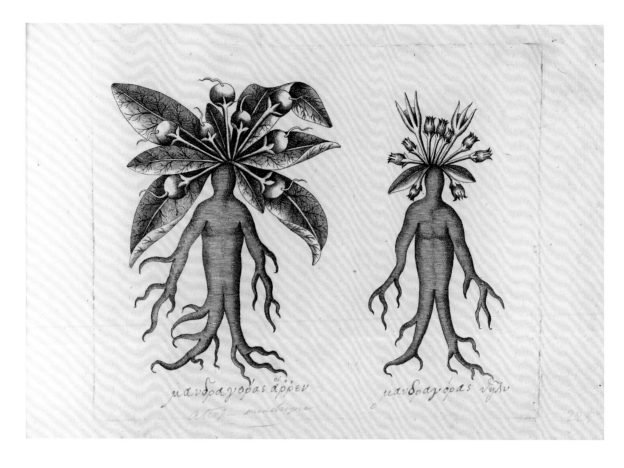

μανδραγόρας ἄρρεν

μανδραγόρας ύγλυ

By the late eighteenth century, however, expectations for botanical illustrations did not align with those of the ancient world. While Bauer's illustration of the mandrake coyly alludes to the notorious anthropomorphism of its root (see, for example, figs. 5.11, 5.12), it also includes closely observed and dissected details—leaves with clear margins, veins, and surfaces, both ab- and adaxial; flowers and floral anatomy; as well as unripe and ripened fruit (fig. 8.15). The Copy of the ancient illustration, by contrast, provides only a few of these details: fruit in the "male" mandrake, flowers in the "female"; leaves with veins, margins, and different surfaces; and both plants' anthropomorphic roots (fig. 8.14). Although considerably less detailed than Bauer's illustration, the ancient depiction does, however, provide a holistic view of the plant, its general habit, and the connections between its parts. Bauer's illustration instead portrays the mandrake as though dismembered on a dissection table. He thus conjures the plant's fabled corporeality but equates it with an anatomized human body. In the process, Bauer dismantles an ancient myth that had gripped the premodern and early modern Mediterranean world. The Enlightenment artist thus lays the "Absurdities" of the past to rest.

FIGURE 8.14 Engraved copy of mandrakes based on the illustrations in the Naples Dioscorides. University of Oxford, Bodleian Libraries, Sherardian Library of Plant Taxonomy, MS Sherard 443, fol. 225r. Courtesy of Bodleian Libraries, University of Oxford.

FIGURE 8.15 Colored engraving of a mandrake based on a painting by Ferdinand Bauer. John Sibthrop and James Edward Smith, *Flora Graeca Sibthorpiana*, vol. 3 (London: Taylor, 1819), plate 232. From Collection of the Lloyd Library and Museum, Cincinnati.

Atropa Mandragora.

Despite the ancient illustrations' idiosyncrasies and general vagueness, Sibthorp evidently held them in high esteem. On at least one occasion, he apparently took the authority of the ancient illustrations over his own reading of Dioscorides' text. In a letter to Alexander MacLeay, James Edward Smith (d. 1828) notes that Sibthorp ultimately based his identification of the plant *aeizōon heteron* on the ancient illustrations, against his own description and study of the plant.[120]

Although Sibthorp died in 1796, his *Flora Graeca Sibthorpiana*, composed of ten massive, illustrated volumes, was eventually completed and published between 1806 and 1840 at great expense through the collective efforts of Bauer, Hawkins, and Smith. Sibthorp's "Copy" of ancient illustrations is today in the Sherardian Library of Plant Taxonomy in Oxford, while Linné's "Copy" resides at the Linnean Society in London. Sibthorp's expedition may represent the last time a botanist seriously considered the illustrations as a plausible window into Dioscorides' botany. In subsequent years, if scholars in Western Europe studied illustrated manuscripts of Dioscorides, it was mainly as objects of historical interest and not as botanical tools.[121]

Conclusion

Ancient illustrations traversed the Mediterranean world in a variety of ways—not just as paintings in deluxe copies of Dioscorides' *De materia medica*, but also in other works, including botanical atlases with little or no text, as well as through memories and sketches. By the end of the thirteenth century, Latin scholars and physicians had access to the ancient illustrations of the Alphabetical Dioscorides. That period of initial reception coincided with the proliferation of new modes of botanical illustration in the Latinate world. Ancient botanical illustrations may have played a role in inspiring some of these changes. The advent of printing meant that the newer humanist forms of botanical illustration could achieve a much wider distribution while ensuring greater regularity and uniformity among editions. During this time, while Latin scholars sought to exclude medieval and Arabic contributions to botany, some nevertheless continued to seek out ancient botanical illustrations as a way to make sense of Dioscorides. By the late eighteenth century, however, these efforts came to an end as well. In the Eastern Mediterranean, Greek and Arabic botanical traditions continued in the markets, monasteries, charitable foundations, and local knowledges that survived and emerged under the Ottomans.

CONCLUSION

The scientific illustration of plants in the Mediterranean Basin constitutes one of the longest unbroken traditions of image-making in the world. More than a thousand years separates the earliest known illustrated herbals from the advent of early modern humanist botanical illustrations in the sixteenth century. During this long period, the makers and readers of illustrated herbals continuously changed and updated them. They modified, simplified, elaborated, discarded, and replaced earlier plant depictions. They compared pictures against each other, against the text, and against actual plants. They made illustrated herbals that reflected their botanical world and responded to their botanical needs. This, the central message of this book, flies in the face of established narratives on illustrated premodern herbals. According to older literature, illustrated herbals were first created in antiquity to a high standard based on the observation of nature only to stagnate in late antiquity and the Middle Ages from repetitive, uncritical copying practices. This book has shown instead that the late antique and medieval Mediterranean was home to dynamic critical traditions of botanical illustration that privileged the utility of illustrated herbals.

The earliest evidence for botanical illustration in Western Eurasia places its initial flourishing in the Hellenistic world. The early illustrated herbal emerged in connection to the profession of root-cutting (*rhizotomia*) and a related genre of herbal texts called *rhizotomika*. Elite Hellenistic and Roman patrons initially supported the illustration of herbals in conjunction with their broader interests in medicine and horticulture. Elite readers may have fostered the production of such texts, but illustrated herbals also circulated through multiple levels of society, particularly among medical practitioners and enthusiasts. These patterns of patronage and readership continued into the Middle Ages and the early modern period. In the Middle Ages, we also encounter greater evidence for scholars copying and illustrating their own herbals. The rise of new medical and academic institutions, including monasteries, hospitals, and medical schools, created new systems of patronage and readership that supported the continued development of botanical illustration.

Over the long period covered in this book, illustrated herbals were closely connected to broader traditions and practices of botanical inquiry. An herbal could refer to a plant through a description, a depiction, or only a name. Ideally, people learned about plants by seeing them in the field. But there were practical limits to learning this way. Consequently, premodern botanical authorities had to draw upon a variety of sources with different ways of referring to plants. While depictions and descriptions both worked by informing the reader of a plant's external appearance, illustrations simulated the experience of seeing a plant. The attitudes and intellectual concerns of Empiricist medicine, which emphasized firsthand knowledge (*autopsia*), while recognizing the importance of earlier written testimonies (*historia*), likely reinforced contemporaries' interest in illustrated herbals as a viable way to learn about plant morphology. The early illustrated herbal further used pictures to stand in for actual specimens. This is confirmed by the fact that the earliest surviving herbals were illustrated prior to the copying of the accompanying text. In this sense, the text was applied to a picture as though it were a tag on a specimen. This practice of picture-first illustration largely fell out of use over the course of sixth and seventh centuries until its reappearance in the thirteenth century.

Premodern botanical illustrations varied considerably over the centuries and from place to place, but also even within a single manuscript. Premodern botanical artists did not aim to make portraits of individual plants, but rather aimed to produce abstracted, idealized images of plant kinds. They often eschewed botanical detail, especially floral anatomy, to focus on plant habit and leaf shape. By the fourth century we find the full articulation of plant parts and the flattening or "pinning" of the botanical subject against a blank, shallow ground well established as conventions. Even when accurate details were lacking, pictures could still help to train visual recognition of plants: they served to acquaint users with a plant's habit, and helped them to establish their own search images, which they could then refine through direct experience in the field. At the same time, the idea that the botanical image simulated a direct, visual encounter with the plant (*autopsia*) meant that illustrations ideally had to retain a level of botanical detail. This set up a tension within traditions of premodern botanical illustration between the rendering of a plant's distinguishing features and the abstracting of its habit. Illustrations could also include references to a plant's habitat, properties, names, and harvesting methods. Subsidiary human figures could sometimes convey or highlight these properties, though they also appear in some manuscripts simply to direct the reader's attention.

We do not find evidence for the combination of depictions and descriptions in illustrated herbals until the creation of illustrated versions of Dioscorides' *De materia medica*, as seen in the Vienna and Naples Dioscorides manuscripts, of the early sixth century and the late sixth or early seventh century, respectively. These versions of Dioscorides may have initially developed over the course of the second or third century CE. Illustrated, alphabetized versions of Dioscorides were likely created by collating pictures from earlier illustrated herbals and by matching names between texts, often with the aid of lists of synonyms. As a result, pictures in the illustrated Dioscorides often retain botanical lore that differs from the text of Dioscorides. The illustrated Dioscorides eventually became the central illustrated text in the Greek and Arabic botanical traditions in the Middle Ages. Collating and combining different pictorial and textual traditions allowed premodern botanical experts to compare and formally discern distinct subtypes of plants, as seen in the Morgan Dioscorides. Gaps and errors in this tradition also meant that illustrators had regular opportunities to create new illustrations, sometimes through the observation of nature or through a close reading of the text. Until the thirteenth century, novel and direct observation of plants is largely found in the critical copying of earlier illustrations and in continual attempts to update and correct the botanical tradition.

The thirteenth century marks a turning point in this book's telling of the history of botanical illustration. In this period we encounter evidence for increased experimentation with new pictorial modes and the development of larger novel pictorial programs for relatively new herbal texts. The earliest known example is the no-longer-extant illustrated herbal of al-Ṣūrī. It is followed by the *Tractatus* herbals in Latin. These developments may have been linked. In the thirteenth century, we find botanical experts, such as Ibn al-Bayṭār, traveling to learn about plants in Anatolia from Greek botanical experts. Word of al-Ṣūrī's and Ibn al-Bayṭār's methods may have reached Southern Italy via Frederick II's contacts in the Levant. Also during this time we encounter classicizing illustrations in the Topkapi Dioscorides and, by the end of the century, the earliest Latinate copy of the illustrated Greek Dioscorides, the Thott Codex.

Botanical images traveled widely at this time. They circulated independently of texts and made their way into the margins of other manuscripts. Specialists studied and sketched from earlier manuscripts. They memorized the images in them. In the thirteenth and fourteenth centuries we again encounter botanical manuscripts that were illustrated before text was added to them. Some manuscripts from this period entirely lack accompanying texts. This emphasis on the transmission of pictorial content parallels the early formation and production of herbals in antiquity.

Despite the preponderance of material evidence for the development of botanical illustration in the thirteenth and fourteenth centuries, these practices of visualization and reference rather represent a continuation or escalation of earlier practices. Manuscripts such as the Morgan Dioscorides, dated to the late ninth or early tenth century, testify to the value that early medieval scholars placed on visual knowledge. We should be mindful that much does not survive from the early Middle Ages. Countless manuscripts have been lost over time, particularly through the devastation of the Latin occupation of Constantinople from 1204 to 1261 and the Mongol siege of Baghdad in 1258. It is therefore potentially misleading to herald the thirteenth and fourteenth centuries as representing entirely "new" developments in medieval botanical traditions. The circulation of botanical knowledge, however, certainly did expand during this period as a result of widening trade networks and the increasing adoption of paper over parchment in the Latin- and Greek-speaking Mediterranean. Books could be copied more cheaply and circulate more widely than before.

This book's account of the history of botanical illustration in the premodern Mediterranean prompts us to adjust our narratives about the advent of critical observation within botanical illustration. It invites us to revisit assumptions about how early medieval people thought about the natural world. The thirteenth-century scholars of Southern Italy were not the first to seek the depiction of nature through observation in the *Tractatus* tradition. This is not to downplay the importance of the *Tractatus* as a milestone in the intertwined histories of art and science. Rather, we should reconsider what kind of change it represents. When compared to novelty within the Greek and Arabic Dioscoridean traditions, the *Tractatus* represents a much larger scale of novel depiction based on observation. It demonstrates a shift from piecemeal revision to wholesale creation, although Al-Ṣūrī's illustrated herbal would have been an earlier instance of such an illustrative overhaul through direct observation. But unlike al-Ṣūrī's herbal, the *Tractatus* survived to be copied many times over. This survival points to the broader historical realities and systems that fostered the continuation of the *Tractatus* tradition in Latinate Western Europe, the most notable among which is the rise of the university.

Many pictorial strategies favored by early modern humanist botanists also find earlier antecedents in premodern herbals. It is consequently harder to view the foundational work of Brunfels and Fuchs as a complete break with the past. Instead, they emulated or reinvented earlier modes of botanical illustration. Early modern botanical authorities, however, also harnessed the medium of print and engaged broader intellectual networks. The labor of botanical illustration, too, increasingly fell to specialized teams of artists, engravers, and printers. Scale and specialization seem to be the primary, initial differences between premodern and early modern botany.

Modern botany became global and universalist in its reach and aspirations. It moved scholars, specimens, and texts across vast distances unthinkable to the botanical authorities of the Hellenistic, Roman, and medieval Mediterranean. Epistemic transformations eventually followed in the wake of this unprecedented academic, imperial, and colonial expansionism. The accumulation of manuscripts and critical editions, novel botanical observations, the circulation of specimens, heated transregional debates among scholars, and the recognition of ever more species required illustrations with increasingly precise, distinguishing details. Visual mastery of local plants according to their growth habit thus gave way to the universal recording of ever more species, in ever more anatomical detail. This rise of botanical detail required an elaboration of botanical terminology and anticipated the florescence of plant anatomy in the seventeenth century.

While the scientific botany of the Western, Latinate world continued to acknowledge its inheritance of the ancient botanical traditions of Theophrastus and Dioscorides, it steadily diverged from the ongoing botanical practices and traditions of the non-Latinate Mediterranean. Many communities living in the Eastern Mediterranean continued to read, copy, and critically engage with earlier premodern botanical traditions well into the present era. These traditions have long since parted ways. As Arthur William Hill, the director of the Royal Botanic Gardens at Kew, observed in an expedition to Mount

Athos in 1937:

> To be a Botanist on the Holy Mountain was not regarded by the Monks as an unusual occupation, for at Karyes there is an official Botanist Monk, who occupies his time in searching for plants of real or supposed medicinal importance. He quickly discovered us and came to call at Pyrgos soon after our arrival—possibly to see that we were not trespassing on his preserves. When he saw the sort of things we were collecting, I fancy his estimation of us was very much lowered and he regarded us as quite harmless.
>
> He was a remarkable old Monk with an extensive knowledge of plants and their properties. Though fully gowned in a long black cassock he travelled very quickly, usually on foot and sometimes on a mule, carrying his "Flora" with him in a large, black, bulky bag. Such a bag was necessary since his "Flora" was nothing less than four manuscript folio volumes of Dioscorides, which apparently he himself had copied out. This Flora he invariably used for determining any plant which he could not name at sight, and he could find his way in his books and identify his plants—to his own satisfaction—with remarkable rapidity.[1]

The director of Kew and this unnamed Botanist Monk were representatives of two different but historically connected botanies. They could recognize each other, but they hardly shared the same concerns. Each tradition was sufficient to address the needs of its adherents. But much has changed since the 1930s. This ever more connected world, transformed by human actions, needs both botanies today. It needs Botanist Monks and others like them to protect their preserves tirelessly. It needs both global and local views of the botanical world. We might yet learn to see and to dwell in each other's botanical worlds.

ACKNOWLEDGMENTS

M any people and places have allowed me to write this book over many years. I initially conceived it at the University of California, Berkeley. I thank Diliana Angelova and Beate Fricke for so many years of enthusiastic support and expert guidance. I also thank Maria Mavroudi for her advice, generosity, and knowledge. Kriszta Kotsis, Linda K. Williams, Zaixin Hong, Gary Shugart, and Peter Wimberger at the University of Puget Sound encouraged me to pursue the intersections of art history and natural history. This book owes much to the guidance and support of Daniela Bleichmar at the University of Southern California. I remain ever grateful for the tireless work of Karen Darling, Caterina MacLean, and Fabiola Enríquez Flores at the University of Chicago Press; the editorial skills of Lys Weiss of Post Hoc Academic Publishing Services; and the critical eyes of the anonymous reviewers.

This project has benefited from the funding of many organizations. Much of the initial research and writing was carried out with generous assistance from the Samuel H. Kress Foundation (2014–16) and the Center for Advanced Studies in the Visual Arts at the National Gallery in Washington, DC (2016–19). Maria Mavroudi hosted me at the Wissenschaftskolleg zu Berlin in November 2016 and advised me through an Arts Research Center Fellowship at UC Berkeley (Spring 2018). I transformed—corrected, pruned, expanded, and polished—the manuscript while holding a fellowship at the Society of Fellows in the Humanities at the Dana and David Dornsife College of Sciences, Arts, and Letters at the University of Southern California (2019–21), and a Getty/ACLS Postdoctoral Fellowship in Art History (2021–22). And I completed the revised manuscript while resident as a postdoctoral fellow at the Seeger Center for Hellenic Studies at Princeton University. I am particularly mindful of the institutional support that allowed me to complete this book manuscript in comfort and safety during a global pandemic. Finally, I am indebted to the College Art Association and to the Medieval Academy of America for grants that have generously supported the publication of this book.

The expertise and assistance of many librarians, curators, and conservators have also sustained this project. I thank Joshua O'Driscoll, Frank Trujillo, Maria Fredericks, John M. McQuillen, John Vincler, Kaitlyn Krieg, and Marilyn Palmieri at the Morgan Library; Yasmin Faghihi at Cambridge University Library; Isabelle Charmantier at the Linnean Society; Oliver Bridle and Stephen Harris of the Department of Plant Sciences, Oxford University; Erin Campbell of the Lloyd Library, Cincinnati; Giovanna Bergantino at the Biblioteca Antica del Seminario Vescovile di Padova; Monsignor Franco Buzzi and Dr. Don Federico Gallo at the Biblioteca Ambrosiana; Father Justin of St. Catherine's of the Sinai; Susan Marshall of the Early Manuscripts Electronic Library at UCLA; as well as the librarians and staff of the Biblioteca Marciana, Venice; Biblioteca Apostolica Vaticana, Rome; Biblioteca Universitaria di Bologna; Bodleian Libraries, Oxford; British Library, London; Kongelige Bibliotek, Copenhagen; Natural History Museum Library, London; Tebtunis Center, UC Berkeley; and Wellcome Library, London.

Finally, this project has benefited from the encouragement, advice, and expertise of many friends, colleagues, and mentors. I thank the directors of the Kunsthistorisches Institut in Florence, Gerhard Wolf and Alessandro Nova, as well as Tomasso Mozzati, Michael Tymkiw, and Didier Milleriot, for their advice and friendship. From my time at the Center for Advanced Studies in the Visual Arts, I thank Dean Elizabeth Cropper and Associate Deans Peter Lukehart and Therese O'Malley, as well as Yuri Long, Danielle Horetsky, Jen Rokoski, Jeannette Shindell, and Helen Tangires. This project gained much from engaging conversations there with current Dean Steven Nelson, Maryan Ainsworth, C. Jean Campbell, Caroline Danforth, Rena Hoisington, J. P. Park, Kimberly Schenk, and Abigail Quandt of the Walters Art Museum, Baltimore. At USC, I thank Isabella Rose Carr, Beth Massari, Tracey Marshall, and Hector Reyes. At Princeton University, I thank Dimitri Gondicas, Beatrice Kitzinger, Charlie

Barber, Alessia Rossi, and Patricia Blessing for their intellectual generosity and kindness. I also thank Kate Baxter, Linda Taylor, Debbie Puskas, Eleni Banis, Monique Joseph, and Nancy Forti for making me feel welcome. I thank Nikolas Bakirtzis, Dimitri Gondicas, and Alessia Rossi for having me in the Mt. Menoikeion Seminar (2023). I owe many thanks to the sisters of the Monastery of St. John the Forerunner, Serres, for their generosity and patience. Many friends, colleagues, and mentors have advised and encouraged me over the years, including Benjamin Anderson, Jess Genevieve Bailey, Armin Bergmeier, Ruthie Ezra, Jack Hartnell, Philip Hollander, Theresa Holler, Derek Krueger, Sarah Kyle, Andrea Mattiello, Kathleen Maxwell, Heba Mostafa, Dominic Olariu, Rachel Patt, Karen Reeds, Rossitza Schroeder, Ma'ayan Sela, Andrew Sears, Ashanti Shih, Fatih Tarhan, Annabel Wharton, Mario Wimmer, and Olga Yunak.

Finally, and most importantly, I am ever grateful to my family— Laura, Louis, Meridith, Forest, Sally, Deb, and Byron—for their enduring love, support, wisdom, and patience.

Land Acknowledgments

As I wrote this book, I often ran into plants known to Dioscorides and Theophrastus far from their homeland. I have seen them in gardens and parks, on hillsides, in empty lots, on the sides of roads, and in the cracks of sidewalks. Many of the Mediterranean plants described and depicted in this book—excluding many that already had cosmopolitan distributions—have contributed to devasting losses of ecosystems and biodiversity. These are consequences of the ongoing legacy of colonialism and an unsustainable, extractive global economy. Naturalized nonnative plants can be harmless and sometimes helpful to existing ecological communities. But many invasive plants are harmful. They can outcompete native plants, hurt native animals, benefit other invasive species, alter soil chemistry, and otherwise rapidly disrupt ecosystems that developed over eons and were historically sustained by ecologically informed, traditional land management practices.

Indigenous communities at present protect most of the biodiversity that remains on this planet.[1] Land acknowledgments are a necessary, if only initial, step in recognizing Indigenous stewardship of the land, supporting Indigenous land claims, acknowledging settler privilege and responsibility, and encouraging more thoughtful and mutually beneficial relationships with the land. This book germinated, took root, and grew up in many lands that were never ceded by their original caretakers and in cities and institutions that have developed through centuries of colonial extraction, genocidal violence, and forced labor. The University of California, Berkeley, where I first conceived this project, sits on the traditional territory of Xučyun (Huichin), the ancestral and unceded land of the Chochenyo-speaking Ohlone people, which continues to be of great importance to the Muwekma Ohlone Tribe and other descendants of the Verona Band of Alameda County. I completed the initial research for the present volume while in Washington, DC, on the traditional territories of the Piscataway and Nacotchtank (Anacostan) peoples. The manuscript was completed while I was in residence in Los Angeles, where I studied, wrote, and taught on the ancestral, unceded territories of the Tongva or Kizh (Gabrieleño) peoples and their neighbors. I completed the revised manuscript while resident in Princeton, New Jersey, on the ancient and traditional homeland of the Lenni-Lenape people. As this book reaches completion, I am an assistant professor at Duke University in Durham, North Carolina, on the original homelands of Tutelo- and Saponi-speaking peoples. The Haliwa-Saponi, Sappony, Occaneechi Band of Saponi, and Lumbee currently live in the area. They make up a flourishing Native American community in the Research Triangle together with tribal members and descendants of Native nations from across the United States.

Land acknowledgments have their limitations. The above statements are summary, and do not mention specific treaties and conflicts, both past and ongoing. Nor have I mentioned the many important ways that Native peoples continue to relate to the land. I encourage readers to find out more about Indigenous peoples and their histories, cultures, lifeways, and visions for a more just and sustainable future.[2]

NOTES

Introduction

1. Werner Greuter, "Botanical Diversity, Endemism, Rarity, and Extinction in the Mediterranean Area: An Analysis Based on the Published Volumes of Med-Checklist," *Botanika Chronika* 10 (1991): 63–79.

2. Pierre Quézel, "Definition of the Mediterranean Region and the Origin of Its Flora," in *Plant Conservation in the Mediterranean Area*, ed. César Gomez-Campo (Dordrecht: Dr. W. Junk Publishers, 1985), 9–24; Greuter, "Botanical diversity," 63–79; Norman Myers, Russell A. Mittermeier, Christina G. Mittermeier et al., "Biodiversity Hotspots for Conservation Priorities," *Nature* 403 (2000): 853–58.

3. Maddalena Rumor, "At the Dawn of Plant Taxonomy: Shared Structural Design of Herbal Descriptions in *Šammu šikinšu* and Theophrastus' *Historia plantarum* IX," in *Mesopotamian Medicine and Magic. Studies in Honor of Markham J. Geller*, ed. Strahil V. Panayotov and Luděk Vacín (Leiden: Brill, 2018), 446–61. I thank Miriam Said for this reference. For a translation of the *Šammu šikinšu*, see Henry Stadhouders, "The Pharmacopoeial Handbook Šammu šikinšu—A Translation," *Le journal des médecines cunéiformes* 19 (2012): 1–21.

4. Gavin Hardy and Laurence Totelin, *Ancient Botany* (New York: Routledge, 2016), 1–3.

5. George H. M. Lawrence, "Herbals, Their History and Significance," in *History of Botany: Two Papers Presented at a Symposium Held at the William Andrews Clark Memorial Library, December 7, 1963* (Los Angeles: University of California Press, 1965), 3.

6. John Scarborough, "Theophrastus on Herbals and Herbal Remedies," *Journal of the History of Biology* 11 (1978): 353–85. More recently, see the introduction in Suzanne Amigues, *Théophraste. Recherches sur les plantes. Tome V. Livre IX. Texte établi et traduit par S. Amigues* (Paris: Les Belles Lettres, 2006).

7. See chapters 1 and 2.

8. John Lowden, "The Transmission of 'Visual Knowledge' in Byzantium through Illuminated Manuscripts: Approaches and Conjectures," in *Literacy, Education and Manuscript Transmission in Byzantium and Beyond*, ed. Catherine Holmes and Judith Waring (Leiden: Brill, 2002), 59–80.

9. Charles Singer, "The Herbal in Antiquity and Its Transmission to Later Ages," *Journal of Hellenic Studies* 47, no. 1 (1927): 1–52.

10. Singer, "The Herbal," 31.

11. Otto Pächt, "Early Italian Nature Studies and the Early Calendar Landscape," *Journal of the Warburg and Courtauld Institutes* 13, nos. 1–2 (1950): 13–47, at 25n1.

12. Pächt, "Early Italian Nature Studies," 25–27.

13. Pächt, "Early Italian Nature Studies," 27–29.

14. Pächt, "Early Italian Nature Studies," 30.

15. Pächt, "Early Italian Nature Studies."

16. See Minta Collins, *Medieval Herbals: The Illustrative Traditions* (Toronto: University of Toronto Press, 2000), 31–114.

17. Jean Givens, *Observation and Image-Making in Gothic Art* (Cambridge: Cambridge University Press, 2005), 34.

18. As of writing, Max Wellmann's edition was the only complete edition of Dioscorides' *De materia medica*. Marie Cronier is currently preparing a new edition of Dioscorides. For Wellmann's edition, see *Pedanii Dioscuridis Anazarbei De materia medica*, 3 vols., ed. Max Wellmann (Berlin: Weidmann, 1907–1914). This book also refers to the first edition of the translation of Dioscorides by Lily Beck, trans. *Pedanius Dioscorides of Anazarbus, De materia medica* (Hildesheim: Olms-Weidmann, 2005). The wording of direct quotations were checked against the text of the fourth edition (2020).

19. Vienna, Österreichisches Nationalbibliothek, Cod. med. Gr. 1, the "Vienna Dioscorides."

20. New York, Morgan Library, MS M 652, the "Morgan Dioscorides."

21. For example, see Collins, *Medieval Herbals*; also Wilfrid Blunt, *The Art of Botanical Illustration* (London: Collins, 1950); Claus Nissen, *Die botanische Buchillustration: Ihre Geschichte und Bibliographie* (Mannsfield, CT: Maurizio Martino, 1951); Alain Touwaide, *A Census of Greek Medical Manuscripts: From Byzantium to the Renaissance* (London: Routledge, 2016).

Chapter 1

1. The author's nephew, Pliny the Younger, recounts his uncle's death in two letters to his friend Tacitus. Pliny the Younger, *Epistulae* 6.16, 6.20.

2. The passage is difficult to translate into idiomatic English. My translation here is based, with modifications, on that in Hardy and Totelin, *Ancient Botany*, 113; and W. H. S. Jones, in Pliny the Elder, *Natural History, Volume VII: Books 24–27*, trans. W. H. S. Jones with A. C. Andrews, Loeb Classical Library (Cambridge, MA: Harvard University Press, 1956). Pliny the Elder, *Naturalis historia* 25.8; all quotations of the original text are based on the edition by Karl F. T. Mayhoff, *Naturalis historiae libri XXVII* (Leipzig: Teubner, 1892–1909): *Praeter hos Graeci auctores prodidere, quos suis locis diximus, ex his Crateuas, Dionysius,*

Metrodorus ratione blandissima, sed qua nihil paene aliud quam difficultas rei intellegatur. pinxere namque effigies herbarum atque ita subscripsere effectus. verum et pictura fallax est coloribus tam numerosis, praesertim in aemulationem naturae, multumque degenerat transcribentium socordia [var. *sors varia, fors varia, sollertia*]. *praeterea parum est singulas earum aetates pingi, cum quadripertitis varietatibus anni faciem mutent.*

3. See, for example, Collins, *Medieval Herbals*, 37–38; Givens, *Observation and Image-Making*, 17–18, 90, 144–45. While Joshua Thomas is more circumspect regarding Pliny's rhetorical concerns, he ultimately upholds Pliny's comments as an accurate reflection on contemporary botanical illustration. See Joshua Thomas, "The Illustrated Dioskourides Codices and the Transmission of Images during Antiquity," *Journal of Roman Studies* 109 (2019): 241–73, at 261.

4. E.g., Guglielmo Cavallo, "Introduction," trans. Salvatore Lilla, in *Dioscorides Neapolitanus. Biblioteca Nazionale di Napoli. Codex ex Vindobonensis Graecus 1. Commentarium*, ed. Carlo Bertelli, Salvatore Lilla, and Giulia Orofino (Rome: Salerno Editrice, 1992), 9–13, at 9–10.

5. The bibliography on this passage is extensive. See Collins, *Medieval Herbals*, 37–38; David Freedberg, "The Failure of Colour," in *Sight and Insight. Essays on Art and Culture in Honour of E. H. Gombrich at 85*, ed. John Onians (London: Phaidon, 1994), 245–62; Givens, *Observation and Image-Making*, 17–18, 90, 144–45; Alfred Stückelberger, *Bild und Wort: Das illustrierte Fachbuch in der antiken Naturwissenschaft, Medizin und Technik* (Munich: P. von Zabern, 1994), 79–81; Thorsten Fögen, *Wissen, Kommunikation und Selbstdarstellung: Zur Struktur und Charakteristik römischer Fachtexte der frühen Kaiserzeit* (Munich: Beck, 2009), 236–38.

6. For the manuscripts, see *Naturalis historia*, ed. Mayhoff, 4:118.

7. Carl Friedrich Wilhelm Müller, "Kritische Bemerkungen zu Plinius' naturalis historia," *Schulnachrichten. 1887–1888*, Programm Nr. 164 (Breslau: Graß, Barth u. Comp. [W. Friedrich], 1888), 20: "ein Wort, das Plinius sehr liebt."

8. Gabriele Marasco, "L'introduction de la médecine grecque à Rome. Une dissension politique et idéologique," in *Ancient Medicine in Its Socio-Cultural Context: Papers Read at the Congress Held at Leiden University, 13–15 April 1992*, ed. Philip van der Eijk, Herman F.J. Horstmanshoff, and Piet H. Schrijvers (Amsterdam: Rodopi, 1995), 1:35–48; Johannes Hahn, "Plinius und die griechischen Ärzte in Rom: Naturkonzeption und Medizinkritik in der *Naturalis Historia*," *Sudhoffs Archiv* 75 (1991): 209–39.

9. E.g., his comments on the titles of Greek scientific works, Pliny, *Naturalis historia* Praef. 24.

10. Pliny, *Naturalis historia* 22.14–15, 24.1, and 29.5. Hahn, "Plinius und die griechischen Ärzte," 231; G. E. R. Lloyd, *Science, Folklore, and Ideology: Studies in the Origin and Development of Greek Science* (Cambridge: Cambridge University Press, 1983), 135–49; Vivian Nutton, "Roman Medicine: Tradition, Confrontation, Assimilation," in *Aufstieg und Niedergang der römischen Welt. Teil II. Principat. Band 37.1*, ed. W. Haase and H. Temporini (New York: De Gruyter, 1993), 49–78; Karin Nijhuis, "Greek Doctors and Roman Patients: A Medical Anthropological Approach," in *Ancient Medicine in Its Socio-Cultural Context*, ed. Van der Eijk, Horstmanshoff, and Schrijvers, 49–67, especially 50–51, 60–61; Heinrich von Staden, "Liminal Perils: Early Roman Receptions of Greek Medicine," in *Tradi-*

tion, Transmission, Transformation: Proceedings of two Conferences on Pre-Modern Science Held at the University of Oklahoma, ed. F. Jamil Ragep, Sally P. Ragep, and S. Livesey (Leiden: Brill, 1996), 369–418.

11. Trevor Murphy, *Pliny the Elder's Natural History: The Empire in the Encyclopedia* (Oxford: Oxford University Press, 2004), 69. On different cultural conceptions of medicine, see Nijhuis, "Greek Doctors," 50–51, 60–61.

12. Case in point is Cato the Elder's (d. 149 BCE) hostility to Greek medicine and physicians. See Plutarch, *Cato the Elder* 23.1, and Pliny, *Naturalis historia* 29.5. Staden, "Liminal Perils." See also Marasco, "L'introduction," 35–48.

13. On Vitruvius, Atticus, and Varro, see Jocelyn Penny Small, *Parallel Worlds of Classical Art and Text* (Cambridge: Cambridge University Press, 2003), 124, 131–33, 135.

14. Galen, *De methodo medendi* 1.7.6 (ed. Kühn, 10:53–54): καὶ τὰ τοιαῦτα ζητεῖν, εἰ ζωγραφία χρήσιμος ἰατροῖς ἐστιν. For translation and commentary, see Galen, *On the Therapeutic Method: Books I and II*, trans. R. J. Hankinson (New York: Oxford University Press, 1991), 27–28, at 145. See also Marie-Hélène Marganne, *Le livre médical dans le monde gréco-romain*, Cahiers du CEDOPAL 3 (Liège: Les Éditions de l'Université de Liège, 2004), 42–43. A passage on *abrotonon*, likely southernwood (*Artemisia abrotonum* L.), in Galen's *De simplicium medicamentum facultatibus* (6.1; ed. Kühn, 11:799) has been understood to present a dim view of illustration; see Sachiko Kusukawa, *Picturing the Book of Nature: Image, Text, and Argument in Sixteenth-Century Human Anatomy and Medical Botany* (Chicago: University of Chicago Press, 2011), 20. I am doubtful of this reading, though. Galen seems not to refer to depiction generally, but rather states that it is unnecessary to describe *abrotonon* as it is so widely known.

15. Cp. Givens, *Observation and Image-Making*, 87, n. 16, and 144–45. This fact is plainly evident from the surviving papyrus fragments of illustrated herbals, see below.

16. Vienna, Österreichische Nationalbibliothek, Cod. med. Gr. 1. This manuscript is sometimes also called the Anicia Juliana Codex, after one of the codex's recipients.

17. The usual dating of the Vienna Dioscorides to 512 or thereabouts is debatable, first, because the codex was assembled in stages, with the frontispieces added later, and second, because the original date of 512 was based on a problematic reading of Theophanes' *Chronographia* to establish the date of the Theotokos church in the Honoratae District, the church the frontispieces apparently allude to. On the former, see Ernst Gamillscheg, "Das Geschenk für Juliana Anicia: Überlegungen zu Struktur und Entstehung des Wiener Dioskurides," in *Byzantina Mediterranea: Festschrift für Johannes Koder zum 65. Geburtstag*, ed. Klaus Belke and Ewald Kislinger (Vienna: Böhlau, 2007), 187–95; on the latter, A. E. Müller, "Ein vermeintlich fester Anker. Das Jahr 512 als zeitlicher Ansatz des 'Wiener Dioskurides,'" *Jahrbuch der österreichischen Byzantinistik* 62 (2012): 103–9.

18. See chapter 5 here.

19. Hans Gerstinger, *Dioscurides, Codex Vindobonensis Med. Gr. 1, Der Österreichischen Nationalbibliothek*, Kommentarband zu der Faksimileausgabe (Graz: Akademische Druck-u. Verlagsanstalt, 1970), 29. The Seven Sages theme is relatively popular in late antique floor mosaics. See David Knipp, "Medieval Visual Images of Plato," in *The Platonic Tradition in the Middle Ages*, ed. Stephen

Gersh and Maarten J. F. M. Hoenen (Berlin: De Gruyter, 2002), 373–416, especially 382–88.

20. That Galen reports using and seeing parchment codices might also lead us to see the red codex on fol. 3v as a reference to a historical reality, perhaps preserved in the tradition of his portraiture. See, e.g., Galen, *De compositione medicamentorum* 1 (ed. Kühn, 12:423): "Our friend Claudianus found this drug [against baldness] written thus in a parchment codex" (τοῦτο τὸ φάρμακον οὕτο γεγραμμένον εὗρε Κλαυδιανὸς ὁ ἑταῖρος ἡμῶν ἐν πυκτίδι διφθέρᾳ). On this passage in connection to the history of the parchment codex, see Colin H. Roberts and T. C. Skeat, *Birth of the Codex* (London: Oxford University Press, 1983), 22. See also Matthew Nicholls, "Parchment Codices in a New Text of Galen," *Greece & Rome*, 2nd ser., 57, no. 2 (2010): 378–86, especially 381–83.

21. Dioscorides does not report this myth about using a dog to extract the mandrake. It is mentioned in the *Herbarius* by Ps.-Apuleius Platonicus, dated to the fourth or fifth century CE. Ps.-Apuleius Platonicus, *Herbarius* 131. For discussion of this passage with English translation, see Laurence Totelin, Karin Tybjerg, Anne-Laurence Caudano, and Catherine Eagleton, "Of Dogs, Mousetraps and Magical Plants: How (Not) to Dig Up a Mandrake," in *Instruments of Mystery*, ed. Patrick Boner and Catherine Eagleton (Cambridge: Cambridge Latin Therapy/ Whipple Museum, 2004), 27–41, at 28–29. This method of extraction also appears in Josephus' account of the *baaras* plant in *De bello judaico* and in Aelian's account of a plant called *aglaophotis* in his *De natura animalium*. Josephus, *De bello Judaico* 7.180–85; Aelian, *De natura animalium* 14.27. Scholars debate when the idea that mandrake killed those who uprooted it was first introduced into the tradition. See Anne Van Arsdall, Helmut W. Klug, and Paul Blanz, "The Mandrake Plant and its Legend: A New Perspective," in *Old Names — New Growth: Proceedings of the 2nd ASPNS Conference, University of Graz, Austria, 6–10 June 2007, and Related Essays*, ed. Peter Bierbaumer and Helmut W. Klug (Frankfurt am Main: Lang, 2009), 285–346.

22. E. C. E. Owen, "Ἐπινοέω, ἐπίνοια and Allied Words," *Journal of Theological Studies* 35 (1934): 368–76; Antonio Orbe, *La Epinoia. Algunos preliminares históricos de la distinción* κατ' ἐπίνοιαν (Rome: Pontificia Universitas Gregoriana, 1955); Richard Vaggione, *Eunomius of Cyzicus and the Nicene Revolution* (Oxford: Oxford University Press, 2000), 241–46; Lewis Ayres, *Nicaea and Its Legacy: An Approach to Fourth-Century Trinitarian Theology* (Oxford: Oxford University Press, 2004), 191–95.

23. Andrew Griebeler, "Production and Design of Early Illustrated Herbals," *Word & Image: A Journal of Verbal/Visual Enquiry* 38, no. 2 (2022): 104–22.

24. The fact that Pliny only mentions three authors does not mean that there were not others. See Collins, *Medieval Herbals*, 37.

25. Other ancient medical and scientific treatises, including works on anatomy, surgery, and bandaging, were probably also illustrated. See Marganne, *Le livre médical*, 35–58; Stückelberger, *Bild und Wort*; Mirko Grmek and Danielle Gourevitch, *Les maladies dans l'art antique* (Paris: Fayard, 1998).

26. I.e., the so-called *Berufsnamen*; see Heikki Solin, "Die sogenannten Berufsnamen antiker Ärzte," in *Ancient Medicine in Its Socio-Cultural Context*, ed. Van der Eijk, Horstmanshoff, and Schrijvers, 1:119–42.

27. See Galen's comments in *De libris propriis* (*On My Own Books*, ed. Kühn, 19:8–48). On the manuscript tradition and problems with the edition, see Véro-

nique Boudon, "Galen's *On My Own Books*: Material from Meshed, Rida, Tibb. 5223," *Bulletin of the Institute of Classical Studies*, Supplement No. 77 (2002): 9–18, and more recently, in light of Galen's *De indolentia* (also known by its Greek title, *Peri alupias*), Peter N. Singer, "New Light and Old Texts: Galen on His Own Books," in *Galen's Treatise "Περὶ Ἀλυπίας" (De Indolentia) in Context: A Tale of Resilience*, ed. Caroline Petit (Leiden: Brill, 2019), 91–132.

28. Max Wellmann published ten fragments of Crateuas preserved in the Vienna Dioscorides; see Max Wellmann, "Krateuas," *Abhandlungen der königlichen Gesellschaft der Wissenschaften zu Göttingen. Philologisch-historische Klasse.* n.s. 2, no. 1 (1897): 2–32; see also fragments and testimonia, *Pedanii Dioscuridis Anazarbei De materia medica*, ed. Max Wellmann (Berlin: Weidmann, 1914), 3:139–46. Dioscorides mentions Crateuas in the preface to his *De materia medica*. Galen mentions Crateuas several times, for example, in the introduction to Book VI of his *De simplicium medicamentum facultatibus*. See also ps.-Hippocrates, *ep.* 16, in *Hippocrates: Pseudoepigraphic Writings*, ed. and trans. Wesley D. Smith (Leiden: Brill, 1990), 70–73.

29. The fragments in the Vienna Dioscorides concern the following: *aristolochia* (fol. 18r); *aristolochios*, i.e., *strongyle* (fol. 19v); *anemōnē hē phoinikē* (fol. 26r); *asphodelos* (fol. 27v); *arnoglōsson* (fol. 29v); *asaron* (fol. 31r); *anagallides amphoterai* (fol. 40v); *argemōnē* (fol. 29r); and *asterion* (fol. 33r).

30. Dioscorides, *De materia medica* 1.1. The term is sometimes translated as "herbalist," although it is better understood more broadly as a "gatherer of simples." See, e.g., Suzanne Amigues, *Théophraste. Recherches sur les plantes. Tome V. Livre IX. Texte établi et traduit par S. Amigues* (Paris: Les Belles Lettres, 2006), xxxiv.

31. Suzanne Amigues, for example, places the *rhizotomoi* at the bottom of a hierarchy, while Riddle suggests that the *rhizotomoi* were above the *pharmakopōlai*. See John Riddle, *Dioscorides on Pharmacy and Medicine* (Austin: University of Texas Press, 1985), 5; and Amigues, *Théophraste. Recherches*, xxxiv.

32. See, e.g., Leanne McNamara, "Conjurers, Purifiers, Vagabonds and Quacks? The Clinical Roles of the Folk and Hippocratic Healers of Classical Greece," *Iris: Journal of the Classical Association of Victoria* 16–17 (2003): 2–25; see also Luciana Repici, "Medici e botanica popolare," in *Medicina e società nel mondo antico: Atti del convegno di Udine, 4–5 ottobre 2005*, ed. A. Marcone (Florence: Le Monnier Università, 2006), 72–90.

33. This fragment is preserved in Macrobius' *Saturnalia*.

34. Macrobius, *Saturnalia* 5.19.10, translated and cited in John Scarborough, "The Pharmacology of Sacred Plants, Herbs, and Roots," in *Magika Hiera: Ancient Greek Magic and Religion*, ed. Christopher A. Faraone and Dirk Obblink (New York: Oxford University Press, 1991), 138–74, at 144.

35. Theophrastus, *Historia plantarum* 9.8.5.

36. Theophrastus, *Historia plantarum* 9.8.6–8.

37. Rebecca Flemming, "Empires of Knowledge: Medicine and Health in the Hellenistic World," in *A Companion to the Hellenistic World*, ed. Andrew Erskine (Oxford: Blackwell, 2003), 449–63, at 459. Guglielmo Cavallo's suggestion that Crateuas' renown as a *rhizotomos* indicates less learning or rigor is unfounded; see Cavallo, "Introduction," 9. That Crateuas was cited by both Galen and Dioscorides should be enough to suggest that his work was taken seriously by physicians, despite his being a mere *rhizotomos*.

38. I.e., Ῥιζοτομικόν. See Wellmann, "Krateuas," 4. The name of Diocles' work has been preserved in a scholion to Nicander's *Theriaca*; see *Scholia in Nicandri* 647a, ed. Annunciata Crugnola, *Scholia in Nicandri Theriaka cum glossis. Testi e documenti per lo studio dell'antichità* (Milan: Cisalpino, 1971), 241. Diocles of Carystus is thought to have been active in Athens and to have written on a broad variety of topics related to medicine. See Daniela Manetti, "Dioklēs of Karustos," in *The Encyclopedia of Ancient Natural Scientists: The Greek Tradition and Its Many Heirs*, ed. P. T. Keyser and G. L. Irby-Massie (London: Routledge, 2008), 255–57.

39. *Scholia in Nicandri* 681a, ed. Crugnola, 252: Ἄλλως· Κρατεύας ἐν τῷ ῥιζοτομικῷ προστίθεται ὅτι πρόβατον ἄρνα εἰ μὴ στέργοι, ἐάν τις κοτυληδόνα τρίψας μεθ'ὕδατος δῷ, στέργει. See also fr. 23 in *Pedanii Dioscuridis Anazarbei De materia medica*, ed. Max Wellmann (Berlin: Weidmann, 1914), 3:139–46.

40. Wellmann, "Krateuas," 5.

41. Wellmann, "Krateuas," 21. See also Friedrich Ernst Kind, "Krateuas," in *Paulys Realencyclopädie der classischen Altertumswissenschaft* (Stuttgart: Alfred Druckenmüller, 1922), 11, 2:1644–46, at 1644.

42. Galen, *In Hippocratis de natura hominis librum commentarii iii*, ed. Kühn, 15:134–35. Dioscorides, *De materia medica* 1.1: "nor did they in fact mention minerals or spices at all. On the other hand, Crateuas, the root cutter, and Andreas, the physician, both of whom are reputed to have addressed themselves to this part of the subject more completely than the rest, left many highly serviceable roots and certain herbs unnoticed" (Lily Y. Beck, *Pedanius Dioscorides of Anazarbus. De Materia Medica*, 3rd ed. [Hildesheim: Olms-Weidmann, 2017], 1–2). Translations of Dioscorides throughout this book (unless otherwise noted) are from Beck's edition, hereafter cited as Beck.

43. Kind, "Krateuas," 1645; Alain Touwaide, "Crateuas," in *Brill's New Pauly: Encyclopedia of the Ancient World*, ed. Hubert Cancik and Helmuth Schneider (Leiden: Brill, 2003) 3:920–21.

44. Pliny, *Naturalis historia* 25.26.

45. Wellmann, "Krateuas," 3; Kind, "Krateuas," 1644; Riddle, *Dioscorides on Pharmacy*, 5; John Scarborough, "Crateuas," in *The Oxford Classical Dictionary*, 4th ed., ed. S. Hornblower, A. Spawforth, and E. Eidinow (Oxford: Oxford University Press, 2012); Natacha Massar, *Soigner et servir. Histoire sociale et culturelle de la médecine grecque à l'époque hellénistique* (Paris: De Boccard, 2005), 227.

46. Touwaide, "Crateuas"; and, more recently, Hardy and Totelin, *Ancient Botany*, 44.

47. Gabriele Marasco has pointed out that Pompey's freedman Lenaeus, who clearly was not in the service of Mithridates, is also supposed to have named a plant after the Pontic king. After Pompey brought Mithridates VI's library to Rome, he tasked his freedman Lenaeus with its translation. Lenaeus subsequently attributed to Mithridates a plant whose properties were evidently first described by him—the proof was a note written in the monarch's own hand. See Pliny, *Naturalis historia* 25.27. Gabriele Marasco, "Les médecins de cour à l'époque hellénistique," *Revue des études grecques* 109 (1996): 435–66 at 457.

48. Pliny the Elder, *Naturalis historia* 25.34–36.

49. Pliny the Elder, *Naturalis historia* 5.16.

50. Georgia Irby-Massie has suggested that Pliny's Metrodorus might be the Hippocratic commentator cited by Erotian (first century CE) or, following an

earlier suggestion by Max Wellmann, an Asclepiadian pharmacist mentioned by Galen. See Georgia Irby-Massie, "Mētrodōros (Pharm.)," in *The Encyclopedia of Ancient Natural Scientists: The Greek Tradition and Its Many Heirs*, ed. P. T. Keyser and G. L. Irby-Massie (London: Routledge, 2008), 553. Also, Erotian, *Epidemics* 5.26; Galen, *De simplicium medicamentum facultatibus* 1.29, 35 (ed. Kühn, 11:432, 442). Natacha Massar has noted that the only physician we know of with this name from the Hellenistic period is Metrodorus of Amphipolis, a physician to Antiochus I Soter (r. 281–261 BCE). As Massar notes, however, the evidence linking Metrodorus to Pliny is meager. See Massar, *Soigner et servir*, 227. We only know of Metrodorus of Amphipolis from a decree, dated sometime c. 275–268/7 BCE, recorded on a marble stele at Ilion. K. M. T. Atkinson, "The Seleucids and the Greek Cities of Western Asia Minor," *Antichthon* 2 (1968): 32–57; Wolfgang Orth, *Könglicher Machtanspruch und städtische Freiheit; Untersuchungen zu den politischen Beziehungen zwischen den ersten Seleukidenherrschern und den Stadten des westlichen Kleinasien* (Munich: C. H. Beck, 1977), 52–54, 73–74. For English translations of the inscription, see Stanley Burstein, ed. and trans., *The Hellenistic Age from the Battle of Ipsos to the Death of Kleopatra VII*, Translated Documents of Greece and Rome (Cambridge: Cambridge University Press, 1985), 26; and Roger S. Bagnall and Peter Derow, *The Hellenistic Period: Historical Sources in Translation* (Malden, MA: Blackwell, 2004), note 79, 138.

51. Pliny the Elder, *Naturalis historia* 1.18.

52. Pliny the Elder, *Naturalis historia* 20.214.

53. Hardy and Totelin, *Ancient Botany*, 113–14.

54. As early as 1897, Max Wellmann identified Dionysius with Cassius Dionysius of Utica (second to first century BCE). Both Marie-Hélène Marganne and Phillip Thibodeau have reiterated this identification. Phillip Thibodeau, "Dionusios of Utica, Cassius," in *The Encyclopedia of Ancient Natural Scientists*, ed. Keyser and Irby-Massie, 265. See also Feliciano Speranza, *Scriptorum romanorum de re rustica reliquiae*, vol. 1 (Messina: Università degli studi, 1974). For Pliny's citation of the name Cassius Dionysius, see Pliny, *Naturalis historia* 11.15; for simply Dionysius, see, e.g., Pliny, *Naturalis historia* 1.1, among the indices of authorities for Books 8, 10, 14, 15, 17, 18.

55. Columella, *De agricultura* 1.1.13. On Mago, see Vilhelm Lundström, "Magostudien," *Eranos* 2 (1897): 60–67; John Pentland Mahaffy, "The Work of Mago on Agriculture," *Hermathena* 7 (1889): 29–35. See also Jacques Heurgon, "L'agronome carthaginois Magon et ses traducteurs en latin et en grec," *Comptes rendus des séances de l'Académie des inscriptions et belles-lettres* 120 (1976): 441–56. Heurgon notes that Mahaffy confused P. Sextilius with C. Sextius Calvinus; see Heurgon, "L'agronome," 444. See also *Naturalis historia* 18.22, 1.18.

56. Varro, *De re rustica* 1.1.10, ed. Georg Goetz (Leipzig: Teubner, 1929), 9: "Hos nobilitate Mago Carthaginiensis praeteriit, poenica lingua qui res dispersas comprendit libris XXIIX, quos Cassius Dionysius Uticensis vertit libris XX ac Graeca lingua Sextilio praetori misit; in quae volumina de Graecis libris eorum quos dixi adiecit non pauca et de Magonis dempsit instar librorum VIII. Hosce ipsos utiliter ad VI libros redegit Diophanes in Bithynia et misit Deiotaro regi." Translation by William Davis Hooper with revisions by Harrison Boyd Ash in *Cato and Varro, On Agriculture*, Loeb Classical Library 283 (Cambridge, MA: Harvard University Press, 1935), 165, 166: "All these [authors] are surpassed in

reputation by Mago of Carthage, who gathered into twenty-eight books, written in the Punic tongue, the subjects they had dealt with separately. These Cassius Dionysius of Utica translated into Greek and published in twenty books, dedicated to the praetor Sextilius. In these volumes he added not a little from the Greek writers whom I have named, taking from Mago's writings an amount equivalent to eight books. Diophanes, in Bithynia, further abridged these in convenient form into six books, dedicated to king Deiotarus." Mago's work was widely known for its length. See, e.g., Cicero, *De oratore* 1.249.

57. Thibodeau, "Dionusios of Utica," 265.

58. Varro, *De re rustica* 1.1.10. Columella also notes that Diophanes of Bithynia abridged Dionysius' work; see Columella, *De agricultura* 1.1.10.

59. A scholiast of Nicander's *Theriaca* cites Dionysius as the author of a work by this name, while Stephen of Byzantium (fl. 528–35 CE) links a work of the same name to an author from Utica. That the scholiast omits the name Cassius could suggest that this *Rhizotomika* circulated simply under the name Dionysius. For the scholion in Nicander's *Theriaka*, see Crugnola, ed., *Scholia in Nicandri* 520a (p. 204). On Stephen of Byzantium see his *Ethnica* (*Epitome*), s.v. Ἰτύκη, ed. A. Meineke, 342, ll. 3–4; and Stephen of Byzantium, *Ethnica*, bk. 9, n. 121, in Margarete Billerbeck and Christian Zubler, *Stephani Byzantii Ethnika (Volumen II: Δ–I)*, Corpus fontium historiae Byzantinae, Series Berolinensis 43, 2 (Berlin: De Gruyter, 2011). As Thibodeau notes, the identification of the Utican author in Stephen of Byzantium with Dionysius of Utica requires accepting an alteration to the text. Thibodeau, "Dionusios," 265. The manuscript names Diocles (c. 400–300 BCE, see above), the author of a similarly titled work. But the Diocles who wrote the *Rhizotomikon* came from Carystus, a city in Euboea. It seems reasonable to revise Stephen of Byzantium's Diocles to Dionysius. See Meineke, ed., *Stephen von Byzanz. Ethnika*, 342.

60. On Mikion, *Scholia in Nicandri Theriaca*, ed. Crugnola, 617; Wilhelm Kroll, "Mikion (5)," in *Paulys Realencyclopädie der classischen Altertumswissenschaft*, ed. Wilhelm Kroll (Munich: Alfred Druckenmüller, 1980), 15, 2:1555; on Amerias of Macedon, see Athenaeus, *Deipnosophistae* 15.681f; on Eumachus of Corcyra, see Giuseppe Squillace, "Tracce del *Rhizotomikon* di Eumaco di Corcira? (Nota ad Ateneo 15,681e)," in *Sulla rotta per la Sicilia: L'Epiro, Corcira e l'Occidente*, ed. Giovanna de Sensi Sestito and Maria Intrieri (Pisa: Edizioni ETS, 2011), 315–27.

61. Oribasius, *Collectiones medicae* 7.26.31.

62. This is evident in Wellmann's discussion of the adaptations to Dioscorides' *De materia medica*; see Wellmann, "Krateuas," 21–25, especially at 21: "Da diese drei illustrierten Herbarien oder wie man sie nennen will in der späteren Fachlitteratur ausser bei Plinius keinerlei Berücksichtigung gefunden haben, so glaube ich annehmen zu dürfen, dass sie in der Art der illustrierten Pflanzenkunden der Humanistenzeit eine mehr für das Bedürfnis des Volkes bestimmte, populäre Form der ριζοτομικά darstellen."

63. Stückelberger, *Bild und Word*, 74–78.

64. Stückelberger, *Bild und Word*, 74–78.

65. Tertullian, *Scorpiace* l. 3. On dating Nicander to the second century, see Jean-Marie Jacques, ed. and trans., *Nicandre: Oeuvres. Tome II: Les Thériaques. Fragments iologiques antérieurs à Nicandre* (Paris: Jacques, 2002), xix; Jean-Marie Jacques, "Nicander," in *The Encyclopedia of Ancient Natural Scientists*, ed. Keyser

and Irby-Massie, 573–74. See also Marco Fantuzzi, "Nicander," in *Brill's New Pauly*, 9: coll. 706–8.

66. Dated to 202/203, 203/204, or 211–13. See Timothy D. Barnes, "Tertullian's 'Scorpiace,'" *Journal of Theological Studies*, n.s. 20, no. 1 (1969): 105–32.

67. For translation and notes, see A. S. F. Gow and A. F. Scholfield, eds. and trans., *Nicander: The Poems and Poetical Fragments* (Cambridge: Cambridge University Press, 1953).

68. Sotera Fornaro, "Eutecnius," in *Brill's New Pauly*, 5:231–32.

69. See Paris, Bibliothèque nationale de France, MS gr. suppl. 247, e.g., fols. 4r–v, 5r, 16r, 17v, 18r, 19r–v, 20r, 21r, 22r, 27v, 28r–v, 29v, 30r–v, 31v, 44r, 45r, 46r (mid-tenth century); see also the paraphrases in the Vienna Dioscorides, fols. 395r–v, 396r–v, 397r–v, 398r (early sixth century); and the Morgan Dioscorides, fols. 339r–v, 340r–v, 341r–v (late ninth or early tenth century).

70. Some plants in the sixth-century paraphrase are identical to those in Dioscorides; some are not. Some plants in the later, mid-tenth-century Paris Nicander (Paris, Bibliothèque nationale de France, MS gr. suppl. 247), moreover, do not appear to be derived from the same source as the paraphrases.

71. See Jacques, *Nicandre*, xix; see also Phillip Thibodeau, "Attalos III of Pergamon, Philomētōr," in *The Encyclopedia of Ancient Natural Scientists*, ed. Keyser and Irby-Massie, 179–80; and John Scarborough, "Attalus III of Pergamon: Research Toxicologist," in *Asklepios: Studies on Ancient Medicine*, ed. L. Cilliers (Bloemfontien: Classical Association of South Africa, 2008), 138–56; and Laurence Totelin, "Botanizing Rulers and Their Herbal Subjects: Plants and Political Power in Greek and Roman Literature," *Phoenix* 66, nos. 1–2 (2012): 122–44.

72. Marganne, *Le livre médical*, 35–36.

73. Stückelberger, *Bild und Wort*; Marganne, *Le livre médical*, 35–58.

74. Marganne, *Le livre médical*, 43: "Ces médecins gravitent dans un milieu de cour, autour de monarques intéressés par les sciences,—spécialement la toxicologie—, et collectionneurs de livres."

75. On anatomy, see Flemming, "Empires of Knowledge," 451–57.

76. Medicine was increasingly exalted as a practical and theoretical discipline, with the result that medical practitioners gained prestige and status. Their elevated status may have resulted in shifting medical practices, such as the rise of compound drugs, which required expertise and access to many often expensive ingredients. In tandem with this expansion of medical knowledge, the cult of the medical deity Asclepius and his many associates (e.g., Hygeia, Telesphorus, Panacea, Iaso, Aceso) became increasingly popular and spread throughout the Hellenistic world. See Flemming, "Empires of Knowledge," 461.

77. Flemming, "Empires of Knowledge," 453–55. Galen, *De antidotis* 2.7; ed. Kühn, 14:150.

78. Plutarch, *Demetrius* 20.3; K. Ziegler, ed., *Plutarchi vitae parallelae*, 2nd ed. (Leipzig: Teubner, 1971), 3:1: Ἄτταλος δ' ὁ Φιλομήτωρ ἐκήπευε τὰς φαρμακώδεις βοτάνας, οὐ μόνον ὑοσκύαμον καὶ ἐλλέβορον, ἀλλὰ καὶ κώνειον καὶ ἀκόνιτον καὶ δορύκνιον, αὐτὸς ἐν τοῖς βασιλείοις σπείρων καὶ φυτεύων, ὀπούς τε καὶ καρπὸν αὐτῶν ἔργον πεποιημένος εἰδέναι καὶ κομίζεσθαι καθ' ὥραν. And translation by Bernadotte Perrin in *Plutarch, Lives, Volume IX: Demetrius and Antony. Pyrrhus and Gaius Marius*, Loeb Classical Library 101 (Cambridge, MA: Harvard University Press, 1920), 47: "And Attalus Philometor used to grow poisonous plants, not only

henbane and hellebore, but also hemlock, aconite, and dorycnium, sowing and planting them himself in the royal gardens, and making it his business to know their juices and fruits, and to collect these at the proper season."

79. Justin, *Phillipic Histories* 36.4.3.

80. Adrienne Mayor, *The Poison King: The Life and Legend of Mithridates, Rome's Deadliest Enemy* (Princeton: Princeton University Press, 2010), 58. Mayor's picture of Attalus is based in large part on Kent J. Rigsby, "Provincia Asia," *Transactions of the American Philological Association* 118 (1988): 123–53.

81. Galen, *De antidotis* 1.1 (ed. Kühn, 14:2): ὁ γάρ τοι Μιθριδάτης οὗτος, ὥσπερ καὶ ὁ καθ' ἡμᾶς Ἄτταλος, ἔσπευσεν ἐμπειρίαν ἔχειν ἁπάντων σχεδὸν τῶν ἁπλῶν φαρμάκων, ὅσα τοῖς ὀλεθρίοις ἀντιτέτακται, πειράζων αὐτῶν τὰς δυνάμεις ἐπὶ πονηρῶν ἀνθρώπων, ὧν θάνατος κατέγνωστο. τινὰ μὲν οὖν αὐτῶν ἀνεῦρεν ἐπὶ φαλαγγίων ἰδίως ἁρμόζοντα, τινὰ δὲ ἐπὶ σκορπίων, ὥσπερ ἐπὶ τῶν ἐχιδνῶν ἄλλα. καὶ ἐπὶ τῶν ἀναιρούντων φαρμάκων τὰ μὲν ἐπὶ ἀκονίτου, τὰ δὲ ἐπὶ λαγωοῦ τοῦ θαλαττίου, τὰ δ' ἐπ' ἄλλου τινὸς ἢ ἄλλου. Πάντα δ' οὖν αὐτὰ μίξας ὁ Μιθριδάτης ἓν ἐποίησε φάρμακον, ἐλπίσας ἕξειν ἀρωγὸν ἐπὶ πᾶσι οἷς ὀλεθρίοις. And translation by Robert Leigh, *On Theriac to Piso, Attributed to Galen: A Critical Edition with Translation and Commentary* (Leiden: Brill, 2016), 24–25, with modification: "For this Mithridates like our countryman Attalus wanted to test the effect of pretty much every single simple drug which is used against poisons, trying their effects on criminals condemned to death. He found some of help against poisonous spiders, some against scorpions and others against vipers. In the case of poisonous drugs he found some effective against aconite, some against the sea hare, and others against other substances. So Mithridates mixed all these together and made one drug hoping to have a defence against all ills."

82. Pliny the Elder, *Naturalis historia* 25.3; see John Scarborough, "Asklepiades of Bithunia," in *The Encyclopedia of Ancient Natural Scientists*, ed. Keyser and Irby-Massie, 170–71.

83. Pliny, *Naturalis historia* 37.60.

84. See, for example, Pliny, *Naturalis historia* 23.77.

85. On mithridatium, see Laurence Totelin, "Mithridates' Antidote: A Pharmacological Ghost," *Early Science and Medicine* 9, no. 1 (2004): 1–19. The expression "mother of all antidotes," taken from the epigraph to Totelin's article, can be attributed to the *Antidotarium Nicolai*. For possible archaeological evidence for mithridatium production, see Marina Ciaraldi, "Drug Preparation in Evidence? An Unusual Plant and Bone Assemblage from the Pompeian Countryside," *Vegetation History and Archaeobotany* 9, no. 2 (2000): 91–98.

86. E.g., Andrew Erskine, "Culture and Power in Ptolemaic Egypt: The Museum and Library of Alexandria," *Greece and Rome* 42, no. 1 (1995): 38–48. On the emergence of libraries, see Thomas Hendrickson, "The Invention of the Greek Library," *Transactions of the American Philological Association* 144 (2014): 371–413. On Eumenes and parchment, see Richard R. Johnson, "Ancient and Medieval Accounts of the 'Invention' of Parchment," *California Studies in Classical Antiquity* 3 (1970): 115–22. On the influence and memory of the Alexandrian library, see Diana Delia, "From Romance to Rhetoric: The Alexandrian Library in Classical and Islamic Traditions," *American Historical Review* 97, no. 5 (1992): 1449–67.

87. Pliny the Elder, *Naturalis historia* 25.7: "is ergo in reliqua ingeni magnitudine medicinae peculiariter curiosus et ab omnibus subiectis, qui fuere magna

pars terrarum, singula exquirens scrinium commentationum harum et exemplaria effectus que in arcanis suis reliquit."

88. Pliny the Elder, *Naturalis historia* 36.5: "Gemmas plures primus omnium Romae habuit—quod peregrino appellant nomine dactyliothecam—privignus Sullae Scaurus, diu que nulla alia fuit, donec Pompeius Magnus eam, quae Mithridatis regis fuerat, inter dona in Capitolio dicaret, ut Varro alii que aetatis eius auctores confirmant, multum praelata Scauri."

89. Bruno Strasser, "Collecting Nature: Practices, Styles, and Narratives," *Osiris* 27, no. 1 (2012): 303–40, at 313.

90. Pliny the Elder, *Naturalis historia* 8.17 (trans. from Jones): "Aristoteles diversa tradit, vir quem in his magna secuturus ex parte praefandum reor. Alexandro Magno rege inflammato cupidine animalium naturas noscendi delegata que hac commentatione Aristoteli, summo in omni doctrina viro, aliquot milia hominum in totius Asiae Graeciae que tractu parere iussa, omnium quos venatus, aucupia piscatus que alebant quibusque vivaria, armenta, alvaria, piscinae, aviaria in cura erant, ne quid usquam genitum ignoraretur ab eo. quos percunctando quinquaginta ferme volumina illa praeclara de animalibus condidit. quae a me collecta in artum cum iis, quae ignoraverat, quaeso ut legentes boni consulant, in universis rerum naturae operibus medio que clarissimi regum omnium desiderio cura nostra breviter peregrinantes."

91. Plutarch, *Life of Alexander* 8.1. On *philiatroi* as "amateur physicians," see Petros Bouras-Valliantos, *Innovation in Byzantine Medicine: The Writings of John Zacharias Aktouarios (c. 1275–c.1330)* (Oxford: Oxford University Press, 2020), 105–10.

92. On this passage, see Flemming, "Empires of Knowledge," 450.

93. Pliny the Elder, *Naturalis historia* 25.7 (trans. from Jones): "Pompeius autem omni praeda regia potitus transferre ea sermone nostro libertum suum Lenaeum grammaticae artis iussit vitae que ita profuit non minus quam reipublicae victoria illa."

94. Heurgon, "L'agronome," 441–56, especially 444; Antoine Pietrobelli, "The Pharmacological Treatise *Περὶ εὐφορβίου* of Juba II, King of Mauretania," in *'Greek' and 'Roman' in Latin Medical Texts*, ed. Brigitte Maire (Leiden: Brill, 2014), 157–82.

95. See Kathryn Gleason, "Porticus Pompeiana: A New Perspective on the First Public Park of Ancient Rome," *Journal of Garden History* 14, no. 1 (1994): 13–27; Gleason, "The Garden of Portico of Pompey the Great: An Ancient Public Park Preserved in the Layers of Rome," *Expedition* 32, no. 2 (1990): 4–13; Ann Kuttner, "Looking Outside Inside: Ancient Roman Garden Rooms," *Studies in the History of Gardens & Designed Landscapes* 19, no. 1 (1999): 7–35; Kuttner, "Culture and History at Pompey's Museum," *Transactions of the American Philological Association* 129 (1999): 347–73, especially 355–56 and 369.

96. Totelin, "Botanizing Rulers," 134.

97. Elizabeth Ann Pollard, "Pliny's Natural History and the Flavian Templum Pacis: Botanical Imperialism in First-Century C.E. Rome," *Journal of World History* 20, no. 3 (2009): 309–88.

98. The library is attested by Aulus Gellius, *Noctes Atticae* 5.21.9–13 and 16.8.1–4. Pier Luigi Tucci, *The Temple of Peace in Rome* (Cambridge: Cambridge University Press, 2017), 174–215. As Tucci notes, the nature and extent of the library's holdings remain unclear.

99. Galen, *De libris propriis* 2 (ed. Kühn, 19:21–22), quoted in Tucci, *Temple of Peace*, 195. Tucci also refutes the common belief that anatomical demonstrations were carried out at the Temple of Peace.

100. E.g., Cicero, *Epistulae ad Atticum* 1.11.3. See Kuttner, "Looking Outside Inside."

101. Kuttner, "Looking Outside Inside," 29.

102. Diliana Angelova, *Sacred Founders: Women, Men, and Gods in the Discourse of Imperial Founding, Rome through Early Byzantium* (Oakland: University of California Press, 2015), 72. See Pliny the Elder, *Naturalis historia* 15.70 and 15.136–37.

103. Angelova, *Sacred Founders*, 300n45. For Livia's laxative, see Pliny, *Naturalis historia* 19.92; for her remedy for sore throat, Marcellus Empiricus, *De medicamentis* 15.6, and for chills, 35.6.

104. See Ovid, *Fasti* 5.149–58; Macrobius, *Saturnalia* 1.12.26. On Livia's restoration of the temple of Bona Dea, see Angelova, *Sacred Founders*, 72; and Hendrik Hubert Jan Brouwer, *Bona Dea: The Sources and a Description of the Cult*, Études préliminaires aux religions orientales dans l'empire romain 110 (Leiden: Brill, 1989), 323–72.

105. Athenaeus, *Deipnosophistae* bk. 15.

106. Aulus Gellius, *Noctes Atticae* 18.10.

107. For a larger study of domestic Roman medical knowledge, see Jane Draycott, *Roman Domestic Medical Practice in Central Italy: From the Middle Republic to the Early Empire* (London: Routledge, 2019).

108. Anton von Premerstein, "Anicia Juliana im Wiener Dioskorides-Kodex," *Jahrbuch der kunsthistorischen Sammlungen des allerhöchsten Kaiserhauses* 24 (1903): 105–24; Leslie Brubaker, "The Vienna Dioskorides, and Anicia Juliana," in *Byzantine Garden Culture*, ed. Antony Robert Littlewood, Henry Maguire, and Joachim Wolschke-Bulmahn (Washington, DC: Dumbarton Oaks, 2002), 189–214; Angelova, *Sacred Founders*, 173–78, especially 177.

109. Angelova, *Sacred Founders*, 177.

110. Angelova, *Sacred Founders*, 173–78, especially 177.

111. Vivian Nutton, "Healers in the Medical Marketplace: Towards a Social History of Graeco-Roman Medicine," in *Medicine in Society: Historical Essays*, ed. Andrew Wear (Cambridge: Cambridge University Press, 1992), 15–52. See also Sarah Pomeroy, "Technikai kai Mousikai: The Education of Women in the Fourth Century and the Hellenistic Period," *American Journal of Ancient History* 2 (1977): 51–68; and Mati Meyer, *An Obscure Portrait: Imaging Women's Reality in Byzantine Art* (London: Pindar Press, 2009), 116–27. On female medical practice in medieval Western Europe, along with a discussion of methodological concerns, see Monica Green, "Women's Medical Practice and Health Care in Medieval Europe," *Signs* 14, no. 2 (1989): 434–73. In Galen's few noted interactions with female health providers, he does not doubt their diagnoses or recommendations; see Susan P. Mattern, *Galen and the Rhetoric of Healing* (Baltimore: Johns Hopkins University Press, 2008), 73–74.

112. Brubaker, "The Vienna Dioskorides," 211–13.

113. Susan Treggiari, "Jobs in the Household of Livia," *Papers of the British School at Rome* 43 (1975): 48–77, at 56; Susan P. Mattern, "Physicians and the Roman Imperial Aristocracy: The Patronage of Therapuetics," *Bulletin of the History of Medicine* 73, no. 1(1999): 1–18, at 9–10.

114. See Mattern, *Galen and the Rhetoric*, 71–80. See also Herman F. J. Horstmanshoff, "Galen and His Patients," in *Ancient Medicine in Its Socio-Cultural Context*, ed. Van der Eijk, Horstmanshoff, and Scrijvers, 1:83–99.

115. P. Tebt. II 679 + P. Tebt. Tait 39–41 + P. Tebt. Tait 39 add. + PSI inv. 4169 a+b = Trismegistos 63596 = MP³2094. The largest group of fragments, now in the Tebtunis Center at the University of California, Berkeley (P.Tebt. II 679 frags. *a–f*), was discovered by the excavation of Bernard Pyne Grenfell and Arthur Surridge Hunt in 1899/1900, and edited and published by J. de M. Johnson in 1913. J. de M. Johnson, "A Botanical Papyrus with Illustrations," *Archiv für die Geschichte der Naturwissenschaften und der Technik* 4 (1912–13): 403–8. Three additional fragments in the Papyrology Rooms of the Sackler Library, Oxford, were edited and published by Wiliam John Tait in 1977 (P.Tebt. Tait 39–41). W. J. Tait, *Papyri from Tebtunis in Egyptian and in Greek* (London: Egyptian Exploration Society, 1977), 94–96. Tait suspected that these fragments might have belonged to the roll published by Johnson, a position echoed by subsequent authors. See Ann Ellis Hanson, "Text and Context for the Illustrated Herbal from Tebtunis," in *Atti del XXII congresso di papirologia, Firenze, 23–29, 1998*, ed. Isabella Andorlini, Guido Bastianini, Manfredo Manfredi, and Giovanna Menci (Florence: Istituto Papirologico G. Vitelli, 2001), 585–604. Kim Ryholt has identified three additional fragments: one in the Papyrology Rooms, Sackler Library, Oxford (Box 20, folder "2/5 FF" / P. Tebt. Tait 39 add.), and two at the Istituto Papirologico G. Vitelli in Florence (PSI inv. 4160 *a–b*). Kim Ryholt, "The Illustrated Herbal from Tebtunis: New Fragments and Archaeological Context," *Zeitschrift für Papyrologie und Epigraphik* 187 (2013): 233–38. See also Riddle, *Dioscorides on Pharmacy*, 177–79; Daniela Fausti, "Erbari illustrati su papiro e tradizione iconografica botanica," in *Testi medici su papiro. Atti del Seminario di studio (Firenze, 3–4 giugno 2002)* (Florence: Istituto papirologico G. Vitelli, 2004), 131–50; Marganne, *Le livre médical*, 37–42; Hardy and Totelin, *Ancient Botany*, 114.

116. It is assumed to have been a roll because the versos are blank. While the codex of the Naples Dioscorides also has largely blank versos, the rarity of codices dating to the second century lends support to the suggestion that the fragments belonged to a roll. On the rarity of codices in the second century, see Roberts and Skeat, *The Birth of the Codex*, 35–37. On dating, see Johnson, "Botanical Papyrus," 403.

117. Ryholt, "The Illustrated Herbal," 233–38. See also Ryholt, "Libraries from Late Period and Graeco-Roman Egypt, c. 800 BCE–250 CE," in *Libraries before Alexandria: Ancient Near Eastern Traditions*, ed. Kim Ryholt and Gojko Barjamovic (Oxford: Oxford University Press, 2019), 388–472, especially 392–400.

118. Ryholt, "Libraries," 395.

119. Peter Van Minnen, "Boorish or Bookish?: Literature in Egyptian Villages in the Fayum in the Graeco-Roman Period," *Journal of Juristic Papyrology* 28 (1998): 99–184, at 168.

120. Ryholt, "Libraries," 400.

121. On the demotic herbal, see W. J. Tait, "P. Carlsberg 230: Eleven Fragments from a Demotic Herbal," in *The Carlsberg Papyri. 1. Demotic Texts from the Collection*, ed. P. J. Frandsen, CNI Publications 15 (Copenhagen: Carsten Niebuhr Institute of Ancient Near Eastern Studies and Museum Tusculanum Press, 1991), 93–101.

122. Philippa Lang, *Medicine and Society in Ptolemaic Egypt* (Leiden: Brill, 2013), 222. See also Van Minnen, "Boorish or Bookish?" 166.

123. Isabella Andorlini, *Trattato di medicina su papiro* (Florence: Istituto papirologico G. Vitelli, 1995), 6–10.

124. Lang, *Medicine and Society*, 222.

125. Hanson, "Text and Context," 585–604. On Greek and Egyptian medical practitioners in Ptolemaic and Roman Egypt, see Lang, *Medicine and Society*, 216–29.

126. Lang, *Medicine and Society*, 177–80; on Pharaonic botany, see the excursus in Jacco Dieleman, *Priests, Tongues, and Rites: The London-Leiden Magical Manuscripts and Translation in Egyptian Ritual (100–300 CE)* (Leiden: Brill, 2005), 195–98.

127. See Marie-Hélène Marganne and Paul Mertens, "Medici et Medica, 2e édition (État au 15 janvier 1997 du fichier MP³ pour les papyrus médicaux littéraires)," in *'Specimina' per il Corpus dei Papiri Greci di Medicina. Atti dell'incontro di studio (Firenze, 28–29 marzo 1996)*, ed. Isabella Andorlini (Florence: Istituto papirologico G. Vitelli, 1997), 3–71. Marie-Hélène Marganne, "Compléments au fichier MP³ pour les papyrus médicaux littéraires (État au 1er décembre 1999)," *Analecta Papyrologica* 12 (2000 [2001]): 151–61.

128. Ann Ellis Hanson, "Greek Medical Papyri from the Fayum Village of Tebtunis: Patient Involvement in a Local Health-Care System?" in *Hippocrates in Context, Papers Read at the XIth International Hippocrates Colloquium, University of Newcastle upon Tyne, 27–31 August 2002*, ed. Philip van der Eijk, Studies in Ancient Medicine 31 (Leiden: Brill, 2005), 387–402.

129. Hanson, "Greek Medical Papyri," 401.

Chapter 2

1. Wellmann, "Krateuas," 20.

2. Translation here with modifications based on that given by Hardy and Totelin, *Ancient Botany*, 38, and W. H. S. Jones in the Loeb series. Pliny, *Naturalis historia* 25.9: "Quare ceteri sermone eas tradidere, aliqui ne effigie quidem indicata et nudis plerumque nominibus defuncti, quoniam satis videbatur potestates vimque demonstrare quaerere volentibus. nec est difficilis cognitio: nobis certe, exceptis admodum paucis, contigit reliquas contemplari scientia Antoni Castoris, cui summa auctoritas erat in ea arte nostro aevo, visendo hortulo eius, in quo plurimas alebat."

3. Wellmann, "Krateuas," 20.

4. See Hanson, "Text and Context," 588–89; and Johnson, "A Botanical Papyrus," 403–8.

5. On the Antinoopolis codex, see David Leith, "The Antinoopolis Illustrated Herbal (PJohnson + PAntin. 3.214 = MP3 2095)," *Zeitschrift für Papyrologie und Epigraphik* 156 (2006): 141–56. For the Sinai Palimpsest, see Giulia Rossetto, Agamemnon Tselikas, and Nigel Wilson, "Sinai Arabic NF 8, *Crateuas?* (Greek)," sinai.library.ucla.edu, a publication of St. Catherine's Monastery of the Sinai in collaboration with EMEL and UCLA. The relevant undertexts are on fols. 16r–17v. At present there is not yet a published transcription and edition of the undertexts from the Sinai illustrated herbal palimpsest. The undertext visible on

fol. 16v (the lower half of fig. 3.5) mentions boiling the plant *philetairion* with oil and applying it to help those struck by scorpions. The undertext on the other side of the folio (fol. 16r) is much harder to read.

6. See Max Wellmann, ed., *Pedanii Dioscoridis Anazarbei De materia medica* (Berlin: Weidmann, 1914), 3:139–46.

7. E.g., frag. *a*, ll. 5–6; also possibly frag. *e*, ll. 4–6. See Johnson, "A Botanical Papyrus," 404–5.

8. *Chondrilla juncea* L. Frag. *a*, ll. 5–6: φύεται [δὲ ἐν] τόποις γεώδεσιν.

9. This example is Leiden, Universiteitsbibliotheek, MS VLQ 9, dated to the second half of the sixth century. For an edition of the text, see Ernst Howald and Henry E. Sigerist, eds. *Antonii Musae De herba Vettonica liber. Pseudoapulei Herbarius. Anonymi de taxone liber. Sexti Placiti Liber medicinae ex animalibus* (Leipzig: Teubner, 1927). For a German translation, see Kai Brodersen, trans., *Heilkräuterbuch. Herbarius. Lateinisch und Deutsch* (Wiesbaden: Marixverlag, 2015).

10. Leith, "Antinoopolis Illustrated Herbal."

11. Andrea Guasparri, "Explicit Nomenclature and Classification in Pliny's *Natural History* XXXII," *Studies in History and Philosophy of Science Part A* 44, no. 3 (2013): 347–53, especially 350.

12. For example, two medicine bottles dated to the first half of the third century CE and found at the *domus del chirurgo* (house of the surgeon) in Rimini bear the letters ΧΑΜΑΙΔΡΥϹ (i.e., *chamaidrys*, germander) and ΑΒΡΟΤΟΝΟΥ (i.e., *abrotonon*, southernwood), accompanied by a Latin abbreviation, "HABR." Jacopo Ortalli, "Rimini. La domus 'del Chirurgo,'" in *Aemilia. La cultura romana in Emilia Romagna dal III secolo a.C. all'età costantiniana*, ed. Mirella Marini Calvani (Venice: Marsilio, 2000), 513–26; Luigi Taborelli, "I contenitori per medicamenti nelle prescrizioni di Scribonio Largo e la diffusione del vetro soffiato," *Latomus* 55 (1996): 148–56; Stefano De Carolis, ed., *Ars Medica. I ferri del mestiere. La domus 'del Chirurgo' di Rimini e la chirurgia nell'antica Roma* (Rimini: Guaraldi, 2009); and Isabella Andorlini, "Gli strumenti perduti di Galeno," *La torre di Babele* 8 (2012): 239–47, at 246–47. On the different purveyors of drugs and simples in ancient Rome, see Jukka Korpela, "Aromatarii, pharmacopolae, thurarii et ceteri. Zur Sozialgeschichte Roms," in *Ancient Medicine in Its Socio-Cultural Context* ed. Van der Eijk, Horstmanshoff, and Schrijvers, 1:101–18.

13. Hardy and Totelin, *Ancient Botany*, 101.

14. On *symphyton*, see, Reinhold Strömberg, *Griechische Pflanzennamen*, Göteborgs Högskolas Årsskrift 46, no. 1 (Goteborg: Elanders Boktryckeri Aktiebolag, 1940), 77, 88. On *aristolocheia*, see Dioscorides, *De materia medica* 3.4; Pliny, *Naturalis historia* 25.95.

15. On parsley's name and relation to rocky habitats, see Hardy and Totelin, *Ancient Botany*, 99. On the meanings of Greek plant names, generally, see Strömberg, *Griechische Pflanzennamen*.

16. Hardy and Totelin, *Ancient Botany*, 71–72.

17. Dioscorides, *De materia medica* 3.4.

18. Hardy and Totelin, *Ancient Botany*, 97–98.

19. R. J. Hankinson, "Usage and Abusage: Galen on Language," in *Language: Volume 3 of Companions of Ancient Thought* (Cambridge: Cambridge University Press, 1994), 166–87, at 170.

20. Quoted and cited in Hankinson, "Usage and Abusage," at 170.

21. Bibliography on *Cratylus* is immense. See, e.g., R. Barney, *Names and Nature in Plato's Cratylus* (London: Routledge, 2001); J. L. Ackrill, "Language and Reality in Plato's *Cratylus*," in *Realtà e ragione*, ed. A. Alberti (Florence: Olschki, 1994), 9–28; reprinted in J. L. Ackrill, *Essays on Plato and Aristotle* (Oxford: Oxford University Press, 1997), 33–52.

22. Hardy and Totelin, *Ancient Botany*, 93–102.

23. Hardy and Totelin, *Ancient Botany*, 93, 101–3.

24. Hardy and Totelin, *Ancient Botany*, 72, 93.

25. Hardy and Totelin, *Ancient Botany*, 93, 103–4.

26. Hardy and Totelin, *Ancient Botany*, 93, 96.

27. Hardy and Totelin, *Ancient Botany*, 98.

28. For example, ἀμάραντον (which means "unfading" or "unwithering") in Pliny and Dioscorides refers to the persistent color of the flowers (Pliny, *Naturalis historia* 21.8; Dioscorides, *De materia medica* 4.57), whereas in later lexica it often refers to the evergreen quality of the plant; see Armand Delatte, "Glossaires de botanique," in *Anecdota Atheniensia et alia* (Paris: Droz, 1939), 2:277–454, at 279, 304, 320, 341, 361, 367, 373, 418.

29. On antiphrasis, see Hardy and Totelin, *Ancient Botany*, 100.

30. On ancient botanical lexica, see Hardy and Totelin, *Ancient Botany*, 102–3.

31. Pliny, *Naturalis historia* 25.9. On Antonius Castor, see P. T. Keyser, "Antonius Castor," in *The Encyclopedia of Ancient Natural Scientists*, ed. Keyser and Irby-Massie, 100.

32. For a discussion of observation in ancient botany, see Hardy and Totelin, *Ancient Botany*, 36–41.

33. Hardy and Totelin, *Ancient Botany*, 36–41.

34. Dioscorides, *De materia medica* Praef. 5 (ed. Wellmann): μετὰ γὰρ πλείστης ἀκριβείας τὰ μὲν πλεῖστα δι' αὐτοψίας γνόντες, τὰ δὲ ἐξ ἱστορίας τῆς πᾶσι συμφώνου καὶ ἀνακρίσεως τῶν παρ' ἑκάστοις ἐπιχωρίων ἀκριβώσαντες πειρασόμεθα. Translation here modified from Beck, 3.

35. On Dioscorides' preface, see John Scarborough and Vivian Nutton, "The Preface of Dioscorides' *De materia medica*: Introduction, Translation and Commentary," *Transactions and Studies of the College of Physicians of Philadelphia* 4, no. 3 (1982): 187–227.

36. Galen, *Subfiguratio empirica*, in *Die griechische Empirikerschule: Sammlung der Fragmente und Darstellung der Lehre*, ed. Karl Deichgräber (Berlin: Weidmann, 1965), 42–90, at 47. For translation, see Galen, *Three Treatises on the Nature of Science: On the Sects for Beginners, An Outline of Empiricism, On Medical Experience*, trans. Richard Walzer and Michael Frede (Indianapolis: Hackett, 1985).

37. See Vivian Nutton, *Ancient Medicine*, 2nd ed. (London: Routledge, 2013), 147–50.

38. On *historia*, see Deichgräber, *Die griechische Empirikerschule*, 298–301; W. J. Slater, "Asklepiades and Historia," *Greek, Roman and Byzantine Studies* 13 (1972): 317–33; Barbara Cassin, "L'histoire chez Sextus Empiricus," in *Le Scepticisme antique. Perspectives historiques et systématiques. Actes du Colloque International sur le Scepticisme Antique; 1–3 juin 1988*, ed. André-Jean Voelke (Lausanne: Université de Lausanne, 1990), 123–38.

39. See Gianna Pomata, "A Word of the Empirics: The Ancient Concept of Observation and Its Recovery in Early Modern Medicine," *Annals of Science* 68, no. 1 (2011): 1–25, at 7–8. See also Nutton, *Ancient Medicine*, 148–49. The term "tripod" is connected to a lost work of Glaucias of Tarentum (fl. c. 175 BCE). Nutton notes that there was some disagreement about reasoning on the basis of similarity.

40. Dioscorides, *De materia medica* Praef. 2, especially οὐ τῇ πείρᾳ τὴν ἐνέργειαν αὐτῶν κανονίζοντες.

41. Dioscorides, *De materia medica* Praef. 7–8: τὸν δὲ βουλόμενον ἐν τούτοις ἐμπειρίαν ἔχειν δεῖ κατά τε τὴν ἀρτιφυῆ βλάστησιν ἐκ τῆς γῆς καὶ ἀκμάζουσι καὶ παρηκμακόσι παρατυγχάνειν. . . . ὁ δὲ πολλάκις ἐντετευχὼς αὐτοῖς καὶ ἐν πολλοῖς τόποις μάλιστα τὴν ἐπίγνωσιν ποιήσεται. Translation with modification from Beck, 4.

42. Dioscorides thus hits upon another common use of *autopsia* from geographical accounts. Loveday Alexander, *The Preface to Luke's Gospel: Literary Convention and Social Context in Luke 1.1–4 and Acts 1.1* (Cambridge: Cambridge University Press, 1993), 36–37.

43. See Scarborough and Nutton, "The Preface," 213–15; and Hardy and Totelin, *Ancient Botany*, 15.

44. Hippocrates, *Aphorisms* 1. The Latin translation, *ars longa, vita brevis*, may be more familiar.

45. Pliny the Elder, *Naturalis historia* 24.150: "id autem, quod Graeci dracontion vocant, *triplici effigie demonstratum mihi est*: foliis betae, non sine thyrso, flore purpureo; hoc est simile aro. alii radice longa veluti signata articulosa que monstravere, ternis omnino cauliculis, foliis, decoqui ex aceto contra serpentium ictus iubentes. *tertia demonstratio* fuit folio maiore quam cornus, radicis harundineae, totidem, ut adfirmabant, geniculatae nodis, quot haberet annos, totidem que esse folia; hi ex vino vel aqua contra serpentes dabant."

46. E.g., Hardy and Totelin, *Ancient Botany*, 114. See also translation given by W. H. S. Jones.

47. Charlton Lewis and Charles Short, eds., *A Latin Dictionary* (New York: Oxford University Press, 1995), s.v. *demonstratio* (hereafter Lewis and Short).

48. Theophrastus, *Historia plantarum* 2.4.1 and 8.7.1; see discussion and further examples in Hardy and Totelin, *Ancient Botany*, 74.

49. Hardy and Totelin, *Ancient Botany*, 141–42; David M. Balme, "Development of Biology in Aristotle and Theophrastus," *Phronesis* 7 (1962): 91–104.

50. On *genos* and *eidos* in Aristotle, see David M. Balme, "*Genos* and *Eidos* in Aristotle's Biology," *Classical Quarterly* 12 (1962): 81–98. See also Balme, "Aristotle's Use of Division and Differentiae," in *Philosophical Issues in Aristotle's Biology*, ed. Allan Gotthelf (Cambridge: Cambridge University Press, 1987), 69–89.

51. On division by *differentiae* in Aristotle, see Balme, "Aristotle's Use of Division."

52. Theophrastus, *Historia plantarum* 1.1.2; trans. Arthur Hort, *Theophrastus. Enquiry into Plants: Books 1–5*, Loeb Classical Library 70 (Cambridge, MA: Harvard University Press, 1916), 5. For more on Theophrastus' approach, see Georg Wöhrle, *Theophrasts Methode in seinen botanischen Schriften* (Amsterdam: Grüner, 1985).

53. See Alain Touwaide, "Art and Science: Private Gardens and Botany in the Early Roman Empire," in *Botanical Progress, Horticultural Innovation and Cultural Change*, ed. Michel Conan and W. John Kress (Washington, DC: Dumbarton Oaks Research Library and Collection, 2004), 37–49, at 42–43.

54. See Hardy and Totelin, *Ancient Botany*, 70.

55. See Kusukawa, *Picturing*, 104–6, 123.

56. Pliny, *Naturalis historia* 25.7: "is ergo in reliqua ingeni magnitudine medicinae peculiariter curiosus et ab omnibus subiectis, qui fuere magna pars terrarum, singula exquirens scrinium commentationum harum et exemplaria effectusque in arcanis suis reliquit."

57. Lewis and Short, s.v. *exemplar*. The term can also refer to images, likenesses, and impressions.

58. Stavros Lazaris, "L'illustration des disciplines médicales dans l'antiquité. Hypothèses, enjeux, nouvelles interprétations," in *La collezione di testi chirurgici di Niceta. Firenze, Biblioteca medicea laurenziana, Plut. 74.7. Tradizione medica classica a Bisanzio*, ed. Massimo Bernabò (Rome: Edizioni di storia e letteratura, 2010), 99–109, at 104–8. And Lazaris, "L'image paradigmatique. Des *Schémas anatomiques* d'Aristote au *De materia medica* de Dioscuride," *Pallas* 93 (2013): 131–64; Thomas, "Illustrated Dioskourides," 262–68.

59. See Andrew Griebeler, "Production and Design of Early Illustrated Herbals," *Word & Image: A Journal of Verbal/Visual Enquiry* 38, no. 2 (2022): 104–22.

60. Leith, "Antinoopolis Codex," 152.

61. Matthew Nicholls, "Parchment Codices in a New Text of Galen," *Greece and Rome* 57, no. 2 (2010): 378–86.

62. Galen, *De indolentia* 32, also often called by its Greek title, *Peri alupias*. See Nicholls, "Parchment Codices," 382.

63. Laurence M. V. Totelin, *Hippocratic Recipes: Oral and Written Transmission of Pharmacological Knowledge in Fifth- and Fourth-Century Greece* (Leiden: Brill, 2009), 228. I thank one of the anonymous reviewers for this suggestion.

64. Pliny, *Naturalis historia* 16.137. The story is adapted from Theophrastus; see Laurence Totelin, "Botanizing Rulers," 135.

65. Dried bundles and garlands of flowers have been found in Roman Egyptian tombs: e.g., Renate Germer, "Ancient Egyptian Plant-Remains in the Manchester Museum," *Journal of Egyptian Archaeology* 73 (1987): 245–46. The remains of a garland with *Celosia argentea* L. was also found in a burial at Hawara. It is now kept at the Liverpool World Museum, no. 56.20.475.

66. Dioscorides, *De materia medica* Praef. 9.

67. Pliny, *Naturalis historia* 37.11.

68. E.g., Pliny, *Naturalis historia* Bk. 37.

69. See Isabella Bonati, "φαρμακοθήκη," published September 8, 2014, in Isabella Andorlini (Principal Investigator), *Medicalia Online*, http://www.papirologia.unipr.it/. I thank the anonymous reviewer for this suggestion.

70. Bruno Latour, *Pandora's Hope. Essays on the Reality of Science Studies* (Cambridge: Harvard University Press, 1999), 56.

71. Latour, *Pandora's Hope*, 30.

72. Nelson Goodman, *Languages of Art: An Approach to a Theory of Symbols* (Indianapolis: Hackett, 1976), 33.

73. The principal exceptions include "tree" forms of shrubs and undershrubs, such as *tithymallos dendritēs*, illustrated in the Vienna Dioscorides, fol. 346v. It may simply have been unnecessary to collect or depict specimens from trees in antiquity, due to the fact that ancient people encountered them differently than they did herbs. Ancient peoples of the classical Mediterranean world seem

better acquainted with trees in their cultivated forms (see Theophrastus, *Historia plantarum* 1.14.4). In a similar way, Dioscorides often dispenses with descriptions of common cultivated trees, presumably on the grounds that they were primarily cultivated, or, as he says, "known to everyone" (e.g., Dioscorides, *De materia medica* 1.126).

74. See, for example, Dioscorides' comments in his preface. Dioscorides, *De materia medica* Praef.

75. E.g., Theophrastus, *Historia plantarum* 9.8.1.

76. See Griebeler, "Production and Design."

77. Collins, *Medieval Herbals*, 37.

78. *Subscripsere* can mean to "write underneath or below" as well as to "write or note down"; see Lewis and Short, s.v. *subscribo*.

79. Fausti, "Erbari illustrati," 131–50.

80. University of California, Berkeley (II 679 a, e, f); Oxford, P. Tebt. Tait 39, frag. 3.

81. Griebeler, "Production and Design."

82. Chrysi Kotsifou, "Books and Book Production in the Monastic Communities of Byzantine Egypt," in *The Early Christian Book*, ed. William E. Klingshirn and Linda Safran (Washington, DC: Catholic University of America Press, 2007), 48–66.

Chapter 3

1. See summary and further support in Thomas, "Illustrated Dioskourides," 253–59.

2. Erich Bethe, *Buch und Bild in Altertum* (Leipzig: Harrassowitz, 1945), especially 22–60, 99–101.

3. Pächt, "Early Italian Nature Studies," 30.

4. Lazaris, "L'image paradigmatique," 131–64.

5. The bibliography on diagrams is vast. Some representative studies on premodern diagrams with further bibliography include Steffen Bogen and Felix Thürlemann, "Jenseits der Opposition von Text und Bild: Überlegungen zu einer Theorie des Diagramms und des Diagrammatischen," in *Die Bildwelt der Diagramme Joachims von Fiore: zur Medialität religiös-politischer Programme im Mittelalter*, ed. Alexander Patschovsky (Ostfildern: Thorbecke, 2003), 1–22; Jeffrey Hamburger, *Diagramming Devotion: Berthold of Nuremberg's Transformation of Hrabanus Maurus's Poems in Praise of the Cross* (Chicago: University of Chicago Press, 2020), especially 13–32; Jeffrey Hamburger, David Roxburgh, and Linda Safran, eds., *The Diagram as Paradigm: Cross-Cultural Approaches* (Baltimore: Dumbarton Oaks, 2022). On diagrams more generally, see especially Birgit Schneider, Christoph Ernst, and Jan Wöpking, eds., *Diagrammatik-Reader: Grundlegende Texte aus Theorie und Geschichte* (Berlin: De Gruyter, 2016); and Sybille Krämer, *Figuration, Anschauung, Erkenntnis: Grundlinien einer Diagrammatologie* (Berlin: Suhrkamp, 2016), especially 59–86.

6. Sybille Krämer, "Trace, Writing, Diagram: Reflections on Spatiality, Intuition, Graphical Practices and Thinking," in *The Power of the Image: Emotion, Expression, Explanation*, ed. András Benedek and Kristóf Nyíri (Frankfurt am Main: Peter Lang, 2014), 3–22, at 3.

7. See also Valeria Giardino, "Diagramming: Connecting Cognitive Systems to Improve Reasoning," in *The Power of the Image*, ed. Benedek and Nyíri, 23–34.

8. See Sybille Krämer and Christina Ljungberg, "Thinking and Diagrams: An Introduction," in *Thinking with Diagrams: The Semiotic Basis of Human Cognition*, ed. Krämer and Ljungberg (Berlin: De Gruyter, 2016) 1–19, at 8, nos. 10–11.

9. Note the tendency in the literature to define diagrams as iconic following Charles Sanders Peirce; e.g., Bogen and Thürlemann, "Jenseits der Opposition," 9.

10. Pächt, "Early Italian Nature Studies," 31.

11. Otto Pächt, "Die früheste abendländische Kopie der Illustrationen des Wiener Dioskurides," *Zeitschrift für Kunstgeschichte* 38, nos. 3–4 (1975): 201–14.

12. Pächt's view echoes here a more general and widespread idea that botanical illustrations are less "artistic" than scientific. See, for example, Joseph Mantuani (Josip Mantuani), "Die Miniaturen im Wiener Kodex Med. Graecus I," in *De codicis Dioscuridei Aniciae Iulianae, nunc Vindobonensis Med. Gr. 1*, ed. Josef Karabacek (Leiden: Sijthoff, 1906), 353–490, at 383.

13. Pächt, "Die früheste abendländische Kopie," 208.

14. Pächt, "Die früheste abendländische Kopie," 210: "Kurz, was die antike Herbarillustration bietet, ist in der Regel manipulierte, vordemonstrierte Natur, eine auf Wiedererkennbarkeit und objektive Bestimmbarkeit gerichtete Empirie, nie der subjektive Eindruck des in spontaner Wahrnehmung Erschauten."

15. Pächt, "Die früheste abendländische Kopie," 208.

16. Collins's notion of plant portraiture may go back to Pächt, "Early Italian Nature Studies," 26. Given the difficulty of defining the term "portrait," I avoid this term. On portraiture, see Shearer West, *Portraiture* (Oxford: Oxford University Press, 2004), 21–41.

17. Collins, *Medieval Herbals*, 28. She defines naturalism in terms of the accuracy of pictorial delineation, as well as the inclusion of pictorial devices emulating three-dimensionality (e.g., overlapping stems and complex modeling) and seemingly "natural" aspects of growth (e.g., curving and scarred stems, fallen petals).

18. Collins, *Medieval Herbals*, 27.

19. Hardy and Totelin, *Ancient Botany*, 120.

20. Givens, *Observation and Image-Making*, 34.

21. Givens, *Observation and Image-Making*, 102.

22. Pächt, "Early Italian Nature Studies," 31.

23. Ernst H. Gombrich, *Art and Illusion: A Study in the Psychology of Pictorial Representation* (London: Phaidon Press, 1984 [1960]), 219.

24. Dominic Lopes, *Understanding Pictures* (Oxford: Clarendon Press, 2004 [1996]), 119.

25. Lopes adapts these categories from Ned Block; see Lopes, *Understanding Pictures*, 118.

26. Lopes, *Understanding Pictures*, 126.

27. Lopes, *Understanding Pictures*, 125: "The reason is simply that not all spatial relations between objects in three-dimensional space can be represented on a two-dimensional surface. Selecting to represent some spatial relations makes other relations unrepresentable."

28. E.g., "it denotes our idea or conception of a thing in contrast to the thing itself" (Archibald Robertson, in *Select Library of Nicene Fathers*, 4:368n, as cited in Owen, "Ἐπινοέω, ἐπίνοια," 372).

29. Ayres, *Nicaea and Its Legacy*, 191.

30. Owen, "Ἐπινοέω, ἐπίνοια," 376.

31. On medieval abstraction, see the contributions in *Abstraction in Medieval Art: Beyond the Ornament*, ed. Elina Gertsman (Amsterdam: Amsterdam University Press, 2021).

32. Ute Mauch, "Pflanzenabbildungen des Wiener Dioskurides und das Habituskonzept: Ein Beitrag zur botanischen Charakterisierung von antiken Pflanzen durch den Habitus," *Antike Naturwissenschaft und ihre Rezeption* 16 (2006): 125–38.

33. Compare with the prominence of perspective and foreshortening in modern botanical illustration. See, for example, Wendy B. Zomlefer, *Flowering Plant Families* (Chapel Hill: University of North Carolina Press, 1994), 20–25. See also Marian Ruff Sheehan, "Illustrating Plants," in *The Guild Handbook of Scientific Illustration*, ed. Elaine R. S. Hodges with Lawrence B. Isham, Martha E. Jessup, and G. Robert Lewis (New York: Van Nostrand Reinhold, 1989), 189–200.

34. See Francis Hallé, "A Life Drawing Trees: Interview with Emmanuele Coccia," trans. Emma Lingwood, in *Trees*, ed. Pierre-Édouard Couton (Paris: Fondation Cartier, 2019), 32–47, at 41. I thank an anonymous reviewer for this suggestion.

35. Hardy and Totelin, *Ancient Botany*, 106.

36. Hardy and Totelin, *Ancient Botany*, 147, 179.

37. On the date palm, see Hardy and Totelin, *Ancient Botany*, 132–33, 171–73.

38. Theophrastus, *Historia plantarum* 1.13.1–5. He distinguishes them according to whether they are "downy" (χνοώδης, probably very small flowers, since he gives the examples of ivy and mulberry); "leafy" (φυλλώδης, i.e., large petals); large (μέγεθος) or small (ἀμέγεθος); monochromatic (μονόχροος) or dichromatic (δίχροος); with a simple corolla or doubled, "twofold flowers" (διᾰνθη); whether they have one fused petal or many petals; and according to the petals' relative location to the seed/fruit case (περικάρπιον). He finally notes differences between "fertile" and "infertile" flowers.

39. Hardy and Totelin, *Ancient Botany*, 111; Touwaide, "Art and Science," 42.

40. Hardy and Totelin, *Ancient Botany*, 167.

41. Compare Collins, *Medieval Herbals*, 304.

42. For overview and further bibliography on search images, see Morten Tønnessen, "The Search Image as Link between Sensation, Perception and Action," *BioSystems* 164 (2018): 138–46. The (re)introduction of the term "search image" into modern ethology can be credited to Luuk Tinbergen, "The Natural Control of Insects in Pinewoods: I. Factors Influencing the Intensity of Predation by Song Birds," *Archives Néerlandaises de Zoologie* 13 (1960): 265–343. The term "search image" first appears in Jacob von Uexküll, *Streifzüge durch die Umwelt von Tieren und Menschen: Ein Bilderbuch unsichtbarer Welten* (Hamburg: Rowohlt, 1956 [1934]), 83–87. For English translation, see Jacob von Uexküll, *A Foray into the Worlds of Animals and Humans with a Theory of Meaning*, trans. Joseph D. O'Neil (Minneapolis: University of Minnesota Press, 2010). Uexküll had a complicated relationship with National Socialism; right-wing ideology may have informed his *Umwelt* theory. See Gottfried Schnödl and Florian Sprenger, *Uexküll's Surroundings: Umwelt Theory and Right-Wing Thought*, trans. Michael Thomas Taylor and Wayne Yung (Lüneberg: Meson Press, 2021).

43. Robin Wall Kimmerer, *Gathering Moss: A Natural and Cultural History of Mosses* (Corvallis: Oregon State University Press, 2003), 8–9.

44. Uexküll, *Streifzüge*, 83: "Das Suchbild vernichtet das Merkbild."

45. Daniela Bleichmar, *Visible Empire: Botanical Expeditions and Visual Culture in the Hispanic Enlightenment* (Chicago: University of Chicago Press, 2012), 151–61.

46. Scholars have had a dim view of the quality of the illustrations in the Tebtunis Roll; e.g., Johnson, "Botanical Papyrus," 404; Singer, "Herbal in Antiquity," 31; Kurt Weitzmann, *Ancient Book Illumination* (Cambridge, MA: Harvard University Press, 1959), 11; and Collins, *Medieval Herbals*, 38. On the contrary, the fine brushstrokes, confident handling of the colors, and the careful pooling of color to suggest space point to the work of a professional painter.

47. Gombrich, *Art and Illusion*, e.g., 105–9, 182–203, 208.

48. See Hardy and Totelin, *Ancient Botany*, 106, 111, 147.

49. On mental rotation, see R. N. Shepard and J. Metzler, "Mental Rotation of Three-Dimensional Objects," *Science* 171 (1971): 701–3. On the role of the motor cortex in mental rotation, not without controversy, see C. Eisenegger, U. Herwig, and J. Jäncke, "The Involvement of Primary Motor Cortex in Mental Rotation Revealed by Transcranial Magnetic Stimulation," *European Journal of Neuroscience* 25 (2007): 1240–44; and S. J. Flusberg and L. Boroditsky, "Are Things That Are Hard to Physically Move Also Hard to Imagine Moving?" *Psychonomic Bulletin and Review* 18 (2011): 158–64.

50. On surface and depth in eighteenth-century botanical illustration, see Michael Gaudio, "Surface and Depth: The Art of Early American Natural History," in *Stuffing Birds, Pressing Plants, Shaping Knowledge: Natural History in North America 1730–1860*, ed. Sue Ann Prince, *Transactions of the American Philosophical Society* 93, no. 4 (2003): 55–74.

51. On the two versions, see Marie Cronier, "L'herbier alphabétique grec de Dioscoride. Quelques remarques sur sa genèse et ses sources textuelles," in *Fitozooterapia antigua y altomedieval. Textos y doctrinas*, ed. Arsenio Ferraces Rodriguez (Coruña: Universidade da Coruña, 2009), 33–59; Cronier, "Un manuscrit méconnu du Περὶ ὕλης ἰατρικῆς de Dioscoride: New York, Pierpont Morgan Library, M. 652," *Revue des études grecques* 125, no. 1 (2012): 95–130; and Thomas, "Illustrated Dioskourides," 247–49, 251–52.

52. On the stylistic differences between the manuscripts, see Giulia Orofino, "The Dioscorides of the Biblioteca Nazionale of Naples: The Miniatures," trans. Linda Lappin, in *Dioscorides Neapolitanus: Biblioteca Nazionale di Napoli Codex ex Vindobonensis Graecus 1, Commentarium*, ed. Carlo Bertelli, Salvatore Lilla, and Giulia Orofino (Rome: Salerno Editrice, 1992), 99–113, at 104–6.

53. *Viola odorata* L. It is depicted in Vienna Dioscorides, fol. 148v; Naples Dioscorides, fol. 142r. Weitzmann, *Ancient Book Illumination*, 12.

54. For similar observations in the Naples Dioscorides, see Orofino, "The Dioscorides of the Biblioteca Nazionale," 105–7.

55. While most depictions of floral anatomy in the Dioscorides manuscripts are rather vague, some detailed depictions of flowers do occur. For example, the depiction of a rose (*rhodon ē rhoda*, *Rosa spp.* L.) shows several unopened flower buds in addition to opened flowers; see Vienna Dioscorides, fol. 282r; Naples Dioscorides, fol. 129r. The open flowers face multiple directions, showing us the base, top, and profile of the flower head. Interestingly, the underside of the rose

flower was not included in the Naples Dioscorides, a detail that may have been simply lost through the process of repeated copying.

56. Theophrastus, *Historia plantarum* 1.1.3.

57. *Rubus ulmifolius* Schott; Vienna Dioscorides, fol. 83r; Naples Dioscorides, fol. 32r.

58. See Riddle, *Dioscorides on Pharmacy*, 212.

59. *Geranium* spp. in Vienna Dioscorides, fol. 85r; Naples Dioscorides, fol. 58r; and *Anemone coronaria* L. in Vienna Dioscorides, fol. 25v; Naples Dioscorides, fol. 12r.

60. Pliny the Elder, *Naturalis historia* 25.4: "praeterea parum est singulas earum aetates pingi, cum quadripertitis varietatibus anni faciem mutent."

61. On diachronic approaches in early modern botanical illustrations, see Kusukawa, *Picturing the Book*, 115–16.

62. See Hardy and Totelin, *Ancient Botany*, 147–49.

63. *Symphytum spp.*, Naples Dioscorides, fol. 132r.

64. Ovid, *Metamorphoses* 10.725–39, trans. Frank Justus Miller, *Ovid. Metamorphoses: Books 9–15*, Loeb Classical Library 43 (Cambridge, MA: Harvard University Press, 1939), 117.

65. On this phenomenon in mostly later examples, see Lina Bolzoni, "The Play of Images: The Art of Memory from Its Origins to the Seicento," in *The Mill of Thought: From the Art of Memory to the Neurosciences*, ed. Pietro Corsi (Milan: Electa, 1989), 17–26; Frances A. Yates, *The Art of Memory* (London: Routledge, 1992 [1966]), 1–26; Kusukawa, *Picturing*, 18.

66. *Eryngium maritimum* L.

67. Naples Dioscorides, fol. 78r.

68. Dioscorides, *De materia medica* 3.21: ἱστορεῖται δ' ὅτι περιαπτομένη διαφορεῖ φύματα.

69. Translation here modified from Henry North Fowler in *Plutarch. Moralia X*, ed. Jeffrey Hendersen, Loeb Classical Library 321 (Cambridge, MA: Harvard University Press, 1936), 33. Plutarch, *Maxime cum principibus philosopho esse disserendum*, Steph. 776f.

70. See also Plutarch, *De sera numinis vindicta* 558e.

71. See comments attributed to Julius Africanus via Michael Psellos: *Cesti: The Extant Fragments*, ed. Martin Wallraff et al. (Berlin: De Gruyter, 2012), 21, 23. For the most recent edition, see Stratis Papaioannou, *Michael Psellus, Epistulae*, 2 vols. (Berlin: De Gruyter, 2019), vol. 1, sec. 30, no. 124. See also Olimpio Musso, ed. and trans, *Michele Psello, Nozioni paradossali. Testo critico, introduzione, traduzione e commentario* (Naples: Università di Napoli, 1977), 47–49n59.

72. We would not expect to find roots, for example, on lichens (λιχὴν ὁ ἐπὶ τῶν πετρῶν, identifications vary, Sprengel says *Pettigera canina* Hoffm., or *Pettigera aphthosa* Hoffm.; Fraas says *Lecanora parella* Ach., e.g., Vienna Dioscorides, fol. 216v; Dioscorides, *De materia medica* 4.53).

73. *Antirrhinum maius* L. or *A. oronteum* L., Vienna Dioscorides, fol. 159v; Naples Dioscorides, fol. 51r. The accompanying text in the Alphabetical Dioscorides does not explain why the illustration of the plant lacks roots and why the plant is shown growing on a substrate. Other authors, however, note simply that *antirrhinon*, a synonym for *kynokephalion*, has no roots (Theophrastus, *Historia plantarum* 9.19; Pliny, *Naturalis historia* 25.129).

74. Vienna Dioscorides, fol. 391v. See Hartmut Böhme, "Koralle und Pfau, Schrift und Bild im *Wiener Dioskurides*," in *Bild/Geschichte. Festschrift für Horst Bredekamp*, ed. Philine Helas, Maren Polte, Claudia Rückert, and Bettina Uppenkamp (Berlin: Akademie Verlag, 2007), 57–72.

75. Laurence M. V. Totelin, "What's a Plant?" in *The Cambridge Companion to Ancient Greek and Roman Science*, ed. L. Taub (Cambridge: Cambridge University Press, 2020), 141–59.

76. The dog-faced *kētos* is a generic iconographic type. Oppian mentions several different *kētea* in his *Halieutika*, one of which is a *kētos*. Oppian, *Halieutika* 1.360–82.

77. Collins suggests a possible identification with Thetis. Collins, *Medieval Herbals*, 97n62. Contemporaries may not have agreed on her identification. For example, various Middle Byzantine sources identified a statue of a sea goddess crowned with crabs in the Forum of Constantine in Constantinople as Thalassa, Thetis, or Amphitrite. Sarah Bassett, *Urban Image of Late Antique Constantinople* (Cambridge: Cambridge University Press, 2004), cat. no. 115, pp. 207–8. Arethas (tenth century) reports that the statue was of Thetis, though his contemporaries called it Thalassa, while Kedrenos (fl. eleventh century) calls the figure Amphitrite. See also R. J. H. Jenkins, "The Bronze Athena at Byzantium," *Journal of Hellenic Studies* 67 (1947): 31–33. The confusion over the statue's identity may even date to the statue's installation in the Forum. Sarah Bassett thinks the statue was originally Amphitrite, but was repurposed as Thetis in order to create a Judgment of Paris statuary group in the Forum. Bassett, *Urban Image*, cat. no. 115, pp. 207–8.

78. Otto Mazal, *Der Wiener Dioskurides: Codex medicus graecus 1 der Österreichischen Nationalbibliothek*, 2 vols. (Graz: Akademische Druck- u. Verlagsanstalt, 1998), 2:47. Mazal favors an identification of the figure as Thalassa. A depiction of Thalassa appears in the Church of the Apostles in Madaba; see Henry Maguire, *Nectar and Illusion: Nature in Byzantine Art and Literature* (Oxford: Oxford University Press, 2012), 16.

79. For example, London, Natural History Museum, Banks Coll. Dio., fol. 399r.

80. See Kurt Weitzmann, "The Greek Sources of Islamic Scientific Illustrations," in *Archaeologica Orientalia in Memoriam Ernst Herzfeld* (Locust Valley, NY: J. Agustin, 1952), 244–66, reprinted in Weitzmann, *Studies in Classical and Byzantine Manuscript Illumination*, ed. Herbert Kessler (Chicago: University of Chicago Press, 1971), 20–44, at 29–30; Riddle, *Dioscorides on Pharmacy*, 201.

81. Grape-Albers, *Spätantike Bilder*, 37–103, and conclusion at 182.

82. For another explanation, see George Saliba and Linda Komaroff, "Illustrated Books May Be Hazardous to Your Health: A New Reading of the Arabic Reception and Rendition of the 'Materia Medica' of Dioscorides," *Ars Orientalis* 35 (2008): 6–65.

Chapter 4

1. For an example of astrological organization, see Leith, "Antinoopolis Illustrated Herbal," 154. Pliny uses a combination of organizational systems; see Hardy and Totelin, *Ancient Botany*, 86–87; also Lloyd W. Daly, *Contributions to*

a History of Alphabetization in Antiquity and the Middle Ages (Brussels: Latomus, 1967), 35–36.

2. William T. Stearn and Erminio Caprotti, *Herbarium Apulei 1481, Herbolario Volgare 1522* (Milan: Polifilo, 1979), 1:xlix.

3. It is not clear when alphabetization became a common organizational scheme, though it could have emerged in the famous libraries of Alexandria. See Daly, *Contributions to a History of Alphabetization*, 24, 36. Crateuas' herbal may have been alphabetically ordered. For the most part, ancient alphabetical organization was based only on the first letter or syllable, and was therefore not absolute. For a rare exception, see Daly, *Contributions to a History of Alphabetization*, 35.

4. Riddle, *Dioscorides on Pharmacy*, 25–93.

5. See Hardy and Totelin, *Ancient Botany*, 76–77.

6. E.g., Theophrastus, *Historia plantarum* 6.6.2. On Theophrastus' ordering of plants, see Hardy and Totelin, *Ancient Botany*, 75–80.

7. Theophrastus, *Historia plantarum* 6.6.

8. For more on this system of organization, see Riddle, *Dioscorides on Pharmacy*, 94–131.

9. Riddle, *Dioscorides on Pharmacy*, 96–97.

10. Dioscorides, *De materia medica* Praef. 3.

11. Hardy and Totelin, *Ancient Botany*, 21–22.

12. On the dating and localization of the Vienna Dioscorides, see Mazal, *Der Wiener Dioskurides*, 1:4–5; and more recently, Müller, "Ein vermeintlich fester Anker," 103–9. On the Naples Dioscorides, see Cavallo, "Introduction," 11–13; also Orofino, "The Dioscorides," 102. Note: the main points in Orofino's essay also appear (in German) in Orofino, "Dioskurides war gegen Pflanzenbilder: Die Illustration der Heilmittellehre des Dioskurides zwischen Spätantike und dem Hochmittelalter," *Die Waage* 30 (1991): 144–49.

13. Anton von Premerstein (Antonius de Premerstein), "De codicis Dioscuridei Aniciae Iulianae, nunc Vindobonensis Med. Gr. I: Historia, Forma, Argumento," in *De codicis Dioscuridei Aniciae Iulianae, nunc Vindobonensis Med. Gr. 1*, ed. Josef Karabacek (Leiden: Sijthoff, 1906), 3–228, at 110. Wellmann, "Krateuas," 24; Singer, "Herbal in Antiquity," 24–26; Collins, *Medieval Herbals*, 48–56. See also Orofino, "The Dioscorides," 100. On the fact that the two codices are not directly related to each other, see Cronier, "Un manuscrit méconnu," 95–130; and Thomas, "Illustrated Dioskourides," 247–49, 251–52. I agree with Cronier's dating and localization of the earliest version of the Alphabetical Dioscorides to Italy of the second or third century CE. See Cronier, "L'herbier alphabétique grec," 52–58.

14. Scholars now generally agree on this point. Most recently, see Thomas, "Illustrated Dioskourides," 242. In addition to the reasons given by Thomas, we can add that Dioscorides would have followed the conventions of illustrated herbals by excluding descriptions had there been illustrations. John Riddle is one of the few scholars to argue that Dioscorides' original work had to include illustrations. See Riddle, *Dioscorides on Pharmacy*, 177.

15. Wellmann, "Krateuas," 25–26.

16. Singer, "Herbal in Antiquity," 5–7.

17. For fragments and testimonia of Crateuas, see Wellmann, ed., *Pedanii Dioscuridis Anazarbei De materia medica*, 3:139–46.

18. The index is on fols. 8r–10v. Premerstein, "De codicis Dioscuridei," 3–228, especially the helpful "codicis conspectus," 193–220.

19. Vienna Dioscorides, fols. 4r, 5r, 6r, 7r. The title appears on fol. 4r: πίναξ ὀρθώτατος τῶν βοτάνων ἅπερ ἔχει τὸ παρὸν βιβλίον ἐνταυθα.

20. E.g., Orofino, "Dioscorides of the Biblioteca Nazionale," 100–101. See discussion (and rejection) of this theory in Collins, *Medieval Herbals*, 47–50.

21. Grape-Albers, *Spätantike Bilder*, 7–10.

22. Orofino, "Dioscorides of the Biblioteca Nazionale," 100–101, especially note 7.

23. Riddle, *Dioscorides on Pharmacy*, 190; Collins, *Medieval Herbals*, 48–49.

24. See Griebeler, "Production Sequence and Page Design."

25. Identifying truly divergent plant depictions depends to a large extent on the eye of the beholder. Giulia Orofino counts 19 different pictures of plants among the 351 that appear in both codices. See Orofino, "The Dioscorides," 104–5. Anton von Premerstein gives fewer (*argemōne, bouglōsson, blēton, gnaphallion, nymphaia*), apparently following an eighteenth-century notice. See Premerstein, "De Codicis Dioscuridei," 100. See also D. Weigel and Ernst Gottfried Baldinger, "Ueber die griechischen Handschriften des Dioscorides in der kaiserl. Bibliothek zu Wien," *Medicinisches und Physiches Journal* 8, no. 32 (1793): 5–12. I agree that *bouglōsson, gnaphallion,* and *nymphaia* differ significantly between the two manuscripts, but the *blēton* and *argemone* in both manuscripts probably go back to a common source, as their main differences lie in their proportions, as well as the number and length of the branches.

26. E.g., Ferdinand Cohn, "Beitrag zur Geschichte der Botanik," *Jahres-Bericht der Schlesischen Gesellschaft für vaterländische Cultur* 59 (1882): 302–12.

27. See Vienna Dioscorides, fol. 76v; Naples Dioscorides, fol. 28r. Compare to Dioscorides, *De materia medica* 4.127.

28. Collins, *Medieval Herbals*, 56.

29. E.g., Mantuani, "Die Miniaturen," 476–77. And against this view, Orofino, "The Dioscorides," 103.

30. For the former view: Orofino, "The Dioscorides," 104–6. For the latter: Cronier, "Un manuscrit méconnu," 109; and Lazaris, "L'illustration des disciplines médicales," 99–109. On critical copying practices as observed in later manuscripts, see chapter 6, below.

31. Cronier, "Un manuscrit méconnu," 109; Cronier, "L'herbier alphabétique grec," 33–59.

32. Nigel Wilson has proposed that this approach to formatting may represent an important early indication for the development of scholia in late antiquity. See Nigel G. Wilson, "Two Notes on Byzantine Scholarship: I. The Vienna Dioscorides and the History of Scholia," *Greek, Roman and Byzantine Studies* 12 (1971): 557–58.

33. See chapter 2, above.

34. See Riddle, *Dioscorides on Pharmacy*, 176–80; Cronier, "L'herbier alphabétique grec."

35. Galen, *De propriis libris* K19, 9, translated with modification, based on Peter N. Singer, *Galen: Selected Works* (Oxford: Oxford University Press, 1997), 3.

36. Oribasius, *Collectiones medicae* (Ἰατρικαὶ συναγωγαί), Bks. 11–13. On Oribasius, see Barry Baldwin, "The Career of Oribasius," *Acta Classica* 18 (1975): 85–97.

37. As Galen does not include interpolations present in the Alphabetical Dioscorides, but which are present in Oribasius, Wellmann thinks that this version of the text must have come between them. Max Wellmann, "Die Pflanzennamen des Dioskurides," *Hermes* 33 (1898): 360–422, at 374–75.

38. Riddle, *Dioscorides on Pharmacy*, 179–80.

39. Riddle, *Dioscorides on Pharmacy*, 180.

40. Riddle, *Dioscorides on Pharmacy*, 180.

41. Cronier, "L'herbier alphabétique grec," 44.

42. See Hardy and Totelin, *Ancient Botany*, 102; Ioana Claudia Popa, "The Lists of Plant Synonyms in *De materia medica* of Dioscorides," *Global Journal of Science Frontier Research* 10, no. 3 (2010): 46–49; and Wellmann, "Pflanzennamen des Dioskurides," 360–422. For examples of entries without synonyms, see *bolbos* and *bettonikē* in the Vienna Dioscorides, fol. 78r–v. See also Premerstein, "De Codicis Dioscuridei," 83–86.

43. For discussion of these different groups, see Hardy and Totelin, *Ancient Botany*, 102; John Scarborough, "More on Dioscorides' Etruscan Herbs," *Etruscan News* 6, no. (2006): 1 and 9; Kyle P. Johnson, "An Etruscan Herbal?" *Etruscan News* 5 (2006): 1 and 8; José Fortes Fortes, "Hispanische Pflanzennamen im Pseudo-Dioskurides und Pseudo-Apuleius," *Glotta* 74 (1997): 1–11; V. L. Bologa, "I sinonimi 'daci' delle piante descritte da Dioscoride possono servire alla ricostruzione della lingua daci?" *Archeion* 12, no. 2 (1930): 166–70; Wellmann, "Pflanzennamen des Dioskurides," 364.

44. See Hardy and Totelin, *Ancient Botany*, 102; Johnson, "An Etruscan Herbal?" 1 and 8.

45. See Wellmann, "Pflanzennamen des Dioskurides," 369–73; and Max Wellmann, "Pamphilos," *Hermes* 51 (1916): 1–64. This lexicographer is mentioned by Galen. See Galen, *De simplicium medicamentum facultatibus*, preface to Book VI (ed. Kühn, 11:793).

46. Hardy and Totelin, *Ancient Botany*, 102.

47. Leiden, Universiteitsbibliotheek, MS VLQ 9. A slight difference is that the main plant names are placed in the same column as the synonyms and are set off by the term "*herba*."

48. The hoary stock (*leukoion thalassion, Matthiola incana* [L.] W. T.Aiton) can be seen in Vienna Dioscorides, fol. 203v; and Naples Dioscorides, fol. 109r. On the identification, see Mazal, *Der Wiener Dioskurides*, 1:79. Dog's cabbage (*kynaia*, also called *kynokrambē, Theligonum cynocrambe* L.) can be found in Vienna Dioscorides, fol. 158v; and Naples Dioscorides, fol. 51r. For the identification, see Mazal, *Der Wiener Dioskurides*, 1:70. Hound's tongue (*kynoglōsson*, either *Cynoglossum columnae* Ten., according to Henry George Liddell, Robert Scott, Henry Stuart Jones, and Roderick McKenzie, *A Greek-English Lexicon*, 9th ed. (Oxford: Clarendon Press, 2007), hereafter *LSJ*; or *Cynoglossum officinale* L., according to Mazal, *Der Wiener Dioskurides*, 1:70) appears in Vienna Dioscorides, fol. 166v; and Naples Dioscorides, fol. 102r. The two delphiniums (*Delphinium* spp.): *delphinion* (Vienna Dioscorides, fol. 96r; Naples Dioscorides, fol. 61r) and *delphinion heteron* (Vienna Dioscorides, fol. 101v [no illustration, just text]; Naples Dioscorides, fol. 61r). The two hawkweeds: *hierakion to mega* (*Urospermum picroides* [L.] F. W.Schmidt), Vienna Dioscorides, fol. 149v; *hierakion to mikron* (perhaps *Hymenonema graecum* DC), Vienna Dioscorides, fol. 150v, and

Naples Dioscorides, fol. 41r. Multiple identifications have been proposed for *zōonychon* (Vienna Dioscorides, fol. 124r; Naples Dioscorides, fol. 76r; see Mazal, *Der Wiener Dioskurides*, 1:59): *Evax pygmaea* (L.) Brot. by Fraas and *Gnaphalium leontopodium* Scop. by Sprengel, both cudweeds; while Penzig gives *Calendula arvensis* L. or field marigold. Cronier identifies two more: *phasiolos* and *sphairitis*. Cronier, "L'herbier alphabétique grec," 47.

49. See Wellmann, ed., *De materia medica* 3.73 RV.

50. Hoary stock, for example, includes a list of synonyms and medicinal uses; see Wellmann, ed., *De materia medica* 3.123 RV; 2:134. For *zōonychon* there is only a list of synonyms; see Wellmann, ed., *De materia medica* 4.133 RV; 2:278.

51. Vienna Dioscorides, fols. 8r–10v. The one exception is that the table of contents notes only one kind of delphinium while there are actually two.

52. See also discussion in Cronier, "L'herbier alphabetique grec."

53. See Vienna Dioscorides, fols. 13v–14v; compare with *De materia medica* 4.88–90. Note that the top corner of fol. 14 is torn and that a later hand mislabeled the text on fol. 14r as *aeizōon to leptophyllon*, when it should be *aeizōon to mikron*. Notably, there is no minuscule transliteration for the chapter on *aeizōon to leptophyllon* on fol. 14v, evidently because the scribe recognized that the text was the same as that on the previous folio.

54. See Vienna Dioscorides, fol. 38r. Compare to Wellmann's *De materia medica* 4.90 and 4.168.

55. *De materia medica* 3.113.

56. Compare *De materia medica* 3.113 with Vienna Dioscorides, fols. 20r–21r, and Naples Dioscorides, fol. 3r. Two of the artemisias could be *Artemisia campestris* L. and *A. arborescens* L. Mazal gives *artemisia monoklōnos* as *A. spicata* Wulfen ex Jacq. For translation, see Beck, 232; and Mazal, *Der Wiener Dioskurides*, 1:34. The third artemisia could be *A. annua* L.

57. For examples of scholarship regarding these as mistakes, see Orofino, "Dioscorides of the Biblioteca Nazionale," 103–4.

58. Another example is *geranion* and *geranion heteron* in Vienna Dioscorides, fols. 84v–87r; compare Naples Dioscorides, fol. 58r.

59. E.g., *Anemōnē hē phoinikē*, and *anemōnē hē agria melaina*. See Vienna Dioscorides, fols. 25v–26r; compare to Naples Dioscorides, fol. 12r. Another example includes *arkeuthis mikra* and *arkeuthis megalē*; see Vienna Dioscorides, fols. 33v–34r, and compare to Naples Dioscorides, fol. 11r. In one case we find an inset illustration apparently so as to save space without the intention to compare two related plants: the illustration of *thymelaia*, Vienna Dioscorides, fol. 134v. This chapter awkwardly appears between the chapters on *thermos* and *thermos agrios* (fols. 133v–34r, and 135r, respectively), which are separated as a result even though they share a text. Compare the more elegant arrangement in Naples Dioscorides, fols. 36r and 38r.

60. See *konyza platyphyllos* and *konyza leptophyllos* (Vienna Dioscorides, fol. 152v, and Naples Dioscorides, fol. 39r); *knēphē ē knidē* and *knēphē hetera* (Vienna Dioscorides, fol. 151v, and Naples Dioscorides, fol. 57r); *kalaminthē orinē* and *kalaminthē* (Vienna Dioscorides, fol. 153v, and Naples Dioscorides, fol. 48r); *linozōstis thēleia* and *linozōstis arrēn* (Vienna Dioscorides, fol. 201v, and Naples Dioscorides, fol. 108r); and *katanankē* and *katanankē hetera* (Vienna Dioscorides, fol. 173v, and Naples Dioscorides, fol. 53r).

61. Cronier, "L'herbier alphabétique grec," 46.

62. ἠρίγγιον, *ēringion*, a kind of eryngo, likely sea holly, *Eryngium maritimum* L. Vienna Dioscorides, fol. 126v; Naples Dioscorides, fol. 78r.

63. Dioscorides, *De materia medica* 3.21.

64. The main chapter heading in the Naples Dioscorides even reads "eryngo or *gorgonion*" (ἠρινγειον η γοργονιο[ν]). *Gorgoneion* also appears as a synonym for *lithospermon*; see Strömberg, *Griechische Pflanzennamen*, 101.

65. See discussion in chapter 3. The accompanying text in the Alphabetical Dioscorides does not explain why the illustration of the plant lacks roots. Other authors, however, note simply that *antirrhinon*, a synonym for *kynokephalion*, has no roots (Theophrastus, *Historia plantarum* 9.19; Pliny the Elder, *Naturalis historia* 25.129).

66. Note: the text in fig. 4.5 is a later addition. *Nymphaia* at Naples Dioscorides, fol. 104r; *pteris*, *pteris hetera* at Naples Dioscorides, fol. 101r, and Vienna Dioscorides, fol. 257r; *thēlypteris* at Naples Dioscorides, fol. 39r, and Vienna Dioscorides, fol. 142r. Beck gives *pteris* as *Polystichum filix-mas* Roth., and *thelypteris* as *Pteris aquilina* L. (Beck, 326). See Dioscorides, *De materia medica* 4.184–85. None matches the picture, which is rather a hart's tongue fern, *Phyllitis scolopendrium* (L.) Newman, also called *Asplenium scolopendrium* L. and *Scolopendrium officinale* DC. Giulia Orofino noted the multiplication of this picture. See Orofino, "The Dioscorides," 103.

67. See *LSJ*, s.v. πτερίς, noted by Dioscorides, *De materia medica* 4.185.

68. The Latin *lingua cervina* follows from the Greek transliteration *lingoua kerbina* (λιγγουα κερβινα) given in the Alphabetical Dioscorides, e.g., Vienna Dioscorides, fol. 141v; Naples Dioscorides, fol. 39r. The name *lingua cervina* appears in list of synonyms for *gladiolum* in Ps.-Apuleius Platonicus, *Herbarius* 79. It is unclear if *lingua cervina* designated hart's tongue at this time. It certainly did so by the late Middle Ages, see R. E. Latham, D. R. Howlett, and Richard Ashdowne, eds., *Dictionary of Medieval Latin from British Sources* (Turnhout: Brepols, 2015), s.v. *lingua* 5.h.

69. Cronier, "L'herbier alphabétique grec," 57–58.

70. Plutarch, *Maxime cum principibus philosopho esse disserendum* Steph. 776 F.

71. Wellmann, "Krateuas," 26–30.

72. Hanson, "Text and Context," 589–90.

73. Grape-Albers, *Spätantike Bilder*, 12.

74. Paris, Bibliothèque nationale de France, MS gr. 2179, the "Old Paris Dioscorides."

75. Edoardo Crisci, *Scrivere greco fuori d'Egitto. Ricerche sui manoscritti greco-orientali di origine non egiztiana dal iv secolo a. C. al ni d. C.*, Papyrologica fiorentina 27 (Florence: Gonnelli, 1996), 95; and Lidia Perria, "Scritture e codici di origine orientale (Palestina, Sinai) dal IX al XIII secolo," in *Tra oriente e occidente. Scritture e libri greci fra le regioni orientali di Bisanzio e l'Italia*, ed. Lidia Perria (Rome: Dipartimento di filologia greca e latina, Sezione bizantino-neoellenica, Università di Roma La Sapienza, 2003), 65–80, at 71–72. See also Guglielmo Cavallo, "Funzione e strutture della maiuscola greca tra i secoli VIII–XI," in *La paléographie grecque et byzantine. Colloques internationaux du Centre national de la recherche scientifique, no. 559, Paris 21–25 October 1974* (Paris: Centre national de la recherche scientifique, 1977), 96–102.

76. Marie Cronier, for example, notes, "The work is treated here as a classic text . . . to be preserved as meticulously as possible." See Cronier, "The Manuscript Tradition," 140.

77. Kurt Weitzmann considered this a holdover of a system of illustration from papyrus rolls. See Kurt Weitzmann, *Illustrations in Roll and Codex: A Study of the Origin and Method of Text Illustration* (Princeton: Princeton University Press, 1975), 71–72. But little evidence suggests that illustrated herbals on papyrus or rolls adopted this method. See also Riddle, *Dioscorides on Pharmacy*, 193. We cannot say if the source text for the Old Paris Dioscorides was a roll or a codex. Compare Collins, *Medieval Herbals*, 88–89.

78. Old Paris Dioscorides, fols. 98r, 28r, 33r, and 33v.

79. Two notable exceptions, however, occur. Two illustrations of plants appear in the chapter on *tēlephion* (τηλέφιον, i.e., τηλεφώνιον, *Andrachne telephioides* L., Dioscorides, *De materia medica* 2.186; Old Paris Dioscorides, fol. 5v). The chapter on mushrooms (Old Paris Dioscorides, fol. 107v) includes multiple mushrooms under one entry.

80. Griebeler, "Production Sequence and Page Design." The illustration of the plants in the Old Paris Dioscorides seems to have followed not only the copying of the text, but also the rubrication and glossing of it. See Collins, *Medieval Herbals*, 85, and 113n333.

81. The illustrations of aloe (ἀλόν i.e., ἀλόη, *Aloe vera* [L.] Burm.; Old Paris Dioscorides, fol. 16r), scammony (*skammōnia, Convolvulus scammonia* L.; Old Paris Dioscorides, fol. 134r), and stinking tutsan (*tragion, Hypericum hircinum* L.; Old Paris Dioscorides, fol. 85v) all depict liquids streaming from them. Dioscorides mentions the medicinal properties of all three plants' exudate, sap, or juice.

82. Old Paris Dioscorides, fol. 16r.

83. See Saliba and Komaroff, "Illustrated Books," 24–38; also the Dioscorides now Athos, Library of the Great Lavra, Cod. Ω 75, on which, see Geōrgios A. Christodoulou, *Σύμμικτα Κριτικά* (Athens: By the author, 1986), 131–92.

84. The violets appear on Old Paris Dioscorides, fol. 119r. See Collins, *Medieval Herbals*, 91; Riddle, *Dioscorides on Pharmacy*, 193; and Singer, "Herbal in Antiquity," 27–28.

85. By my own count, only 42 illustrations, or roughly 12 percent, bear a clear resemblance to the pictures in the Alphabetical Dioscorides. Determining the relationship between the pictures of the Old Paris Dioscorides and those of the Alphabetical Dioscorides, however, remains difficult due to stylistic differences. Most pictures have little (n = 75, or 21%) or no (n = 244, 68%) visible resemblance to those in the Alphabetical Dioscorides.

86. Marie Cronier, "Transcrire l'arabe en grec. À propos des annotations du Parisinus gr. 2179," in *Manuscripta graeca et orientalia. Mélanges monastiques et patristiques en l'honneur de Paul Géhin* (Leuven: Peeters, 2016), 247–65, at 255–59.

87. χελιδόνιον, *Chelidonium majus* L.; see Dioscorides, *De materia medica* 2.180; Old Paris Dioscorides, fol. 3v.

88. ἰσάτις, woad, *Isatis tinctoria* L.; see Dioscorides, *De materia medica* 2.184; Old Paris Dioscorides, fol. 5r.

89. λογχίτης / λογχῖτης, possibly *Serapias lingua* L.; see Dioscorides, *De materia medica* 3.144; Old Paris Dioscorides, fol. 65r. Orofino notes the relationship as well. Orofino, "Dioscorides in the Biblioteca Nazionale," 112. She also points out

that Charles Singer had earlier confused *lonchitis* with *lonchitis hetera*. Compare Singer, "Herbal in Antiquity," 28–29.

90. Trans. Beck, 244, with modification; Dioscorides, *De materia medica* 3.144 (ed. Wellmann, 2:153–54): φύλλα ἔχει πράσῳ καρτῷ ὅμοια, πλατύτερα δὲ καὶ ὑπέρυθρα, πλεῖστα πρὸς τῇ ῥίζῃ, περικλώμενα ὡς ἐπὶ τὴν γῆν· ἔχει δὲ καὶ περὶ τὸν καυλὸν ὀλίγα, ἐφ' οὗ ἄνθη ὅμοια πιλίσκοις, τῷ τύπῳ δὲ κωμικοῖς προσωπείοις κεχηνόσι, μέλανα, λευκὸν δέ τι ἐξ αὐτῶν ἐξέχει ἀπὸ τοῦ χάσματος πρὸς τῷ κάτω χείλει ὥσπερ γλωσσάριον· τὸ σπέρμα δὲ ὅμοιον λόγχῃ, τρίγωνον, ἐν περικαρπίοις, ὅθεν καὶ τῆς ἐπωνυμίας ἠξιώθη, ῥίζα ὁμοία δαύκῳ. φύεται ἐν τραχέσι καὶ ἀνίκμοις τόποις. ταύτης ἡ ῥίζα διουρητικὴ πινομένη σὺν οἴνῳ.

91. The picture of *lonchitis* in the Naples Dioscorides shows yellow dog-shaped flowers, contradicting Dioscorides' description of them as black. Naples Dioscorides, fol. 113r.

92. See Dioscorides, *De materia medica* 2.93.

93. Beck's identification of *tragos* could better signal it is a spelt product. See Krzystztof Jagusiak and Maciej Kokoszko, "Spelt (*ólyra*)," in *Cereals of Antiquity and Early Byzantine Times: Wheat and Barley in Medical Sources (Second to Seventh Centuries AD)*, ed. Maciej Kokoszko, Krzysztof Jagusiak, and Zofia Rzeźnicka; trans. Karolina Wodarczyk, Maciej Zakrzewski, and Michał Zytka (Łódź: Łódź University Press, 2014), 293–310, especially 296, 298, 306–8.

94. They appear in the Old Paris Dioscorides, fols. 2r, 3v, 4v, 5r, 5v, and 7v. Although the figures were probably executed after the plant illustrations and are painted in an entirely different way, their inclusion in the codex seems to have been intended from the beginning. Sketches in faint red pigment clearly preceded the execution of both the illustrations of the plants as well as the figures. The illustrators clearly left space to the figures. They may have been executed later or by a different miniaturist.

95. Weitzmann, "Greek Sources," 29.

96. For example, Paris, Bibliothèque nationale de France, MS syr. 341. Both the Old Paris Dioscorides and the earlier Syriac illustrations render figures with thick, oblong limbs, and similar proportions. By contrast, the outlines in the plant illustrations tend to be of uniform thickness and coincide with the lower layers of pigment.

97. Weitzmann, "Greek Sources," 29; Riddle, *Dioscorides on Pharmacy*, 198–203; Alain Touwaide, "Le traité de matière médicale de Dioscoride en Italie depuis la fin de l'Empire romain jusqu'aux débuts de l'école de Salerne. Essai de synthèse," in *From Epidaurus to Salerno*, ed. A. Krug (Rixensart: PACT Belgium, 1992), 275–305, at 300.

98. Μυὸς ὦτα, literally, "mouse-ears."

99. Dioscorides, *De materia medica* 2.183: ταύτης ἡ ῥίζα καταπλασθεῖσα αἰγιλώπια ἰᾶται. Weitzmann, "Greek Sources," 29–30; see also Riddle, "Dioscorides," 198–99; and Touwaide, "Traité," 291–92.

100. Collins, *Medieval Herbals*, 85.

101. The figure accompanying the illustration of two varieties of pimpernel (*anagallis*) points toward the one with blue flower (*Lysimachia arvensis* [L.] U.Manns. & Anderb. and *L. foemina* Mill. [U.Manns. & Anderb.]; Dioscorides, *De materia medica* 2.178; Old Paris Dioscorides, fol. 2r). This gesture echoes a warning given in the text: "Some say that the one that has the dark-blue flowers

stems prolapses of the anus . . . and that the red-flowered aggravates them" (trans. Beck, 169). The skin-clad figure beside the unidentified *othonna* plant might indicate an ethnic group, the "troglodytes" mentioned in the text (Old Paris Dioscorides, fol. 4v; Dioscorides, *De materia medica* 2.182). On the verso of the following folio (fol. 5v), we find a figure pointing to the leaves of *tēlephion* (i.e., τηλεφώνιον, likely *Andrachne telephioides* L.), which the text suggests can be used as a poultice for treating a skin condition called *leukē* (Dioscorides, *De materia medica* 2.186). The figure crawling toward the gentian plant (*gentianē*) grasps its lower leaves, which are singled out in the description of the plant in the text (Old Paris Dioscorides, fol. 7v). The gentian here is perhaps *Gentiana lutea* L. or *G. purpurea* L.; see Dioscorides, *De materia medica* 3.3.

102. The slanted ogival uncial hand dates the fragment to the ninth century. Riddle, *Dioscorides on Pharmacy*, 193. Yerevan, Matenadaran, MS arm. 141. F. C. Conybeare took photographs and copies in 1888 and donated them to the Bodleian Library in 1892. It can now be found in Oxford, Bodleian Library, MS gr. Class. E. 19. See Collins, *Medieval Herbals*, 112n322. See also Touwaide, *Census of Greek Medical Manuscripts*, no. 0803; also Falconer Madan and Herbert Henry Edmund Craster, *A Summary Catalogue of Western Manuscripts in the Bodleian Library at Oxford with References to the Oriental Manuscripts and Papyri*, vol. 6: Accessions, 1890–1915, Nos. 31001–37299 (Oxford: Clarendon Press, 1924), 62, no. 31528.

103. *Androsaimon, Hypericum perforatum* L.; Dioscorides, *De materia medica* 3.156; *koris, Hypericum empetrifolium* Willd.; Dioscorides, *De materia medica* 3.157; and *chamaipitys, Ajuga chamaepitys* (L.) Schreb.; Dioscorides, *De materia medica* 3.158.

104. Collins, *Medieval Herbals*, 112n322.

Chapter 5

1. Cassiodorus, *Institutiones* 31.2; Roger A. B. Mynors, ed., *The Institutiones of Cassiodorus*, 2nd ed. (Oxford: Clarendon Press, 1961): "Quod si vobis non fuerit Graecarum litterarum nota facundia, in primis habetis Herbarium Dioscoridis, qui herbas agrorum mirabili proprietate disseruit atque depinxit. post haec legite Hippocratem atque Galienum Latina lingua conversos, id est Tharapeutica Galieni ad philosophum Glauconem destinata, et anonymum quendam, qui ex diversis auctoribus probatur esse collectus. deinde Caeli Aureli de Medicina et Hippocratis de Herbis et Curis diversos que alios medendi arte compositos, quos vobis in bibliothecae nostrae sinibus reconditos Deo auxiliante dereliqui." For another translation, see James W. Halporn, trans., *"Institutions of Divine and Secular Learning" and "On the Soul"* (Liverpool: Liverpool University Press, 2004). For discussion of this passage with earlier bibliography, see Collins, *Medieval Herbals*, 163–65.

2. Ultimately, we do not know what version of Dioscorides Cassiodorus recommends. He could refer to a rare Latin translation of Dioscorides, such as that preserved in a manuscript produced in Southern Italy in the tenth century, now in Munich (Munich, Bayerische Staatsbibliothek, Cod. Clm. 337; on this codex, see Collins, *Medieval Herbals*, 149–54). The earliest translations of Dioscorides into Latin, however, appear to have been unillustated; see John

Riddle, "Pseudo-Dioscorides' *Ex herbis femininis* and Early Medieval Medical Botany," *Journal of the History of Biology* 14 (1981): 43–81, at 44–45. John Riddle suggests that Cassiodorus refers to a Latin text called *Ex herbis femininis*, once attributed to Dioscorides (56–57). But Riddle's argument is shaky by his own admission. It remains possible that Cassiodorus refers to a Greek Dioscorides. Dioscorides writes in a plain, straightforward style (as Dioscorides himself notes; Dioscorides, *De materia medica* Praef. 5). His language is eminently more comprehensible with limited Greek than the writings of Galen or Hippocrates. At the same time, we can expect various degrees of functional Greek literacy among the monks at the Vivarium monastery, given the large Greek-speaking population in Southern Italy. At the same time, I suspect that Giulia Orofino may go too far when she concludes that the "technical education of these monks was primarily based on iconographical transmission" through illustrated Greek manuscripts of Dioscorides. See Orofino, "Dioscorides of the Biblioteca Nazionale," 111.

3. For a more detailed overview with bibliography, see Nutton, *Ancient Medicine*, 296–309.

4. Nutton, *Ancient Medicine*, 294.

5. Nutton, *Ancient Medicine*, 299.

6. I thank Joshua Allbright for his suggestions on my translation of this text. Photios, *Epistulae et Amphilochia* ep. 223, Ζαχαρίᾳ μητροπολίτῃ Χαλκηδόνος, ed. V. Laourdas and L. G. Westerink (Leipzig: Teubner, 1984), 2:135: Εὗρον, οἶμαι, τὴν τοῦ αἵματος ἀφαίρεσιν καὶ θερίας οὔσης τῇ τοῦ σώματός σου διαθέσει συμφέρουσαν. εἰ δὲ παρὰ δόξαν ἐστὶν τῶν νῦν ἐπιπολαζόντων ἰατρῶν, παράδοξον οὐδέν· οἷς γάρ, ἵνα μὴ νῦν τὰς ἄλλας αὐτῶν περὶ τὴν τέχνην ἁμαρτάδας, οἴμοι, λέγω, ἀλλ᾽ ἵππουρις μὲν νομίζεται τὸ πολύγονον, σέσελις δὲ τὸ Ἡράκλειον πάνακες, καὶ οἷα δὴ βατράχιον τὸ μικρὸν κρίνεται χελιδόνιον, καὶ ἀντὶ μὲν πεπλίου τιθύμαλλος, ὡς ποταμογείτων δὲ τὸ λειμώνιον, καὶ ὁ μὲν μέλας χαμαιλέων ὡς λευκὸς παραλαμβάνεται, ἀντὶ δὲ τῆς ἀνεμώνης ἡ ἀργεμώνη, καὶ μυρίων ἄλλων φύσεις βοτανῶν ξένοις ὑπηρετοῦσι καὶ καιροῖς καὶ χρείαις καὶ ὀνόμασι (ταῦτα δή, ταῦτα τὰ πρόχειρα, καὶ ἃ μηδὲ τοῖς ἐν προθύροις ἰατρικῆς προσῆκον ἦν ἐν ἀμφισβητήσει καθίστασθαι), τί θαυμαστὸν κἂν ἡ φλεβοτομία, χρειώδης οὖσά σοι, τούτοις νομισθείη ξενίζουσα; αὐτὸς δὲ τὴν συμβουλὴν δεχόμενος, σὺν θεῷ σωτῆρι φάναι, καὶ τὸν ἔλεγχον ἐκείνων τῇ πείρᾳ καὶ τὸ σὸν εὑρήσεις κέρδος.

7. Photios' enthusiasm for bloodletting echoes evidence for bloodletting practices in the early medieval Latin West, as evident, for example, in the presence of a bloodletting house in the St. Gall plan (St. Gall, MS 1092). On connections between Byzantine and Latin medicine, see Gerhard Baader, "Early Medieval Latin Adaptations of Byzantine Medicine in Western Europe," *Dumbarton Oaks Papers* 38 (1984): 251–59; on bloodletting in Western Europe, see Linda Voigts and Michael McVaugh, "A Latin Technical Phlebotomy and Its Middle English Translation," *Transactions of the American Philosophical Society* 74, no. 2 (1984): 1–69. On bloodletting in Galen, see Peter Brain, *Galen on Bloodletting: A Study of the Origins, Development and Validity of His Opinions, with a Translation of Three Works* (Cambridge: Cambridge University Press, 1986).

8. Riddle, *Dioscorides on Pharmacy*, 168–217.

9. E.g., σέσελι (*De materia medica* 3.53) is equated with τὸ Ἡράκλειον πάνακες (*De materia medica* 3.48); βατράχιον τὸ μικρὸν (*De materia medica* 2.175) with χελιδόνιον (*De materia medica* 2.180–81; no βατράχιον τὸ μικρὸν is mentioned, but there is a χελιδόνιον τὸ μικρόν); πεπλίον (*De materia medica* 4.168) with τιθύμαλλος

(*De materia medica* 4.164); μέλας χαμαιλέων (*De materia medica* 3.9) with λευκὸς χαμαιλέων (*De materia medica* 3.8); ἀνεμώνη (*De materia medica* 2.176) with ἀργεμώνη (*De materia medica* 2.176). Two exceptions: ἵππουρις (*De materia medica* 4.46) is equated with πολύγονον (*De materia medica* 4.4–5), and ποταμογείτων (*De materia medica* 4.100) with τὸ λειμώνιον (*De materia medica* 4.16).

10. I thank Joshua Allbright for this insight.

11. See Maria Mavroudi, "The Naples Dioscorides," in *Byzantium and Islam: Age of Transition*, ed. Helen Evans and Brandie Ratliff (New Haven: Yale University Press, 2012), 22–26.

12. See Riddle, *Dioscorides on Pharmacy*, 168–217.

13. Photios, *Bibliotheca* Cod. 178, ll. 23–25 (ed. René Henry, *Photius. Bibliothèque. Tome II: Codices 84–185* [Paris: Les belles lettres, 1960], 182): χρήσιμον δὲ τὸ βιβλίον οὐ πρὸς ἰατρικὴν φιλοπονίαν μόνον, ἀλλὰ καὶ πρὸς ἐμφιλόσοφον καὶ φυσικὴν θεωρίαν.

14. Dimitri Gutas, *Greek Thought, Arabic Culture: The Graeco-Arabic Translation Movement in Baghdad and Early Abbasid Society, 2nd–4th/8th–10th Centuries* (New York: Routledge, 1998), especially 175–86. On the translation of Dioscorides into Arabic, see Saliba and Komaroff, "Illustrated Books," 6–65; and Manfred Ullmann, *Untersuchungen zur arabischen Überlieferung der Materia medica des Dioskurides* (Wiesbaden: Harrassowitz, 2009).

15. Until recently, most of what was known about the translation of Dioscorides' *De materia medica* into Arabic was based on historical accounts, such as the *Ṭabaqāt al-aṭibbāʾ wa-l-hukamāʾ* by Ibn Juljul, the *Fihrist* by Ibn al-Nadīm, or the *Uyūn ul-anbāʾ fī ṭabaqāt al-aṭibbāʾ* by Ibn Abī Uṣaybiʿa.

16. E.g., John W. Watt, "The Syriac Translation of Ḥunayn ibn Isḥāq and Their Precursors," in *Geschichte, Theologie und Kultur des syrischen Christentums, Beiträge zum 7. Deutschen Syrologie-Symposium in Göttingen, Dezember 2011*, ed. Martin Tamcke and Sven Grebenstein (Wiesbaden: Harrassowitz, 2014), 423–45, at 423–26.

17. Grigory Kessel, "Syriac Medicine," in *The Syriac World*, ed. Daniel King (Routledge: London, 2019), 438–59, at 449–51. On the pharmacological terminology in Sergius' translation of Galen, see Siam Bhayro, "Syrian Medical Terminology: Sergius and Galen's Pharmacopia," *Aramaic Studies* 3, no. 2 (2005): 147–65. On the role of Syriac in the transmission of scientific and technical knowledge, see Hidemi Takahashi, "Syriac as a Vehicle for Transmission of Knowledge across Borders of Empires," *Horizons* 5, no. 1 (2014): 29–52. See also Peter E. Pormann, "The Development of Translation Techniques from Greek into Syriac and Arabic: The Case of Galen's *On the Faculties and Powers of Simple Drugs*, Book Six," in *Medieval Arabic Thought: Essays in Honour of Fritz Zimmermann*, ed. R. Hansberger et al., Warburg Institute Studies and Texts 4 (London: Warburg Institute, 2012), 143–62.

18. On the competitive atmosphere of the Abbasid court at this time, see George Saliba, "Competition and the Transmission of the Foreign Sciences: Ḥunayn at the Abbasid Court," *Bulletin of the Royal Institute for Inter-Faith Studies* 2 (2000): 85–101.

19. On this, see Saliba and Komaroff, "Illustrated Books," 8–9, and Alain Touwaide, "Translation and Transliteration of Plant Names in Ḥunayn b. Isḥaq's and Isṭifan b. Bāsil's Arabic Version of Dioscorides, *De materia medica*," *Al-Qanṭara*

30, no. 2 (2009): 557–80. I am unable to confirm that Sergius translated Dioscorides (at 559). On transliterating Greek words in Arabic script, see Nikolaj Serikoff, "Mistakes and Defences: Foreign (Greek) Words in Arabic and Their Visual Recognition," in *Fremde, Feinde und Kurioses: Innen- und aussenansichten unseres muslimischen Nachbarn*, ed. Benjamin Jokisch, Ulrich Rebstock, and Lawrence I. Conrad (Berlin: De Gruyter, 2009), 1–10.

20. This early Arabic botanical tradition was rich with synonyms and descriptions with extensive quotations of literary sources, particularly verse. On Abū Ḥanīfa, see Thomas Bauer, *Das Pflanzenbuch des Abū Ḥanīfa ad-Dīnāwarī: Inhalt, Aufbau, Quellen* (Wiesbaden: Harrassowitz, 1988). See also Remke Kruk, "Nabāt," in *Encyclopaedia of Islam*, ed. Bearman et al.

21. Saliba and Komaroff, "Illustrated Books," 9.

22. In the thirteenth century, the Artūqid emir Fakhr al-Dīn Qara Ar-slan (r. 1144–67) commissioned a translation of Ḥunayn's Syriac rendition of Dioscorides' *De materia medica* from Abū Sālim al-Malṭī. This translation, however, was rejected by Fakhr al-Dīn's cousin Najm al-Dīn Alpi. Manuscript evidence indicates more complex transmission. See Saliba and Komaroff, "Illustrated Books," 6–65. Manfred Ullmann has identified that several different Greek versions of Dioscorides were translated into Arabic earlier. See Ullmann, *Untersuchungen zur arabischen Überlieferung*.

23. Ullmann, *Untersuchungen zur arabischen Überlieferung*.

24. Saliba and Komaroff, "Illustrated Books."

25. B. Lewin, "Adwiya," in *Encyclopaedia of Islam*, ed. Bearman et al.

26. J. Vernet, "Ibn al-Bayṭār," in *Encyclopaedia of Islam*, ed. Bearman et al.

27. For example, see Albert Dietrich, *Dioscurides Triumphans: Ein anonymer arabischer Kommentar (Ende 12. Jahrh. n. Chr.) zur Materia medica*, 2 vols. (Göttingen: Vandenhoeck and Ruprecht, 1988).

28. On Byzantine Greek translations of Arabic, including broader ideological factors, see Dimitri Gutas, "Arabic into Byzantine Greek: Introducing a Survey of Translation," in *Knotenpunkt Byzanz: Wissensformen und kulturelle Wechselbe-zeihungen*, ed. Philipp Steinkruger and Andreas Speer (Berlin: De Gruyter, 2012), 246–62. For a thorough case study, see Maria Mavroudi, *A Byzantine Book on Dream Interpretation: The Oneirocriticon of Achmet and Its Arabic Source* (Leiden: Brill, 2002). On Arabic into Greek translation of medical texts, specifically, see Marie-Hélène Congourdeau, "Medical Art, Erudition, and Practice at the Byzantine Capital," in *Life Is Short, Art Long: The Art of Healing in Byzantium* (Istanbul: Pera Müsezi, 2015), 91–103, at 101–3; Congourdeau, "La médecine byzantine à la croisée de l'Orient et de l'Occident," in *Knotenpunkt Byzanz*, ed. Steinkruger and Speer, 223–31; Alain Touwaide, "Arabic Medicine in Greek Trans-lation: A Preliminary Report," *Journal of the International Society for the History of Islamic Medicine* 1 (2002): 45–53; Touwaide, "Arabic into Greek: The Rise of an International Lexicon of Medicine in the Medieval Eastern Mediterranean," in *Vehicles of Transmission, Translation, and Transformation in Medieval Textual Culture* (Turnhout: Brepols, 2011), 195–222; Petros Bouras-Vallianatos, "Contex-tualizing the Art of Healing by Byzantine Physicians," in *Life Is Short, Art Long*, 108–9; Antoine Pietrobelli, "Les traductions byzantines des textes médicaux arabes," *Horizons maghrébins* 63 (2010): 57–63. On the Greek translation of the *Ephodia*, see Herbert Hunger, *Die hochsprachliche profane Literatur der Byzantiner*

(Munich: Beck, 1978), 2:306. On the translation of Al-Rāzī, see Marie-Hélène Congourdeau, "Le traducteur grec du traité de Rhazès sur la variole," in *Storia ecdotica dei Testi Medici Greci. Atti del II Convegno Internazionale Parigi 24–26 maggio 1994*, ed. Antonio Garzya (Naples: M. D'Auria, 1996), 99–111. For an English translation of the *Zād al-musāfir*, see Michael R. McVaugh, Gerrit Bos, and Fabian Käs, eds., *Ibn Jazzār's Zād al-musāfir wa-qūt al-ḥāḍir. Provisions for the Traveller and Nourishment for the Sedentary* (Leiden: Brill, 2022).

29. H. J. L. Drossaart Lulofs and E. L. J. Poortman, *De plantis: Five Translations* (New York: North Holland, 1989).

30. E.g., see examples given in Gutas, "Arabic into Byzantine Greek," 252–54.

31. E.g., the so-called "Saracenic" lexicon in Margaret H. Thomson, *Textes grecs inédits relatifs aux plantes* (Paris: Les belles lettres, 1955). For a more skeptical view of Byzantine lexica, see Nikolaj Serikoff, "'Syriac' Plant Names in a Fifteenth Century Greek Glossary (from the Wellcome Library Books and Manuscripts," in *Medical Books in the Byzantine World*, ed. Barbara Zipser (Bologna: Eikasmós Online II, 2013), 97–123, with bibliography. See also Alain Touwaide, "Lexica medico-botanica byzantina: prolégomènes à une étude," in *Tès filiès ta'de dôra: Miscelánea léxica en memoria de Conchita Serrano*, Manuales y Anejos de Emerita 41 (Madrid: Consejo Superior de Investigaciones Científicas, 1999), 211–28.

32. On the medical texts associated with Isidore of Seville, see William D. Sharpe, "Isidore of Seville: The Medical Writings, An English Translation with an Introduction and Commentary," *Transactions of American Philosophical Society*, n.s. 54, part 2 (1964): 1–75.

33. This translation appears in three manuscripts. The earliest is a fragmentary eighth- or ninth-century text preserved in the codex Paris, Bibliothèque nationale de France, MS lat. 9332. The second copy occurs in a ninth-century codex in Paris, Bibliothèque nationale de France, MS lat. 12995. The third copy of this translation is the manuscript in Munich, Bayerische Staatsbibliothek, Cod. Clm. 337. It is a small-format codex written in a late tenth-century Beneventan hand. See Collins, *Medieval Herbals*, 149–54.

34. E.g., Collins, *Medieval Herbals*, 154–65.

35. See Patricia Skinner, *Health and Medicine in Early Medieval Southern Italy* (Leiden: Brill, 1997), 127–31.

36. Skinner, *Health and Medicine*, 138–39.

37. On Gariopontus, see Florence Eliza Glaze, "Gariopontus and the Salernitans: Textual Traditions in the Eleventh and Twelfth Centuries," in *La "Collectio Salernitana" di Salvatore De Renzi. Convegno internazionale, Università degli studi di Salerno, 18–19 giugno 2008*, ed. Danielle Jacquart and Agostino Paravicini Bagliani (Florence: SISMEL, 2009), 149–90.

38. On Constantine the African's translation of ʿAlī ibn al-ʿAbbās al-Majūsī's *Kitāb kāmil al-ṣināʿa al-ṭibbiyya*, see the contributions in *Constantine the African and ʿAlī Ibn al-ʿAbbās al-Maǧūsī: The Pantegni and Related Texts*, ed. Danielle Jacquart and Charles Burnett (Leiden: Brill, 1994). The traditional historical narrative on Constantine the African is presented in Peter the Deacon's *On Famous Men (De viris illustribus 23)*. See Herbert Bloch, *Monte Cassino in the Middle Ages*, 3 vols. (Cambridge, MA: Harvard University Press, 1986), 1:127–29. For a translation, see Monica H. Green, "Medicine in Southern Italy, Twelfth–Fourteenth Centuries: Six Texts," in *Medieval Italy: Texts in Translation*, ed. Katherine L.

Jansen, Joanna Drell, and Frances Andrews (Philadelphia: University of Pennsylvania Press, 2009), 311–25, at 312–13.

39. Paul Oskar Kristeller, "The School of Salerno: Its Development and Its Contribution to the History of Learning," *Bulletin of the History of Medicine* 17 (1945): 138–94, at 154. For an example of glossing practices in Salernitan texts, see Florence Eliza Glaze, "Speaking in Tongues: Medical Wisdom and Glossing Practices in and around Salerno, c. 1040–1200," in *Herbs and Healers from the Ancient Mediterranean through the Medieval West: Essays in Honor of John M. Riddle*, ed. Anne Van Arsdall and Timothy Graham (Farnham, UK: Ashgate, 2012), 63–106.

40. E.g., Iolanda Ventura, "Per una storia del *Circa Instans*. I *Secreta Salernitana* ed il testo del manoscritto London, British Library, Egerton 747. Note a margine di un'edizione," *Schola Salernitana: Annali 7–8* (2002–3): 39–109; Ventura, "Un manuale di farmacologia medievale ed i suoi lettori. Il *Circa instans*, la sua diffusione, la sua ricezione dal XIII al XV secolo," in *La scuola medica salernitana. Gli autori e i testi*, ed. Danielle Jacquart and Agostino Paravicini Bagliani (Florence: SISMEL, 2007), 465–533; Ventura, "Une oeuvre et ses lecteurs. La diffusion du *Circa Instans* salernitain," in *Florilegium mediaevale. Études offertes à Jacqueline Hamesse à l'occasion de son éméritat* (Louvain: Institut d'études médiévales, 2009), 585–607; Ventura, "Il *Circa Instans* attribuito a Platearius. Trasmissione manoscritta, redazioni, criteri di costruzione di un'edizione critica," *Revue d'histoire des textes* 10 (2015): 249–362.

41. Bruce P. Flood, "The Medieval Herbal Tradition of Macer Floridus," *Pharmacy in History* 18, no. 2 (1976): 62–66. Relatively little is known about Serapion the Younger (Sirāfiyūn or Sīrabīyūn al-Ṣaghīr). See Lucien Leclerc, *Histoire de la médecine arabe* (Paris: Ernest Leroux, 1876), 2:152–56.

42. See Felix Baumann, *Das Erbario Carrarese und die Bildtradition des Tractatus de herbis. Ein Beitrag zur Geschichte der Pflanzendarstellung im Übergang von Spätmittelalter zu Frührenaissance* (Bern: Benteli, 1974). More recently, on the intellectual and political context for this manuscript, see Sarah R. Kyle, *Medicine and Humanism in Late Medieval Italy: The Carrara Herbal in Padua* (Abingdon, UK: Routledge, 2017).

43. Kristeller, "School," 138–94; Paul Oskar Kristeller, "Bartholomaeus, Musandinus and Maurus of Salerno and Other Early Commentators on the Articella, with a Tentative List of Texts and Manuscripts," *Italia medioevale e humanistica* 19 (1976): 57–87; Mark D. Jordan, "The Construction of a Philosophical Medicine: Exegesis and Argument in Salernitan Teaching on the Soul," in "Renaissance Medical Learning: Evolution of a Tradition," ed. Michael R. McVaugh and Nancy G. Siraisi, *Osiris*, 2nd ser., 6 (1990): 42–46; and Jordan, "Medicine as Science in the Early Commentaries on 'Johannitius,'" *Traditio* 43 (1987): 121–45.

44. E.g., Peter Murray Jones, *Medieval Medicine in Illuminated Manuscripts* (London: British Library, 1998), 58.

45. John of Damascus, *Orationes de imaginibus tres* 1.11. Here, trans. David Anderson, *On the Divine Images: Three Apologies against Those Who Attack the Divine Images* (Crestwood, NY: St. Vladimir's Seminary Press, 1980), 20.

46. See Paul J. Alexander, *The Patriarch Nicephorus of Constantinople; Ecclesiastical Policy and Image Worship in the Byzantine Empire* (Oxford: Clarendon Press, 1958), 196.

47. Paul Magdalino, *L'orthodoxie des astrologues. La science entre le dogme et*

la divination à Byzance (VIIe–XIVe siècle) (Paris: Lethielleux, 2006), especially 12, 55–89; see also Magdalino, "Science and Imperial Power in Byzantine History and Historiography (9th–12th Centuries)," in *The Occult Sciences in Byzantium*, ed. Paul Magdalino and Maria Mavroudi (Geneva: La Pomme d'Or, 2006), 119–62, especially 133, 135.

48. Maguire, *Nectar and Illusion*.

49. New York, Morgan Library, MS M 652, the "Morgan Dioscorides." See chapters 6 and 7, below. See also Andrew Griebeler, "Botanical Illustration and Byzantine Visual Inquiry in the Morgan Dioscorides," *Art Bulletin 105, no. 1* (2023): 93–116.

50. Photios, *Bibliotheca* Cod. 178, ll. 23–25.

51. Beatrice Gruendler, *The Rise of the Arabic Book* (Cambridge, MA: Harvard University Press, 2020), 12–15, 23–26.

52. Eva R. Hoffman, "The Beginnings of the Illustrated Arabic Book: An Intersection between Art and Scholarship," *Muqarnas* 17 (2000): 37–52.

53. Hoffman, "The Beginnings."

54. See Bayard Dodge, trans., *The Fihrist of Al-Nadim: A Tenth-Century Survey of Muslim Culture* (New York: Columbia University Press, 1970), 690. Also Mahmoud Sadek, trans., *The Arabic Materia Medica of Dioscorides* (Queebec: Éditions du Sphinx, 1983), 2. See Muḥammad ibn Isḥāq ibn al-Nadīm, *Kitāb al-fihrist*, ed. Gustav Flügel, 2 vols. (Leipzig: Vogel, 1871–72; reprinted, Beirut: Khayyāt, 1964), 293.

55. Ibn Abī Uṣaybiʿa, *ʿUyūn al-anbāʾ fī ṭabaqāt al-aṭibbāʾ* 4.1.11.1.

56. Cronier, "Transcrire l'arabe en grec," 247–65, especially 250–51. These annotations do not correspond to the plant names given in the Arabic translations of Dioscorides. For related phenomena, see Barbara Zipser, "Griechische Schrift, arabische Sprache und graeco-arabische Medizin: eine neues Fragment aus dem mittelalterlichen Sizilien," *Mediterranean Language Review* 15 (2003–4): 154–66; also Maria Mavroudi, "Arabic Words in Greek Letters: The Violet Fragment and More," in *Moyen arabe et variétés mixtes de l'arabe à travers l'histoire*, ed. Jérôme Lentin and Jacques Grand'Henry (Louvain: Peeters, 2008), 321–54.

57. Dieter Harlfinger, "Weitere Beispiele frühester Minuskel," in *I manoscritti greci tra riflessione e dibattito. Atti del V Colloquio Internazionale di Paleografia greca (Cremona, 4–10 ottobre 1998)*, ed. Giancarlo Pratto (Florence: Gonnelli, 2000), 153–56.

58. Cronier, "Transcrire," 255.

59. Cronier, "Transcrire," 257–59.

60. Cronier, "Transcrire," 255–56.

61. E.g., the Vienna Dioscorides, the Old Paris Dioscorides, the Morgan Dioscorides, and Padua, Biblioteca del Seminario, Cod. 194 (the "Padua Dioscorides").

62. See Saliba and Komaroff, "Illustrated Books," 17. Leiden, Universiteits-bibliotheek, MS Or. 289. See also Mahmoud M. Sadek, "Notes on the Introduction and Colophon of the Leiden Manuscript of Dioscorides' 'De Materia Medica,'" *International Journal of Middle East Studies* 10, no. 3 (1979): 345–54. On the identification of al-Nātilī, including the argument against it, see Sadek, "Notes on the Introduction," at 347.

63. Cronier, "The Manuscript Tradition," 149.

64. *Kitāb al-filāḥa al-nabaṭiyya*, p. 1127, trans. Jaako Hämeen-Antilla, *The Last Pagans of Iraq: Ibn Waḥshiyya and His Nabatean Agriculture* (Leiden: Brill, 2006), text 30, 255–56, at 256.

65. *Kitāb al-filāḥa al-nabaṭiyya*, p. 1127, in Hämeen-Antilla, *The Last Pagans of Iraq*, 256.

66. D. Fairchild Ruggles, *Gardens, Landscape, and Vision in the Palaces of Islamic Spain* (University Park: Pennsylvania State University Press, 2000), 42–45.

67. *Theophanes Continuatus* 328.23–329.2, ed. Bekker: "Παράδεισον . . . παντοῖς κομῶντα φυτοῖς,"; trans. Cyril Mango, *Art of the Byzantine Empire, 312–1453* (Englewood Cliffs, NJ: Prentence-Hall, 1972), 195.

68. On the *Nabataean Agriculture*, see Hämeen-Antilla, *The Last Pagans of Iraq*. See also A. Alves Carrara, "*Geoponica* and *Nabatean Agriculture*: A New Approach Into Their Sources and Authorship," *Arabic Sciences and Philosophy* 16 (2006): 103–32. On agricultural works from Al-Andalus, see J. M. Carabaza Bravo, "La *Filāḥa yūnāniyya* et les traités agricoles arabo-andalous," *Arabic Sciences and Philosophy* 12 (2002): 155–78. On the *Geoponica*, see Andrew Dalby, trans., *Geoponica. Farm Work. A Modern Translation of the Roman and Byzantine Farming Handbook* (Totnes, Devon, UK: Prospect Books, 2011). See also Carlo Scardino, "Editing the *Geoponica*: The Arabic Evidence and Its Importance," *Greek, Roman, and Byzantine Studies* 58 (2018): 102–25; Robert Rodgers, "Κηποποΐα: Garden Making and Garden Culture in the *Geoponika*," in *Byzantine Garden Culture*, ed. Littlewood, Maguire, and Wolschke-Bulmahn, 159–76.

69. For an edition and translation of this passage, see Ibn Abī Uṣaybiʿa, ʿ*Uyūn al-anbāʾ fī ṭabaqāt al-aṭibbāʾ*, 13.36.2.1–3, trans. Ignacio Sánchez, from *A Literary History of Medicine*, ed. E. Savage-Smith, S. Swain, and G. J. van Gelder (Leiden: Brill, 2020). Perhaps Romanos I Lekapenos (r. 920–44) or Romanos II (r. 959–63). See Franz Rosenthal, *The Classical Heritage of Islam* (Berkeley: University of California Press, 1975), 194–97; and Mavroudi, *Byzantine Book*, 416.

70. Rosenthal, *Classical Heritage*, 194–97; and Mavroudi, *Byzantine Book*, 415–17.

71. Perhaps the Isṭifan-Ḥunayn translation.

72. Yvette Hunt, "Bang for His Buck: Dioscorides as a Gift of the Tenth-Century Byzantine Court," in *Byzantine Culture in Translation*, ed. Amelia Robertson Brown and Bronwen Neil (Leiden: Brill, 2017), 73–94, at 80.

73. Mavroudi, *Byzantine Book*, 416.

74. Ibn Abī Uṣaybiʿa, ʿ*Uyūn al-anbāʾ fī ṭabaqāt al-aṭibbāʾ* 13.36.2.3, trans. Ignacio Sánchez, from *A Literary History of Medicine*, ed. Savage-Smith, Swain, and Van Gelder.

75. The possibility that this passage refers to pictures of plants is raised in a note in the translation, with a further citation to Reinhart Dozy, *Supplément aux dictionnaires arabes*, vol. 1 (Leiden: Brill, 1881), s.v. *sh-k-l*. Ibn Abī Uṣaybiʿa, ʿ*Uyūn al-anbāʾ fī ṭabaqāt al-aṭibbāʾ* 13.36.2.3, trans. Ignacio Sánchez, from *A Literary History of Medicine*, ed. Savage-Smith, Swain, and Van Gelder.

76. Weitzmann, "Greek Sources."

77. Paris, Bibliothèque nationale de France, MS ar. 4947, the "Parchment Arabic Dioscorides." On the fact that MS ar. 4947 is technically a redaction, see Saliba and Komaroff, "Illustrated Books," 15. On the relationship of the pictures between the Old Paris Dioscorides and the Parchment Arabic Dioscorides, see

Edmond Bonnet, "Étude sur les figures de plantes et d'animaux peintes dans une version arabe manuscrite de la *Matière médicale* de Dioscoride conservée à la BN de Paris," *Janus* 14 (1909): 294–303. See also Collins, *Medieval Herbals*, 88–91.

78. Bonnet, "Étude."

79. Parchment Arabic Dioscorides, fol. 23v.

80. Ibn al-Bayṭar, *Tafsīr kitāb Diyusqūrīdūs* 2.78, spelled *ṭraghīs*, in Albert Dietrich, *Die Dioskurides-Erklärung des Ibn al-Baiṭār: Ein Beitrag zur arabischen Pflanzensynonymik des Mittelalters* (Göttingen: Vandenhoeck and Ruprecht, 1991), 113. See also Dietrich, *Dioscorides Triumphans*, 2:245, with a note and additional bibliography on the difficulties of identifying the plant.

81. Parchment Arabic Dioscorides, fol. 28v, lupine (*turmus*), *Lupinus spp*. L. This plant is mentioned in the second book of Dioscorides' *De materia medica* and therefore belongs to part of the text that is no longer part of the codex of the Old Paris Dioscorides. There are two lupines in the Alphabetical Dioscorides, *thermos agrios* and *thermos hēmeros*; see Vienna Dioscorides, fols. 135r and 134r, respectively.

82. See Jaclynne J. Kerner, "The Illustrated *Herbal* of al-Ghāfiqī: An Art Historical Introduction," in *The Herbal of al-Ghāfiqī: A Facsimile Edition with Critical Essays*, ed. F. Jamil Ragep and Faith Wallis with Pamela Miller and Adam Gacek (Montreal: McGill-Queen's University Press, 2014), 121–56, at 128.

83. Cronier, "Transcrire," 259–60.

84. Collins, *Medieval Herbals*, 148–220.

85. Copenhagen, Kongelige Bibliotek, MS Thott 190 4° (the "Thott Codex"). On this manuscript, see Alain Touwaide, "Latin Crusaders, Byzantine Herbals," in *Visualizing Medieval Medicine and Natural History, 1200–1550*, ed. Jean A. Givens, Karen M. Reeds, and Alain Touwaide (New York: Routledge, 2016), 25–50.

86. Munich, Bayerische Staatsbibliothek, Cod. Clm. 337 (the "Munich Dioscorides"), fol. 54r.

87. Fabio Troncarelli, "Una pietà più profonda. Scienza e medicina nella cultura monastica medievale italiana," in *Dall'eremo al cenobio. La civiltà monastica in Italia dalle origini all'età di Dante* (Milan: Libri Scheiwiller, 1987), 703–27.

88. I suspect the image may show the purple flowers of *Stachys officinalis* L., though Marianna Shreve Simpson identifies the plant as *Betonica alopcurus* (*sic*) (i.e., *Betonica alopecuros* L.); see Marianna Shreve Simpson, *Arab and Persian Painting in the Fogg Art Museum* (Cambridge, MA: Fogg Art Museum, 1980), 16.

89. The original manuscript to which it belonged is now Istanbul, Süleymaniye Library, MS Aya Sofya 3703.

90. Saliba and Komaroff, "Illustrated Books," 31–32.

91. Saliba and Komaroff, "Illustrated Books," 35.

92. See Timothy S. Miller, *The Birth of the Hospital in the Byzantine Empire* (Baltimore: Johns Hopkins University Press, 1985), critically reviewed by Vivian Nutton in *Medical History* 30 (1986): 218–21; and further examined by Peregrine Horden, "The Byzantine Welfare State: Image and Reality," *Bulletin of the Society of the Social History of Medicine* 37 (1985): 7–10; Horden, "The Earliest Hospitals in Byzantium, Western Europe, and Islam," *Journal of Interdisciplinary History* 35 (2005): 361–89; and Horden, "How Medicalised Were the Byzantine Hospitals?" *Medicina e storia* 10 (2006): 45–74, reprinted in Horden, *Hospitals and Healing from Antiquity to the Later Middle Ages* (Burlington, VT: Ashgate Variorum, 2008).

93. Ahmed Ragab, *Medieval Islamic Hospital: Medicine, Religion, and Charity* (Cambridge: Cambridge University Press, 2015), 21–42. See discussion in Michael Dols, "The Origins of the Islamic Hospital: Myth and Reality," *Bulletin of the History of Medicine* 61, no. 3 (1987): 367–90, especially 371–78.

94. Ragab, *Medieval Islamic Hospital*, 21–26. The Umayyad caliph Al-Walid I (ruled 705–15) is said to have built the first Islamic *bīmāristān*. But most scholars doubt that Al-Walid I's foundation was as medicalized as later *bīmāristānāt*. Michael Dols supposes that Al-Walid merely followed preexisting Byzantine precedents in segregating the leper population and providing for them. Michael Dols, "Origins of the Islamic Hospital," 378. See also Ahmed Issa, *Histoire des bimaristans (hopitaux) à l'époque islamique* (Cairo: Paul Barbey, 1928), 188–89.

95. Ragab, *Medieval Islamic Hospital*, 26–33. Traditional accounts trace the birth of the *bīmāristān* back to the Sassanian medical center of Jundi-Shapur in Khuzistan. There, Greek Hippocratic medicine fruitfully mingled with Indian and Persian medicine. See Cyril Elgood, *A Medical History of Persia and the Eastern Caliphate: From the Earliest Times until the Year A.D. 1932* (Cambridge: Cambridge University Press, 1951), especially 173. Researchers now doubt this narrative. Ragab has called it an "ideal image" (Ragab, *Medieval Islamic Hospital*, 27). According to Michael Dols, the Sassanian king surrounded himself with Eastern Christian doctors, who were already actively involved in the translation of Greek medical texts into Syriac at medical schools such as that at Nisibis. Dols hypothesized that the legend of Jundi-Shapur arose as reaction to Syrian Christians' early medical monopoly in Baghdad. See Dols, "Origins of the Islamic Hospital," 369–81.

96. See Elgood, *Medical History*, 172.

97. See Dunlop, "Bimaristan"; and Issa, *Histoire*, 89–90.

98. Issa, *Histoire*, 111.

99. Ragab, *Medieval Islamic Hospital*, 49–59; and Issa, *Histoire*, 191. On the Nūrī Bīmāristān and its relation to Nūr al-Dīn's other endowments, see Nikita Elisséeff, "Les monuments de Nūr ad-Dīn: Inventaire, notes archéologiques et bibliographiques," *Bulletin d'études orientales* 13 (1949–51): 5–43, especially 13 (the *bīmāristān* in Aleppo), and 19–20 (the *bīmāristān* in Damascus); see also Issa, *Histoire*, 190–221; and Dieter Jetter, "Zur Architektur islamischer Krankenhäuser," *Sudhoffs Archiv für Geschichte der Medizin und der Naturwissenschaften* 45, no. 3 (1961): 261–73, at 264–67.

100. E.g., Ibn Abī Uṣaybiʻa, *ʻUyūn al-anbāʼ fī ṭabaqāt al-aṭibbāʼ* 15.50.5–6; Issa, *Histoire*, 191; Jetter, "Zur Architektur," 267.

101. Ibn Abī Uṣaybiʻa, *ʻUyūn al-anbāʼ fī ṭabaqāt al-aṭibbāʼ* 15.50.5, trans. N. Peter Joosse, from *A Literary History of Medicine*, ed. Savage-Smith, Swain, and van Gelder.

102. See, for example, K. A. C. Creswell, "The Origin of the Cruciform Plan of Cairene Madrasas," *Extrait du bulletin de l'institut français d'archéologie orientale* 21 (Cairo: L'institut français d'archéologie orientale, 1922). On the antique lintel, see Terry Allen, *A Classical Revival in Islamic Architecture* (Wiesbaden: Reichert, 1986), 57–71. On the inscriptions, see Julian Raby, "Nur al-Din, the Qastal al-Shuʼaybiyya, and the 'Classical Revival,'" *Muqarnas* 21 (2004): 289–310, at 300. Ragab, *Medieval Islamic Hospital*, 56. See also Yasser Tabbaa, "The Architectural Patronage of Nur Al-Din, 1146–1174," PhD diss., New York University, 1982, 228;

Ernst Herzfeld, "Damascus: Studies in Architecture I," *Ars Islamica* 9 (1942): 1–53, at 5.

103. Terry Allen sees this as a plausible reading of the lintel (though he does not mention the muqarnas in relation to this reading), see Allen, *Classical Revival*, 62–63. While Julian Raby is probably right to suppose that the classical lintel helped to link the building to the region, there is no reason to suppose that it did not also denote the "pagan past," especially the "pagan past" of the medical tradition. See Raby, "Nur al-Din," 300. The spoliated lintel (and the reused altar) might have had talismanic functions; see Finnbar Barry Flood, "Image against Nature: Spolia as Apotropaia in Byzantium and the Dar al-Islam," in "Mapping the Gaze: Vision and Visuality in Classical Arab Civilisation," special issue, *Medieval History Journal* 9, no. 1 (2006): 143–66.

104. Istanbul, Topkapi Library, MS Sultanahmet III 2127 (the "Topkapi Dioscorides"), fols. 1v, 2r–v. Here: fol. 2v.

105. See Richard Ettinghausen, *Arab Painting* (Geneva: Skira, 1962), 67.

106. Ettinghausen, *Arab Painting*, 67. Mahmoud Sadek translates the colophon: "The five maqalat of the book of Dioscorides have been transcribed by the weak slave desiring the mercy of the kind God, Abu Yusuf Bihnam ibn Musa al-Mawsili, educated in the profession of medicine. This work was completed on the evening of Thursday the 27th of the Safar in the year 626 A.H. that is to say, the 25th day of the Kanun Thani, in the year 1540 after Alexander and thanks to the almighty God." This date based on the death of Alexander the Great along with a Syriac blessing at the end of the colophon. See Sadek, *Arabic Materia Medica*, 47. See also Ettinghausen, *Arab Painting*, 67–74.

107. Linda Komaroff credits this point to George Saliba. See Linda Komaroff, "*De materia medica* by Dioskorides," in *The Glory of Byzantium: Art and Culture of the Middle Byzantine Era, A.D. 843–1261* (New York: Metropolitan Museum, 1997), 429–33, at 433.

108. Sergio Toresella, "Il Dioscoride di Istanbul e le prime figurazioni naturalistiche botaniche," *Atti e memorie della Accademia italiana di storia della farmacia* 13, no. 1 (1996): 21–40, at 29; and Ettinghausen, *Arab Painting*, 67–74. Richard Ettinghausen, "Interaction and Integration in Islamic Art," in *Unity and Variety in Muslim Civilization*, ed. Gustave Edmund von Grünebaum (Chicago: University of Chicago Press, 1955), 107–31, at 119–20; and Eva R. Hoffmann, "The Author Portrait in Thirteenth-Century Arabic Manuscripts: A New Islamic Context for a Late Antique Tradition," *Muqarnas* 10 (1993): 6–17, especially 8, 12.

109. For more on contrasting elements, see also Hoffmann, "Author Portrait," 12.

110. See Ibn Waḥshiyya, *Kitāb al-filāḥa al-nabaṭiyya* 155–57, trans. Hämeen-Antilla, *Last Pagans*, 222–24.

111. On the relationships between Islamic and Byzantine hospitals, see Peregrine Horden, "Medieval Hospital Formularies: Byzantium and Islam Compared," in *Medical Books in the Byzantine World*, ed. Barbara Zipser (Bologna: Eikasmos Online, 2013), 145–64.

112. See Tyler Wolford, "'To Each according to Their Need': The Various Medical and Charitable Institutions of the Pantokrator Monastery," in *Piroska and the Pantokrator: Dynastic Memory, Healing and Salvation in Komnenian*

Constantinople, ed. Marianne Sághy and Robert Ousterhout (New York: Central European University Press, 2019), 195–224.

113. See the translation of the *typikon* (no. 28, at sec. 55) by Robert Jordan in *Byzantine Monastic Foundation Documents: A Complete Translation of the Surviving Founders' Typika and Testaments*, ed. John Thomas and Angela Constantinides Hero (Washington, DC: Dumbarton Oaks Library and Collection, 2000), 2:765. For an edition of the Greek text, see Paul Gautier, "Le typikon du Christ Sauveur Pantocrator," *Revue des études byzantines* 32 (1974): 1–145, at 106–7.

114. Alexandre Philipsborn, "Der Fortschritt in der Entwicklung des byzantinischen Krankenhauswesens," *Byzantinischen Zeitschrift* 54, no. 2 (1961): 338–65, at 355.

115. See Miller, *Birth of the Hospital*, 158–59, 195–96; also M. Živojinović, "Bolnica Kralja Milutina u Carigradu," *Zbornik radova Visantoloshkog Instituta* 16 (1975): 105–17.

116. Scholars have attributed the current fifteenth-century bindings of the Morgan Dioscorides to the Petra monastery. See Nadezhda Kavrus-Hoffmann, "Catalogue of Greek Medieval and Renaissance Manuscripts in the Collections of the United States of America, Part IV.2: The Morgan Library and Museum," *Manuscripta* 52, no. 2 (2008): 207–324, especially 212–30, at 225–26; Annaclara Cataldi Palau, "Legature costantinopolitane del monastero di Prodromo Petra tra i manoscritti di Giovanni di Ragusa (+ 1443)," *Codices manuscripti* 37–38 (2001): 11–50. On manuscript production at the Prodromos monastery, see Annaclara Cataldi Palau, "The Manuscript Production in the Monastery of Prodromos Petra (twelfth–fifteenth centuries)," in *Studies in Greek Manuscripts* (Spoleto: Fondazione Centro Italiano di Studi sull'alto Medioevo, 2008), 197–208, at 203, see also 203–6; and Michel Cacouros, "Marginalia de Chortasménos dans un opuscule logique dû à Prodromènos (Vatican gr. 1018)," *Revue des études byzantines* 53 (1995): 271–78, at 274.

117. One monk from this monastery, named Neophytos Prodromenos, authored several texts that testify to the research output of the Petra monastery. In a manuscript now in Paris, copied entirely in his hand, we find his botanical lexicon and a large number of individual entries on specific plants copied from Dioscorides (Paris, Bibliothèque nationale de France, MS gr. 2286, fols. 83r–88v). He included several sketchy illustrations of mandrakes (fols. 52r–v, and 53v, which should precede fol. 52. The illustrations are on fol. 52v). Neophytos wrote on a variety of topics, especially philosophy, theology, and medicine. Hunger, *Hochsprachliche profane Literatur*, 308–9; E. D. Kakoulidē, "Η βιβλιοθήκη τῆς μονῆς Προδρόμου-Πέτρας στὴν Κωνσταντινουπόλη," *Ἑλληνικά* 21 (1968): 3–39, especially 24–26; John Duffy, "Michael Psellos, Neophytos Prodromenos, and Memory Words for Logic," in *Gonimos: Neoplatonic and Byzantine Studies Presented to L. G. Westerink*, ed. J. Duffy and J. Peradotto (Buffalo: Arethusa, 1988), 207–16; Michel Cacouros, "Néophytos Prodromènos copiste et responsable (?) de l'édition quadrivium-corpus aristotelicum du 14e siècle," *Revue des études byzantines* 56 (1998): 193–212. Cacouros, "Jean Chortasménos katholikos didaskalos, annotateur du Corpus logicum dû à Néophytos Prodromènos," *Bollettino della Badia greca di Grottaferrata* 52 (1998): 185–225; Brigitte Mondrain, "La constitution de corpus d'Aristote et de ses commentateurs aux XIIIe–XIVe siècles," *Codices manuscripti* 29 (2000): 11–33. See also Charles Barber, "Neophytus

Prodromenus on Epigraphy," in *Legitimation des Bildes: Festschrift Martin Büchsel* (Berlin: Mann Verlag, 2015), 211–25.

118. See Andrew Griebeler, "How to Illustrate a Scientific Treatise in the Palaiologan Period," in *Late Byzantium Reconsidered: The Arts of the Palaiologan Era in the Mediterranean*, ed. Maria Alessia Rossi and Andrea Mattiello, (London: Routledge, 2019), 85–103, at 87–89.

119. A note on fol. 1r records this restoration: "John Chortasmenos restored this book of Dioscorides, which had become quite old and in danger of falling completely into ruin, at the behest and cost of the most venerable monk, Lord Nathanael, then nurse (*nosokomos*) in the hospital of the Kral in the year 6914 [i.e., 1406], of the 14th indiction." Τὸ παρὸν βιβλίον τὸν Διοσκουρίδην παντάπασι παλαιωθέντα καὶ | κινδυνεύοντα τελείως διαφθαρῆναι ἐστάχωσεν ὁ Χορτασμένος Ἰωάννης | προτροπῇ καὶ ἐξόδῳ τοῦ τιμιωτάτου ἐν μοναχοῖς κυροῦ Ναθαναὴλ νοσοκ|όμου τηνικαῦτα τυγχάνοντος ἐν τῷ ξενῶνι τοῦ Κράλη ἔτους ͵ϛϡΙ͞Δου ιδου | ἰν[δικτιῶν]ος ιδη.

120. Additional repairs may have also been made at this time. For example, a thirteenth-century paper gathering (fols. 287r–289v) was added to make up for a missing entry on mandrake. See Mazal, *Wiener Dioskurides*, 2:18.

121. E.g., John Meyendorff, "Wisdom–Sophia: Contrasting Approaches to a Complex Theme," *Dumbarton Oaks Papers* 41 (1987): 391–401.

122. Paris, Bibliothèque nationale de France, MS gr. 2144, fols. 10v–11r.

123. Paris, Bibliothèque nationale de France, MS gr. 2144, fol. 10v, l. 1; in Joseph A. Munitiz, "Dedicating a Volume: Apokaukos and Hippocrates (Paris. gr. 2144)," in *Φιλέλλην, Studies in Honour of Robert Browning*, ed. Costas N. Constantinides, Nikolaos M. Panagiotakis, Elizabeth Jeffreys, and Athanasios D. Angelou, Istituto Ellenico di Studi Bizantini e Postbizantini di Venezia Bibliotheke 17 (Venice: Istituto Ellenico, 1996), 267–80, at 268: ἰατρικῆς μὲν τῆς κρατίστης ἐν τέχναις.

124. Paris, Bibliothèque nationale de France, MS gr. 2144, fol. 11r, l. 4; Munitiz, "Dedicating," 270: μαθεῖν πάρεστιν ἐνθέους λόγους.

125. Munitiz, "Dedicating," 278.

126. For an example from the Arabic tradition, see the arguments in favor of medicine in the first chapter of Ibn Jumayʿ, *Al-maqāla al-Ṣalāḥiya* (*Treatise to Ṣalāḥ al-Dīn*), ed. and trans. in Hartmut Fähndrich, *Treatise to Ṣalāḥ ad-Dīn on the Revival of the Art of Medicine* (Wiesbaden: Franz Steiner, 1983), 8–16.

127. Paris, Bibliothèque nationale de France, MS lat. 6823, fol. 1r. On this frontispiece, see Theresa Holler, "Naturmaß, künstlerisches Maß und die Maßlosigkeit Ihrer Anwendung. Simiplicia in zwei 'Tractatus de herbis'-Handschriften des 13. und 14. Jahrhunderts," *Das Mittelalter* 13, no. 1 (2018): 1–25.

128. On the identification of the man as Manfredus or as Adam, see Holler, "Naturmaß," 25. On the plants, see Baumann, *Das Erbario Carrarese*, 103.

129. Baumann, *Das Erbario Carrarese*, 103.

130. "*Prima et ultima medicina propter corpus et animam est abstinentia.*"

131. "*Omnia probate q(uo)d bonu(m) est tenete / Eligite ergo q(uo)d bonu(m) est et reprobate malu(m).*"

132. The frontispiece gives a Christian reinterpretation of a scene from earlier herbals in the Latin West that depict the pre-Christian origins of herbal medicine: the presentation of *artemisia* to Chiron the Centaur by Artemis. See Holler, "Naturmaß," 22–25.

133. Kotsifou, "Books and Book Production," 50.

134. On Roman libraries, see George W. Houston, *Inside Roman Libraries: Book Collections and Their Management in Antiquity* (Chapel Hill: University of North Carolina Press, 2014); and Houston, "The Slave and Freedman Personnel of Public Libraries in Ancient Rome," *Transactions of the American Philological Association* 132, nos. 1–2 (2002): 139–76.

135. Kotsifou, "Books and Book Production," 52.

136. E.g., Guglielmo Cavallo, "Libro e pubblico alla fine del mondo," in *Libri, editori e pubblico nel mondo antico. Guida storica e critica*, ed. Cavallo (Rome: Laterza, 1984), 83–132.

137. See Griebeler, "Production and Design."

138. The copying and binding of texts often formed an important part of monastic craft industry. See Anne Boud'hors, "Copie et circulation des livres dans la région thébaine (viiᵉ–viiiᵉ siècles)," in *"Et maintenant ce ne sont plus que des villages . . ." Thèbes et sa région aux époques hellénistique, romaine et byzantine* (Brussels: Association égyptologique Reine Elisabeth, 2008), 149–61.

139. On the differences between Western and Byzantine book production and illustration, see Irmgard Hutter, "Decorative Systems in Byzantine Manuscripts, and the Scribe as Artist: Evidence from Manuscripts in Oxford," *Word & Image* 12, no. 1 (1996): 4–22.

140. Niels Gaul, "12.4. Culture of Writing and Books II: Books and Libraries—Writing and Reading in Byzantium," in *Brill's New Pauly Supplements II, Vol. 10: History and Culture of Byzantium*, English edition by John N. Dillon, trans. Duncan A. Smart (Leiden: Brill, 2019).

141. Anthony Cutler has estimated that monks accounted for only around half of the scribes active in tenth- and eleventh-century Byzantium. See Anthony Cutler, "The Social Status of Byzantine Scribes, 800–1500: A Statistical Analysis Based on Vogel-Gardthausen," *Byzantinische Zeitschrift* 74 (1981): 328–34.

142. See Griebeler, "How to Illustrate," 89. Manfredus provides an example of this phenomenon in the case of a Latin herbal.

143. See Griebeler, "How to Illustrate," 85–87.

144. S.v. Scribes and Copyists, in Adam Gacek, *Arabic Manuscripts. A Vademecum for Readers* (Leiden: Brill, 2009), 238–40. See also Gruendler, *Rise of the Arabic Book*, 103–39.

145. M. A. J. Beg, "Warrāḳ," and C. Huart and A. Grohmann, "Ḳāghad," both in *Encyclopaedia of Islam*, ed. Bearman et al.

146. W. Heffening and J. D. Pearson, "Maktaba," in *Encyclopaedia of Islam*, ed. Bearman et al.

147. Oxford, Bodleian Library, MS Arab. d. 138. The scribe is named as al-Ḥasan ibn Aḥmad ibn Muḥammad al-Nashawī. Jaclynne Kerner is skeptical of identification of this madrasa with the Niẓāmiyya madrasa in Baghdad. On the difficulty of locating this madrasa, see Kerner, "Illustrated Herbal," 126–27.

148. Etan Kohlberg, *A Medieval Muslim Scholar at Work: Ibn Ṭāwūs and His Library* (Leiden: Brill, 1992), 86.

149. See Hutter, "Decorative Systems." On the division of labor among subcontracted painters in a deluxe illustrated Byzantine manuscript, see Ihor Ševčenko, "The Illuminators of the Menologion of Basil II," *Dumbarton Oaks Papers* 16 (1962): 245–76.

150. See Saliba and Komaroff, "Illustrated Books," 17. Leiden, Universiteits-bibliotheek, MS Or. 289. See also Sadek, "Notes on the Introduction."

151. The Thott Codex and Padua Dioscorides, respectively. On the Thott Codex, see Touwaide, "Latin Crusaders, Byzantine Herbals." On the Padua Dioscorides, see Elpidio Mioni, "Un ignoto Dioscoride miniatio," in *Libri e stampatori in Padova. Miscellanea di studi storici in onore di Mons. G. Bellini* (Padua: Tipografia Antoniana, 1959), 345–76.

152. Oleg Grabar, "About an Arabic Dioskorides Manuscript," in *Byzantine East, Latin West: Art Historical Studies in Honor of Kurt Weitzmann*, ed. Doula Mouriki, C. Moss, and K. Kiefer (Princeton: Department of Art and Archaeology, Princeton University, 1995), 361–64; and Saliba and Komaroff, "Illustrated Books," 21.

153. Translation quoted here is from Saliba and Komaroff, "Illustrated Books," 21.

154. Toronto, Aga Khan Museum, AKM3, dated to the 1200s.

155. Griebeler, "How to Illustrate," 90–91. For a Latin example, see New York, Morgan Library, MS M 873, c. 1350–75. And in Greek: Vatican, Biblioteca Apostolica Vaticana, MS Chigi F.VII.159. On this codex see Miguel Ángel González Manjarrés and María Cruz Herrero Ingelmo, *El Dioscórides Grecolatino del Papa Alejandro VII. Manuscrito Vat. Chigi 53 (F. VII 159)* (Madrid: Testimonio, 2001); and Collins, *Medieval Herbals*, 77–82.

156. On late Byzantine lexicography and botanical illustration, see Andrew Griebeler, "*Aeizōon to amaranton*: Intercultural Collaboration in a Late Byzantine Nature Study," *Convivium. Exchanges and Interactions in the Arts of Medieval Europe, Byzantium, and the Mediterranean. Seminarium Kondakovianum*, n.s. 2, no. 6 (2019): 16–29.

157. Paris, Bibliothèque nationale de France, MS ar. 2964. See Jaclynne J. Kerner, "Art in the Name of Science: The *Kitāb al-diryāq* in Text and Image," in *Text and Image in Illustrated Arabic Manuscripts*, ed. Anna Contadini (Leiden: Brill, 2007), 25–39, at 35; Oya Pancaroğlu, "Socializing Medicine: Illustrations of the Kitāb al-diryāq," *Muqarnas* 18 (2001): 155–72, at 155 and 157.

158. Pancaroğlu, "Socializing Medicine," 157.

159. Vienna, Österreichische Nationalbibliothek, Cod. A.F. 10.

160. Ibn Jumayʿ, *Treatise to Ṣalāḥ ad-Dīn*, ch. 3, sec. 119 [231B]; trans. Fähndrich, 28.

Chapter 6

1. Pliny the Elder, *Naturalis historia* 25.4.

2. See discussion of Pliny's comments in chapter 1.

3. See, for example, the discussion of Anglo-Saxon herb gardens and climate in Linda E. Voigts, "Anglo-Saxon Plant Remedies and the Anglo-Saxons," *Isis* 70, no. 2 (1979): 250–68, especially 261–66.

4. Some material in this chapter, mainly text concerning the Morgan Dioscorides, has been previously published in Griebeler, "Botanical Illustration."

5. New York, Morgan Library, MS M 652. Earlier studies dated the codex to the mid-tenth century. Aletta, Kavrus-Hoffmann, and Cronier all characterize the script of the codex as an intermediate between *bouletée* and *minuscola antica*

oblunga, thus pushing the date of the codex to the late ninth or early tenth century. See Alessia A. Aletta, "Per una puntualizzazione cronologica del Morgan 652 (Dioscoride)," in *Praktika tou 6' Diethnous Symposiou Ellenikes Palaiographias (Drama, 21–27 Septembriou 2003)*, ed. Basiles Atsalos and Nike Tsirone, 3 vols. (Athens: Société hellénique de reliure, 2008), 2:771–87; Kavrus-Hoffmann, "Catalogue of Greek Medieval and Renaissance Manuscripts," 218; and Marie Cronier, "Un manuscrit méconnu."

6. This dating puts the production of the manuscript either at the end of the reign of Basil I (r. 867–86), or during the reigns of his sons, Leo VI "the Wise" (r. 886–912) or Alexander (r. 912–13). Alessia Aletta suggests that Photios might have commissioned the manuscript, but she also acknowledges there is not much evidence to support such a claim. See Aletta, "Per una puntualizzazione," 787. Nadezhda Kavrus-Hoffmann instead suggests Leo VI may have commissioned the codex for his physician or for a hospital, although she, too, notes that another elite person could have done so. See Kavrus-Hoffmann, "Catalogue of Greek Medieval and Renaissance Manuscripts," 226–27.

7. Cronier, "Un manuscrit," 116.

8. For example, see *melissophyllon* (fols. 1v and 102v), *bettonikē* (fol. 22r), *erigerōn* (*sic*, i.e., ἠριγέρων, fol. 42v); *erysimon* (fol. 46v); *ēryngion* (fol. 57r); *thlaspi hetera* (fol. 61v); *krinon basilikon* (fol. 84r); *lychnis stephanōmatikē* (fol. 93v); *polygonon arren* (fol. 130r); *skolymos* (fol. 154v); and *sinēpi agrion* (fol. 157v).

9. Likely *Eryngium spp.*; Morgan Dioscorides, fol. 57r.

10. The sequence of production was confirmed through microscopy. I thank Frank Trujillo of the Morgan Library, New York, for his work photographing and assessing the layering of colors in the manuscript, and Joshua O'Driscoll for supporting this work. See Griebeler, "Botanical Illustration and Byzantine Visual Inquiry."

11. The Padua Dioscorides; it is possible (though unlikely) that the pictures were added to both manuscripts at about the same time.

12. Likely *Sisymbrium spp.*; Morgan Dioscorides, fol. 103v.

13. ἐρύσιμον ἕτερον.

14. *Eryngium campestre* L. *E. creticum* Lam. grows locally: D. A. Webb, "Flora of European Turkey," *Proceedings of the Royal Irish Academy* Section B, 65 (1966/1967): 1–100, at 47.

15. *Eryngium maritimum* L. *E. planum* L. also grows locally: Webb, "Flora of European Turkey," 47.

16. In particular (fol. 57v), with differences from Wellmann's edition indicated in brackets: αὐξόμενα δὲ [ἀκανθοῦται] κατα πλείονας ἐξονας καυλὸν [καυλῶν] ἐφ' ὧν κατὰ τὰ ἄκρα κεφαλια [κεφάλιά] [ἐστι] σφερωδη [σφαιροειδῆ]· ἀκάνθας περικείμενα ὥσπερ ἀστὴρ κύκλω [κύκλῳ] ὀξυτατα [ὀξυτάτας], σκληρας ὧν τὸ χρῶμα [χλωρὸν ἢ] λευκόν [ἢ χλωρὸν], η κοθανουν [ἐνίοτε δὲ καὶ κυανοῦν] εὑρίσκεται. κοθανουν is likely an error for κυανοῦν indicating a dark blue color. Thus the text does not indicate the possibility (ἐνίοτε) that the thorny bracts (ἀκάνθας περικείμενα ὥσπερ ἀστὴρ κύκλω [κύκλῳ] ὀξυτατα [ὀξυτάτας]) of the plant are a dark blue. Consequently, the text mentions only white and green, which describes field eryngo.

17. Vatican, Biblioteca Apostolica Vaticana, MS Chigi F.VII.159, the "Chigi Dioscorides." Anton von Premerstein suggested that the codex was copied after 1406, because it reflects John Chortasmenos' rearrangement of the Vienna

Dioscorides. Premerstein, "De codicis Dioscuridei," 171–72. On this codex see González Manjarrés and Herrero Ingelmo, *El Dioscórides Grecolatino*, especially 46–53; and Collins, *Medieval Herbals*, 77–82.

18. The Chigi codex lacks text except for Greek titles and later notes, most of which are in Latin.

19. His hand appears in tiny inscriptions at the top edges of many of the folios (fols. 13–219), as well as in the red Greek titles for the animals (fols. 221–33). Another scribe wrote the Greek plant names in larger letters closer to the illustrations. These inscriptions often contain spelling errors that probably derive from Isidore's tiny inscriptions, which suggests that the Vienna and Morgan Dioscorides manuscripts were no longer present. Isidore's unceremonious inscriptions seem to tell the painter what pictures to copy. Giovanni Mercati, *Scritti d'Isidoro il cardinal Ruteno*, Studi e testi 46 (Rome: Biblioteca Apostolica Vaticana, 1926), 93. Three other hands are associated with Latin inscriptions. On the sequence of production, see González Manjarrés and Herrero Ingelmo, *El Dioscórides Grecolatino*, 51. See also Premerstein, "De codicis Dioscuridei," 11, 89. It is difficult to know exactly when the codex was produced. Isidore could have been involved in the production of the manuscript in the 1430s, while he was *hegoumenos* of the monastery of St. Demetrios in Constantinople, or perhaps during a later stay in the city in 1450 or 1452–53. See Joseph Gill, *Personalities of the Council of Florence* (New York: Barnes and Noble, 1964), 65–78; and more recently, Marios Philippides and Walter K. Hanak, *Cardinal Isidore, c. 1390–1462: A Late Byzantine Scholar, Warlord, and Prelate* (Abingdon, UK: Routledge, 2008). In favor of a later date of c. 1450 for the Chigi Dioscorides is the fact that Isidore owned a manuscript copy of Church Councils copied in two parts in 1445 and 1446 by the scribe Athanasios (Munich, Bayerische Staatsbibliothek, Cod. Gr. 186). On fol. 298v, Isidore notes that the second part of this codex was copied from an exemplar in the library of the Petra Prodromos monastery, also the location of the exemplars of the Chigi Dioscorides. See Kerstin Hajdú, ed., *Katalog der griechischen Handschriften der Bayerischen Staatsbibliothek München*, 4: *Codices graeci Monacenses 181–265* (Wiesbaden: Harrassowitz, 2012), 4:53. On Isidore's copying of manuscripts, see Mercati, *Scritti*; and Philippides and Hanak, *Cardinal Isidore*, 250. That Isidore also owned an unillustrated Dioscorides has led Minta Collins to suppose that he might have had the Vatican codex made to accompany it. Collins, *Medieval Herbals*, 82. The unillustrated Dioscorides is Vatican, Biblioteca Apostolica Vaticana, gr. 289. See also Alain Touwaide, "Un recueil grec de pharmacologie du Xe siècle illustré au XIVe siècle. Le Vaticanus gr. 284," *Scriptorium* 39, no. 1 (1985): 13–56, at 49. It remains possible, too, that the book was used in conjunction with other medical works.

20. "The narrowleaf eryngo" (ἤριγγιον τὸ λεπτόφυλλον) at Chigi Dioscorides, fol. 68v, and "the big eryngo" (ἠρύγγιον τὸ μέγα) at fol. 191v.

21. Likely *Isatis tinctoria* L.; Morgan Dioscorides, fol. 68v.

22. See Dioscorides, *De materia medica* 2.185.

23. Pliny, *Naturalis historia* 20.25.

24. E.g., Yunus Doğan, Süleyman Başlar, Hasan Hüseyin Mert and Güngör Ay, "Plants Used as Natural Dye Sources in Turkey," *Economic Botany* 57, no. 4 (2003): 442–53.

25. See John Riddle, "Byzantine Commentaries on Dioscorides," in "Sympo-

sium on Byzantine Medicine," *Dumbarton Oaks Papers* 38 (1984): 95–102, at 102.

26. Riddle adds that this scholium "reveals excellent attention to botanical detail rarely equaled in ancient or medieval herbals. . . . The unknown Byzantine commentator has trusted his observation of the plant in nature in a way that his western counterpart would not have done." Riddle, "Byzantine Commentaries," 102.

27. Evidence for the recognition of disagreement between picture and text can also be found in a marginal annotation beside a picture on fol. 79v in the chapter on *krambē thalassion* (perhaps sea bindweed, *Convolvulus soldanella* L.). The note (ἔχει τὰ φύλλα ὅμοια ἀριστολοχι[ᾷ] στρογγύλη) paraphrases accompanying text comparing the leaves of *krambē thalassion* to those of birthwort (*aristolochia strongylē*). This description notably disagrees with the picture included. This note was apparently written to highlight the disagreement.

28. On the relationship between late Byzantine lexicography and botanical illustration, see Griebeler, "*Aeizōon to amaranton*," 16–29.

29. Likely *Melissa officinalis* L.; Morgan Dioscorides, fols. 1v and 102v.

30. See, for example, the inclusion of the second balm, labeled *melissophyllon heteron* (μελισσόφυλλον ἕτερον) in the Padua Dioscorides, fol. 189v.

31. *Nelumbo nucifera* Gaertn.; Morgan Dioscorides, fol. 75r.

32. *Asparagus officinalis* L., at Morgan Dioscorides, fol. 12v, and *Vicia faba* L., at fol. 74v.

33. The note beside the picture on fol. 74v reads: μὴ ζωγραφήσῃς.

34. Other corrections include the "whiting out" of hairs in the depiction of *bounion* (earthnut? fol. 16r), evidently so that the stem would appear smooth; the washing off and repainting of elder (*chamaiaktē, Sambucus ebulus* L., fol. 195v); and the enlargement and addition of leaves in "another geranium" (*geranion heteron*, fol. 31r).

35. Morgan Dioscorides, fol. 30v.

36. *G. robertianum* has clearly documented medicinal usages going back to the thirteenth century. See Geoffrey Grigson, *An Englishman's Flora* (London: Phoenix House, 1955); Peter F. Yeo, *Hardy Geraniums*, 2nd ed. (Portland, OR: Timber Press, 2002), 9. On the medicinal usages, see H. Gams, "Geraniaceae," 1656–1725, in G. Hegi, *Illustrierte Flora von Mittel-Europa*, 1st ed. (Munich: Carl Hanser, 1923–24), vol. 4, part 3. For modern ethnobotanical uses of the herb (for treating male sterility), see, e.g., Sulejman S. Redzić, "The Ecological Aspect of Ethnobotanny and Ethnopharmacology of Population in Bosnia and Herzegovina," *Collegium Antropologicum* 31, no. 3 (2007): 869–90, at 887. On the anti-inflammatory properties of the herb, see Marcelo D. Catarino, Artur M. S. Silva, Maria Teresa Cruz, and Susana M. Cardoso, "Antioxidant and Anti-Inflammatory Activities of *Geranium robertianum* L. Decoctions," *Food & Function* 8, no. 9 (2017): 3355–65.

37. *Spartium junceum* L.

38. Vienna Dioscorides, fols. 327v–28r.

39. Other illustrations of this plant in the same tradition lack these loops (e.g., Naples Dioscorides, fol. 150r).

40. Vienna Dioscorides, fol. 328r.

41. I thank Maria Mavroudi for help reading this inscription. The last part of the inscription is unclear, and has been reconstructed here. Τοιοῦτον σχῆμα |

ἔχει ὁ σπάρτος τὸ ἄνθος αὐτοῦ ὅμοιον | φασιόλου ἢ κυάμου | πλὴν ξανθὸν κίτρινον | τὸ σπέρμα αὐτοῦ μιά[ζει?] (=μοιάζει?) | βίκον ἢ φακήν. *Phasiolos* appears to be φάσηλος or calavance (*Vigna sinensis* [L.] Savi ex Hassk.); *kyamos*, κύαμος, likely fava (*Vicia faba* L.); βίκον, i.e., βικίον (*Vicia sativa* L.); *phakēn*, φακήν (*Ervum lens* L., i.e., *Lens culinaris* Medik.).

42. Padua Dioscorides, fol. 155r. See Mioni, "Un ignoto Dioscoride miniato."

43. Padua Dioscorides, fol. 155r.

44. Chigi Dioscorides, fol. 157r.

45. *Potentilla reptans* L. Chigi Dioscorides, fol. 135r; compare with Vienna Dioscorides, fol. 273r; Naples Dioscorides, fol. 118r; and Morgan Dioscorides, fol. 126r.

46. The following illustrations appear in the Chigi Dioscorides, but in neither the Vienna nor the Morgan Dioscorides: *artemistra* (*sic*, i.e., ἀρτεμισία, fol. 95r), *krambē nea* (fol. 99r), *krambē megalē* (fol. 100r), *kardamon* (fol. 100v), *libanōteis hetera* (fol. 108r), *nymphaia* (fol. 120r), *nymphaia hetera* (fol. 120v), *seutlon hēmeron leukon* (fol. 148r), *seutlon kokkeinon agrion* (fol. 148v), *cotilidon cimbalaria* (fol. 150, Greek name is missing), *to phagedaikon* (fol. 172v), *ōkimon heteron* (fol. 183v), *sikyos pepōn* and *sikyos* (fol. 203v), *heteros sikyos pepōn* and *sikyos kolokynthi* (fol. 204r), *karpin* (fol. 209r), and *kynosbatos* (fol. 214v). The illustration of *artemistra* (fol. 95r), likely a misspelling for *artemisia*, seems to depict feverfew, *Tanacetum parthenium* (L.) Sch. Bip.

47. Morgan Dioscorides, fol. 252v.

48. The text on fol. 252v of the Morgan Dioscorides is similar to that given in Wellmann's edition (*De materia medica* 1.94).

49. Compare Givens, *Observation and Image-Making*, 17–18, 90, 144–45.

50. Paul Lemerle, *Le premier humanisme byzantin. Notes et remarques sur enseignement et culture à Byzance des origines au Xe siècle* (Paris: Presses universitaires de France, 1971), 266. Translated into English: *Byzantine Humanism: The First Phase: Notes and Remarks on Education and Culture in Byzantium from Its Origins to the 10th Century*, trans. Helen Lindsay and Ann Moffatt (Canberra: Australian Association for Byzantine Studies, 1986). And criticism: Paolo Odorico, "La cultura della Συλλογή," *Byzantinische Zeitschrift* 83 (1990): 1–21.

51. George Kubler, *The Shape of Time: Remarks on the History of Things* (New Haven: Yale University Press, 1962), 74–75.

52. Kubler, *Shape of Time*, especially 53–61, 96–123.

53. Hans Jörg Rheinberger, "Experimental Systems: Historiality, Narration, and Deconstruction," *Science in Context* 7, no. 1 (1994): 65–81, at 74–75.

54. Pomata, "A Word of the Empirics."

55. Katharine Park, "Observations in the Margins, 500–1500," in *Histories of Scientific Observation*, ed. Lorraine Daston (Chicago: University of Chicago Press, 2011), 15–44, at 30.

56. Pomata, "A Word of the Empirics," 14.

57. Alain Touwaide, "The Development of Paleologan Renaissance: An Analysis Based on Dioscorides' *De materia medica*," in *Philosophie et sciences à Byzance de 1204 à 1453. Actes de la Table Ronde organisée au xxe Congrès International d'Études Byzantines (Paris, 2001)*, ed. Michel Cacouros and Marie-Hélène Congourdeau (Leuven: Peeters, 2006), 189–224. On the bindings linking these manuscripts to the Prodromos monastery, see Kavrus-Hoffmann, "Catalogue of

Greek Medieval and Renaissance Manuscripts," 212–30, at 225–26; Cataldi Palau, "Legature costantinopolitane."

58. Cataldi Palau, "Manuscript Production"; Annaclara Cataldi Palau, "The Library of the Monastery of Prodromos Petra in the Fifteenth Century (to 1453)," in *Studies in Greek Manuscripts* (Spoleto: Fondazione Centro Italiano di Studi sull'alto Medioevo, 2008), 209–18. On the hospital, see Urs Benno Birchler-Argyros, "Die Quellen zum Kral-Spital in Konstantinopel," *Gesnerus* 45 (1988): 419–44; Vassilios Kidonopoulos, *Bauten in Konstantinopel 1204–1328. Verfall und Zerstörung, Restaurierung, Umbau und Neubau von Profan- und Sakralbauten* (Wiesbaden: Harrassowitz, 1994), 218–21.

59. Other hospitals may have existed in Byzantium prior to the fourteenth century, though most care facilities would have been fairly humble institutions. A foundation document (*typikon*) for the Pantokrator monastery gives specifications for medical lectures and instruction. On the Pantokrator Typikon, see *Byzantine Monastic Foundation Documents*, 725–81.

Chapter 7

1. London, British Library, MS Egerton 747 (hereafter "Egerton 747"). See Collins, *Medieval Herbals*, 240, 245, and 249. For a facsimile of the manuscript (fols. 1–109), see *A Medieval Herbal: A Facsimile of British Library Egerton MS 747*, with introduction by Minta Collins and plant list by Sandra Raphael (London: British Library, 2003).

2. Pächt, "Early Italian Nature."

3. See discussion in Collins, *Medieval Herbals*, 239, 254–55; and Jean Givens, "Reading and Writing the Illustrated *Tractatus de herbis*, 1280–1526," in *Visualizing Medieval Medicine and Natural History, 1200–1550*, ed. Jean Givens and Alain Touwaide (Aldershot, UK: Ashgate, 2006), 115–45, at 118.

4. Collins, *Medieval Herbals*, 254.

5. In Egerton 747, the *Tractatus de herbis* (fols. 1r–106v) is followed by several folios with depictions of plants not included in the *Tractatus de herbis* (fols. 106v–9r), a few folios with texts on forecasting and a lunar calendar in a mid-fourteenth-century hand (fols. 109v–11r), the *Antidotarium Nicolai* (fols. 112r–24r), *De dosibus medicinarum* by Nicolaus (fol. 124v), a substitution list or *quid pro quo* (fols. 125v–27v), three short texts on weights and measures (fols. 127v–28v), a list of synonyms for plant names (fols. 128v–46r), and an incomplete supplement to the *Antidotario Nicolai* (fols. 146v–47v). For the text of the *Tractatus de herbis*, see Ps.-Bartholomaeus Mini de Senis, *Tractatus de herbis (MS London, British Library, Egerton 747)*, ed. Iolanda Ventura, Edizione nazionale "La scuola medica salernitana" 5 (Florence: SISMEL, 2009). See also Givens, "Reading and Writing," 119; Collins, *Medieval Herbals*, 243–45.

6. See Givens, "Reading and Writing," 118.

7. On the *Circa instans*, see Ventura, "Per una storia del *Circa Instans*"; Ventura, "Un manuale di farmacologia medievale"; and Ventura, "Une oeuvre et ses lecteurs." On the manuscript transmission of the text, see Ventura, "Il *Circa instans*." On vernacular translations, see Iolanda Ventura, "Il *Circa instans* nella cultura medica del Tardo Medioevo. Le traduzioni volgari in francese, tedesco ed olandese," in *Le civiltà e la medicina*, ed. Gerardo Sangermano (Salerno: Gaia

editrice, 2011), 138–219.

8. Givens, "Reading and Writing," 118–21.

9. Collins, *Medieval Herbals*, 240, 284n7.

10. Collins, *Medieval Herbals*, 242.

11. Collins, *Medieval Herbals*, 250–52.

12. Egerton 747, fol. 16v. Fig. 69 in Collins, *Medieval Herbals*, 248.

13. See discussion in Givens, *Observation and Image-Making*, 16–36. Givens does not use the term "accuracy" but speaks of the correspondence between artworks and plants known today. The term "accuracy" would be consistent with Byzantine understandings of faithful representations; see Henry Maguire, *The Icons of the Their Bodies: Saints and Their Images in Byzantium* (Princeton: Princeton University Press, 2000), 16, 22.

14. Pächt, "Early Italian Nature Studies," 29–30.

15. Egerton 747, fol. 74v.

16. See Baumann, *Das Erbario Carrarese*, 102. For a discussion of textual evidence supporting the possible use of pressed specimens, see Givens, *Observation and Image-Making*, 90–91; for an argument based on close visual analysis, see Holler, "Naturmaß," 7–8.

17. See, e.g., Karen Reeds, *Botany in Medieval and Renaissance Universities* (New York: Garland, 1991), 10, 24–25.

18. Givens, "Reading and Writing," 116; and Collins, "Medieval Herbals," 242, 245. On Salerno, see Paul Oskar Kristeller, *Studi sulla cuola medica salernitana* (Naples: Istituto italiano per gli studi filosofici, 1986); Andrea Cuna, *Per una bibliografia della scuola medica salernitana (secoli XI–XIII)* (Milan: Guerini, 1993). More recently, see the contributions in *La scuola medica salernitana. Gli autori e i testi*, ed. Jacquart and Bagliani; and *Salerno nel XII secolo. Istituzioni, società, cultura, Atti del Convegno internazionale (Raito di Vietri sul Mare, 16–20 giugno 1999)*, ed. Paolo Delogu and Paolo Peduto.

19. Some of this section has been previously published in Griebeler, "Botanical Illustration."

20. An eleventh-century illustrated Dioscorides in Greek, now in the library of the Great Lavra on Mount Athos (Athos, Library of the Great Lavra, Cod. Ω 75), includes hundreds of illustrations of plants and people harvesting them. On this manuscript, see Christodoulou, *Σύμμικτα Κριτικά*, 131–99.

21. Some of the losses in the Old Paris Dioscorides can be tentatively reconstructed or hypothesized on the basis of Arabic manuscripts that reproduce this tradition, such as the Parchment Arabic Dioscorides.

22. *Vitis vinifera* L. or *Vitis vinifera* subsp. *sylvestris* (C.C.Gmel.) Hegi; compare to the Naples Dioscorides, fol. 26r.

23. Cronier connects these to a hypothetical source, "Mc." See Cronier, "Un manuscrit méconnu."

24. These divisions reflect depictive approaches and not groupings by their hypothetical source. It does not include one example from the Alphabetical Dioscorides, as well as nontrees, including mushrooms (fol. 260v), papyrus (fol. 263v) and lotus (fol. 256v), and tree "products," such as oak galls (fol. 253r) and various tools made of reeds (fol. 252r), all of which were originally excluded from the Alphabetical Dioscorides. Minta Collins makes a similar division of the trees. See Collins, *Medieval Herbals*, 66.

25. Juniper berries, fol. 246r, and juniper, fol. 251v. The text for juniper berries is part of the chapter on juniper in the "original," Dioscorides, *De materia medica* 1.77.

26. Laurel at Dioscorides, *De materia medica* 1.78, and juniper at 1.77.

27. An index may have also been used in the compilation of the later *Fragmenta Constantiniana*, see András Németh, "Compilation Methods of the Excerpta Constantiniana Revisited: From One Compiler to the Three-Stage Model of Teamwork," *Byzantinoslavica: Revue international des études byzantines* 1–2 (2017): 265–90, at 287. See also András Németh, *The Excerpta Constantiniana and the Byzantine Appropriation of the Past* (Cambridge: Cambridge University Press, 2018), especially 88–120.

28. Cronier calls this hypothetical source "Ma" and identifies it with the chapters in the Morgan Dioscorides that contain τοῦ ἀναζαρβέως in their title. See Cronier, "Un manuscrit méconnu," 109–18. These chapters are not well represented in Book IV, and tend to be concentrated at the beginning. The chapters that include τοῦ ἀναζαρβέως in their titles are as follows: *aigēros* (fol. 243r), *akakia* (fol. 244v), *brathy* (fol. 244v), *fegos etoi prinos* (fol. 245r), *daphnē hetera* (fol. 246r), *eirēkē* (f. 246r), *kisthos* (fol. 248r), *heteron eidos kisthou* (fol. 248r), *heteron eidos kisthou* (fol. 248v), *krania* (fol. 249r), *peri tou mēlizontas karpou kranias* (fol. 249r), *kinnamōmon* (fol. 249v), *xylokinnamōmon* (fol. 250r), *pseudokinnamōmon* (fol. 250v), *melimēlea ēpeirōtikē* (fol. 258r), *myrsinē* (fol. 258v), *myrtidanon* (fol. 259r).

29. These include *aigeiros* (black poplar, *Populus nigra* L., fol. 243r), *brathy* (savin juniper, *Juniperus sabina* L., or stinking juniper, *Juniperus foetidissima* Willd., fol. 244v), *phēgos* (Valonia oak or Holm oak, *Quercus ithaburensis* subsp. *macrolepis* [Kotschy] Hedge & Yalt. or *Quercus ilex* L., fol. 245r), *eirēkē* (*sic*, i.e., ἐρείκη, alternate spelling in the title, heath tree, *Erica arborea* L., fol. 246r), and finally *melimēlea ēpeirōtikē* ("apples of Epiros," fol. 258r). Only two pictures associated with Ma texts in Book IV, that is, the pictures of myrtle (*myrsinē*, fol. 258v) and Cornelian cherry (*krania*, fol. 249r), can be said to include naturalistic or accurate details. Two other miniatures—those of the Shittah tree (*akakia*, *Acacia* sp., fol. 244v) and *peri tou mēlizontos karpou kranias* (fol. 249r)—do not match any of the groups discussed here or below, but may also come from the Ma source.

30. *Juniperus deltoides* R. P. Adams= *J. oxycedrus* L. subsp. *oxycedrus*, and subsp. *deltoides* R. P. Adams. The fruit of *J. oxycedrus* var. *oxycedrus* is today noted to have stimulant and diuretic properties, which may be related to the properties named by Dioscorides (*De materia medica* 1.79). Cade oil, which is used as an antiparasitic (antihelminthic) and for dermatitis, is also extracted from the wood of these two species. See Robert P. Adams, *Junipers of the World: The Genus Juniperus*, 2nd ed. (Vancouver, BC: Trafford Publishing, 2008), 159–61, 237–39. See also Robert P. Adams, "*Juniperus deltoides*, a New Species and Nomenclatural Notes on *Juniperus polycarpos* and *J. turcomanica* (*Cupressaceae*)," *Phytologia* 86, no. 2 (2004): 49–53; Robert P. Adams, Julie A. Morris, Ram N. Pandey, Andrea E. Schwarzbach, "Cryptic Speciation between *Juniperus deltoides* and *Juniperus oxycedrus* (Cupressaceae) in the Mediterranean," *Biochemical Systematics and Ecology* 33, no. 8 (2005): 771–87; Robert P. Adams, "Morphological Comparison and Key to *Juniperus deltoides* and *J. oxycedrus*," *Phytologia* 96, no. 2 (2014): 58–62.

31. Morgan Dioscorides, fol. 245v.

32. *Amygdalis pikra* (bitter almond, *Prunus amygdalus* [L.] Batsch or *Prunus dulcis* var. *amara* DC, fol. 243v), *daphnē* (bay laurel, *Laurus nobilis* L., fol. 245v), *eitea* (willow, *Salix alba* L., fol. 246v), *elaia agria* (oleaster, *Olea europaea* L., var. *sylvestris*, fol. 247r), which may have been repaired at a later point, *kerasea* (cherry, *Prunus avium* L., fol. 253v), *karya basileika* (walnut, *Juglans regia* L., fol. 254v), *karya pontika* (hazelnut, *Corylus avellana* L., fol. 255r), *leukē* (white poplar, *Populus alba* L., fol. 255r), *mēlea* (apple, *Malus* spp., fol. 257v), *myrsinē* (myrtle, *Myrtus communis* L., fol. 258v), *morea* (mulberry, *Morus nigra* L., fol. 259v).

33. Cherry (*Prunus avium* L., fol. 253v), hazelnut (*Corylus avellana* L., fol. 255r), mulberry (*Morus nigra* L., fol. 259v).

34. Aristotle, *De generatione animalium* 1.23.731a25–26.

35. See Hardy and Totelin, *Ancient Botany*, 147; Theophrastus, *De causis* 1.21.1. Theophrastus' works on plants were known in ninth- and tenth-century Byzantium. One of the earliest extant manuscripts with Theophrastus' *Historia plantarum* and his *De causis plantarum* (Vatican, Biblioteca Apostolica Vaticana, Cod. Urb. gr. 61) is roughly contemporary with the Morgan Dioscorides and is written in a similar script.

36. See further discussion and citations to specific authors in Hardy and Totelin, *Ancient Botany*, 147–48.

37. Photios, *Bibliotheca* Cod. 178, ll. 23–25 (ed. Henry): χρήσιμον δὲ τὸ βιβλίον οὐ πρὸς ἰατρικὴν φιλοπονίαν μόνον, ἀλλὰ καὶ πρὸς ἐμφιλόσοφον καὶ φυσικὴν θεωρίαν.

38. Michael Psellos, *Epistulae* sec. 17, ep. 70; ed. Stratis Papaioannou, *Michael Psellus Epistulae* (Berlin: De Gruyter, 2019): Τοὺς γοῦν λόγους ἡμῖν θεωρητέον τῶν ἐκείνης μηχανημάτων, ἵν' ἔχοιμεν λόγων λόγῳ κατατρυφᾶν.

39. See summaries and bibliography in Michael Jeffreys and Mark D. Lauxtermann, *The Letters of Psellos: Cultural Networks and Historical Realities* (Oxford: Oxford University Press, 2017): KD236, p. 279; KD239, p. 278; KD 235, p. 279; and G26, p. 167; KD237, p. 280.

40. For mulberry, fol. 259v, in the first line: Μορέα ἡ συκάμινος δένδρον ἐστὶν πᾶσιν γνώριμον.

41. Among the second group of illustrations, the following foodstuffs would have been familiar to Constantinopolitans: *mēlea* (apple, *Malus* spp., fol. 257v), *amygdalis pikra* (bitter almond, *Prunus amygdalus* (L.) Batsch or *Prunus dulcis* var. *amara* DC, fol. 243v), *morea* (mulberry, *Morus nigra* L., fol. 259v), *karya basileika* (walnut, *Juglans regia* L., fol. 254v), *karya pontika* (hazelnut, *Corylus avellana* L., fol. 255r), *kerasea* (cherry, *Prunus avium* L., fol. 253v). Monasteries in Greece frequently planted apples, mulberries, and cherries. See Alice Mary Talbot, "Byzantine Monastic Horticulture: The Textual Evidence," in *Byzantine Garden Culture*, ed. Littlewood, Maguire, and Wolschke-Bulmahn, 37–67. Walnut is considered an anthropogenic indicator in palynological studies of the Balkans and Anatolia; see Adam Izdebski, Grzegorz Koloch, and Tymon Słoczyński, "Exploring Byzantine and Ottoman Economic History with the Use of Palynological Data: A Quantitative Approach," *Jahrbuch der österreichischen Byzantinistik* 65 (2015): 67–109, especially 83, 86. Oleaster or wild olive (*elaia agria*, fol. 247r) may have also been familiar, but not cultivated locally or widely. See B. Geyer, "Physical Factors in the Evolution of the Landscape and Land Use," in *Economic History of Byzantium: From the Seventh through the Fifteenth Century*, ed. Angeliki E. Laiou (Washington, DC: Dumbarton Oaks, 2002), 1: 31–45. Note olives do

grow locally: Webb, "Flora of European Turkey," 61. Wild olives may have been a common sight, as many olive orchards went feral in the seventh century. See Alexander Olson, *Environment and Society in Byzantium, 650–1150: Between the Oak and the Olive* (Cham: Palgrave Macmillan, 2020), 95–135; also Lin Foxhall, *Olive Cultivation in Ancient Greece: Seeking the Ancient Economy* (Oxford: Oxford University Press, 2007). Some plants, such as the laurel (*daphnē, Laurus nobilis* L., fol. 245v), myrtle (*myrsinē, Myrtus communis* L., fol. 258v), and cypress (*kyparissos, Cupressus sempervirens* L., fol. 251r), were planted as ornamentals. For evidence of these plants in gardens from later literature, see Mary-Lyon Dolezal and Maria Mavroudi, "Theodore Hyrtakenos' *Description of the Garden of St. Anna* and the Ekphrasis of Gardens," in *Byzantine Garden Culture*, ed. Littlewood, Maguire, and Wolschke-Bulmahn, 105–58, at 117. *Strobylea* (stone pine, *Pinus pinea* L., fol. 269v) was cultivated for its nuts.

42. Collins, *Medieval Herbals*, 66. Collins's assertion that these depictions "are not true to nature" and are "identifiable only with the help of the title and text" are misleading. Collins refers here to the illustration of the cherry, *kerasea*, which also compares favorably against the depictions of trees in Egerton 747.

43. *Pinus pinea* L.; Morgan Dioscorides, fol. 269v.

44. See Egerton 747, fol. 74v.

45. Morgan Dioscorides, fol. 251v.

46. The beginning of the text (first five lines of the main text, excluding the rubric title) on fol. 251 can be transcribed as follows: κέδρος | δένδρον ἐστὶν μέγα ἐξ οὗ λεγομένη κεδρία συνάγεται. καρπὸν | δὲ ἔχει ὥσπερ κυπαρίσσου· μακρότερον μέντοι παρα πολυ| γενεται [γεννᾶται] δὲ καὶ ἄλλη κέδρος μικρὰ ἀκανθώδης ὥσπερ ἄρ|κευθος φέρουσα μέγεθος μύρτου περιφερή.

47. The Greek *kedros* (κέδρος) can apply not only to several different junipers, but also to cedars. *LSJ*, s.v. κέδρος. In this case, it is likely to be cedar of Lebanon, *Cedrus libani* A. Rich.

48. Dioscorides, *De materia medica* 1.77: καρπὸν δὲ ἔχει ὥσπερ κυπάρισσος, μικρότερον μέντοι παρὰ πολύ.

49. Fol. 251: καρπὸν | δὲ ἔχει ὥσπὲρ κυπαρίσσου· μακρότερον μέντοι παρα πολυ| γενεται [γεννᾶται] δὲ καὶ ἄλλη κέδρος μικρὰ ἀκανθώδης ὥσπερ ἄρ|κευθος φέρουσα μέγεθος μύρτου περιφερή.

50. *Juniperus phoenicea* L. Compare Lily Beck's translation of the *De materia medica*, which identifies the smaller *kedros* as *Juniperus communis* L.: Beck, 59. Dioscorides does not in fact mention the color of this juniper's fruit, but rather compares them to the berries of a myrtle (*myrtos*, i.e., *Myrtus communis* L.), and to those of another juniper, called *arkeuthos*. Dioscorides elsewhere describes the *arkeuthos* as having yellow fruit, and it seems possible that the depiction here is influenced by that comparison or was meant to actually depict *arkeuthos* (Dioscorides, *De materia medica* 1.75).

51. *Kynosbaton* (fol. 254r), and dyer's buckthorn (fol. 255v).

52. *Ebenos* (ebony or persimmon, *Diospyrus* spp., fol. 264), *krania* (Cornelian cherry, *Cornus mas* L., fol. 249r), *kedros* (cedar and juniper, likely *Cedrus libani* A. Rich. and *Juniperus phoenicea* L., fol. 251v), *kynosbaton* (*sic*, i.e., κυνόσβατος, evergreen rose, *Rosa sempervirens* L., fol. 252r), *kyprion dendron* (i.e., κύπρος, henna, *Lawsonia inermis* L., fol. 252v), *kerataia* (carob, *Ceratonia siliqua* L., fol. 253v), *kokkymēlea* (plum tree, *Prunus domestica* L., fol. 254r), *komaros* (strawberry-tree,

Arbutus unedo L., fol. 254r), *lykion* (perhaps dyer's buckthorn, *Rhamnus* spp., fol. 255v), *lōtos* (nettle tree, *Celtis australis* L., fol. 256v), *melia* (manna or flowering ash, *Fraxinus ornus* L., fol. 257r), *myrikē* (tamarisk, *Tamarix tetrandra* Pall. ex M. Bieb., fol. 257r), *melimela* (jenneting or summer apple, *Pyrus praecox* [Pall.] Borkh., or apple grafted on quince [*LSJ*, s.v. μελίμελα], fol. 257r), *mespēlaia* (i.e., μέσπλιον, medlar, *Mespilus germanica* L., fol. 259v), *mēlakydōnia* (quince, *Cydonia oblonga* Mill., fol. 260r), *oxyakantha* (firethorn, *Pyracantha coccinea* M. Roem., fol. 261r), *paliouros* (Christ's thorn, *Paliurus australis* Gaertn. or *P. spina-christi* Mill., fol. 262r), *persika mēla* (peach, *Prunus persica* [L.] Stokes, fol. 262v), *pitys* (some kind of pine, *Pinus* sp., fol. 263v), *rhous dendron* (sumac, *Rhus coriaria* L., fol. 264v), *sykomorea* (sycamore fig, *Ficus sycomorus* L., fol. 266r), *schinos* (mastic, *Pistacia lentiscus* L., fol. 266v), *terebinthos* (terebinth, *Pistacia terebinthus* L., fol. 267v), *phillyrea* (i.e., φιλυρέα, mock privet, *Phillyrea medi* L., fol. 268r), *phoinix ho aigyptios* ("Egyptian" date palm, *Phoenix dactylifera* L., fol. 268v), *phoinix thibaikos* ("Theban" date palm, *P. dactylifera* L., fol. 269r).

53. The depictions of carob (fol. 253v), the nettle tree (fol. 256v), medlar (fol. 259v), quince (fol. 260r), peach (fol. 262v), and plum (fol. 254r).

54. For example, Demetrios Kydones sent the empress Helena Kantakouzene Palaiologina medlar fruits from his garden (letter dated c. 1374–75); see Frances Kianka, "The Letters of Demetrius Kydones to the Empress Helena Kantakouzene Palaiologina," *Dumbarton Oaks Papers* 46 (1992): 155–64, at 160. Plum and nettle trees also grow in the region.

55. κερ_ataία, i.e., κεράτια [=κερατωνία], *Ceratonia siliqua* L., fol. 253v. On the distribution of the carob, see Güven Şahin and Nuran Taşlıgil, "Agricultural Geography Analysis of Carob Tree (*Ceratonia siliqua* L.) from Turkey," *Turkish Journal of Agriculture: Food Science and Technology* 12, n. 4 (2016): 1192–1200. On carob trees in Near Eastern monasteries, see Talbot, "Byzantine Monastic Horticulture," at 52.

56. ῥοῦς δένδρον, *Rhus coriaria* L., fol. 264v.

57. Dioscorides, *De materia medica* 1.108; trans. Beck, 77, with modification. The relevant text from fol. 264v in the Morgan Dioscorides reads: καρπὸς δὲ βοτρυδίοις ἐοικώς, πυκνός, κατὰ μέγεθος τερμίνθου, ὑπόπλατυς.

58. *Theophanes Continuatus* 328.23–329.2, ed. Immanuel Bekker, *Theophanes Continuatus* (Bonn: Weber, 1838): "Παράδεισον . . . παντοῖς κομῶντα φυτοῖς"; trans. Mango, *Art of the Byzantine Empire*, 195.

59. See Henry Maguire, "Gardens and Parks in Constantinople," *Dumbarton Oaks Papers* 54 (2000): 251–64, at 259; see also Antony R. Littlewood, "Gardens of Byzantium," *Journal of Garden History* 12, no. 2 (1992): 126–53; and Antony R. Littlewood, ed., *The Progymnasmata of Ioannes Geometres* (Amsterdam: Hakkert, 1972).

60. Leslie Brubaker has similarly pointed out that the more accurate illustrations in the Vienna Dioscorides tend to be those from Anatolia and Thrace. See Brubaker, "Vienna Dioskorides," 207.

61. Olson, *Environment and Society in Byzantium*, 49–94.

62. On the history of representing and collecting *Aloe vera*, see Urs Eggli, Andrew Griebeler, Anastasia Stefanaki, Marie Cronier, and Louise Isager Ahl, "Flowers of *Aloe vera* from Medieval Manuscripts to Renaissance Printed Books" (in preparation).

63. Sadek, *Arabic Materia Medica*, 206.

64. Zohar Amar, Efraim Lev, and Yaron Serri, "On Ibn Juljul and the Meaning and Importance of the List of Medicinal Substances Not Mentioned by Dioscorides," *Royal Asiatic Society*, ser. 3, 24, no. 4 (2014): 529–55, with English translation; for Arabic text, German translation, and commentary, see Albert Dietrich, *Die Ergänzung Ibn Ğulğul's zur "Materia Medica" des Dioskurides* (Göttingen: Vandenhoeck and Ruprecht, 1993).

65. See Adam Gacek, "The Palaeographical and Codicological Features of the Osler Manuscript in the Context of the Manuscript Transmission of al-Ghāfiqī's *Herbal*," in *The Herbal of al-Ghāfiqī. A Facsimile Edition of MS 7508 in the Osler Library of the History of Medicine, McGill, with Critical Essays*, ed. F. Jamil Ragep and Faith Wallis (Montreal: McGill-Queen's University Press, 2014), 18–34. On the later manuscript, see Kerner, "The Illustrated *Herbal* of al-Ghāfiqī," 140–43.

66. Kerner, "Illustrated *Herbal*," 121–56; and Alain Touwaide, "Al-Ghāfiqī's *Kitāb fī l-adwiya al-mufrada*, Dioscorides' *De materia medica*, and Mediterranean Herbal Traditions," in *The Herbal of al-Ghāfiqī. A Facsimile Edition of MS 7508 in the Osler Library of the History of Medicine, McGill, with Critical Essays*, ed. F. Jamil Ragep and Faith Wallis (Montreal: McGill-Queen's University Press, 2014), 84–120.

67. Montreal, McGill University, Osler Library, MS 7508, fol. 73r.

68. As Andrew Dalby has noted, the absence of a classical Greek name for the eggplant led the eleventh-century Byzantine author Simeon Seth to call it "garden nightshade [*strychnos kēpeutos*]," that is, a different, but related plant. See Andrew Dalby, *Tastes of Byzantium: The Cuisine of a Legendary Empire* (London: I. B. Tauris, 2010), 75.

69. Ibn Juljul, *Maqaāla fī dhikr al-adwiya allatī lam yadhkurhā Diyasqūrīdus fī kitābihi*, 56; Amar, Lev, and Serri, "On Ibn Juljul," 553–54; and Dietrich, *Ergänzung*, 72–73.

70. Ibn Abī Uṣaybiʿa, *ʿUyūn al-anbāʾ fī ṭabaqāt al-aṭibbāʾ* 15.45.4, trans. N. Peter Joosse from *A Literary History of Medicine*, ed. Savage-Smith, Swain, and van Gelder, with some modification:

وكان يستصحب مصوراً ومعه الأصباغ والليق على اختلافها وتنوعها فكان يتوجه رشيد الدين بن الصوري إلى المواضع التي بها النبات مثل جبل لبنان وغيره من المواضع التي قد اختص كل منها بشيء من النبات فيشاهد النبات ويحققه ويريه للمصور فيعتبر لونه ومقدار ورقه وأغصانه وأصوله ويصور بحسبها ويجتهد في محاكاتها ثم إنه سلك أيضاً في تصوير النبات مسلكاً مفيداً وذلك أنه كان يري النبات للمصور في إبان نباته وطراوته فيصوره ثم يريه إياه أيضاً وقت كماله وظهور بزره فيصوره تلو ذلك ثم يريه إياه أيضاً في وقت ذواه ويبسه فيصوره فيكون الدواء الواحد يشاهده الناظر إليه في الكتاب و هو على أنحاء ما يمكن أن يراه في الأرض فيكون تحقيقه له أتم ومعرفيه له أبين.

71. Ibn Abī Uṣaybiʾa, *ʾUyūn al-anbāʾ fī ṭabaqāt al-aṭibbāʿ*, 4.1.11.1 (ed. Savage-Smith, Swain, van Gelder):

كان معتزلاً عن قومه متعلقاً بالجبل ومواضع النبات مقيماً بها في كل الأزمنة

72. See ʿAbd al-Laṭīf al-Baghdādī, *A Physician on the Nile and A Description of Egypt and Journal of the Famine Years*, ed. and trans. Tim MacKintosh-Smith (New York: New York University Press, 2021), xxiii.

73. Ibn Abī Uṣaybiʿa, *ʿUyūn al-anbāʾ fī ṭabaqāt al-aṭibbāʾ* 13.86, trans. Ignacio Sánchez from *A Literary History of Medicine*, ed. Savage-Smith, Swain, and van Gelder. See Zohar Amar and Yaron Serri, "Ibn al-Suri, Physician and Botanist of

al-Sham," *Palestine Exploration Quarterly* 135, no. 2 (2003): 124–30, at 126. Abū l-ʿAbbās al-Nabātī ("the herbalist") came from an area near Seville. The name "al-Jayyānī" refers to Jaen.

74. Ibn Abī Uṣaybiʿa, *ʿUyūn al-anbāʾ fī ṭabaqāt al-aṭibbāʾ* 14.58.2, trans. Franak Hilloowala and Emilie Savage-Smith from *A Literary History of Medicine*, ed. Savage-Smith, Swain, and van Gelder.

75. P. E. Pormann, "Medical Methodology and Hospital Practice: The Case of Fourth/Tenth-Century Baghdad," in *In the Age of al-Fārābī: Arabic Philosophy in the Fourth/Tenth Century* (London: Warburg Institute, 2008), 95–118.

76. Ragab, *Medieval Islamic Hospital*, 156–58.

77. Ibn Abī Uṣaybiʿa, *ʿUyūn al-anbāʾ fī ṭabaqāt al-aṭibbāʾ* 4.1.11.1.

78. André Miquel, *La géographie humaine du monde musulman jusqu'au milieu du 11e siècle. Géographie et géographie humaine dans la littérature arabe des origins à 1050* (Paris: Mouton, 1967), 267–330.

79. ʿAbd al-Laṭīf al-Baghdādī, *A Physician on the Nile*, 2–3.

80. These excerpts are preserved in a treatise written by ʿAlī ibn ʿAbd al-ʿAzīm al-Anṣārī in 1268, now in a manuscript in Bethesda, MD, National Library of Medicine, MS A 64. See Amar and Serri, "Ibn al-Suri."

81. Ibn Abī Uṣaybiʿa, *ʿUyūn al-anbāʾ fī ṭabaqāt al-aṭibbāʾ* 15.45.2. Trans. Joosse, *A Literary History of Medicine*, ed. Savage-Smith, Swain, and van Gelder.

82. Istanbul, Topkapi, MS Sultanahmet III 2127 (the "Topkapi Dioscorides").

83. On the identification of the text as a redaction (*taḥrīr*) and its connection to the Parchment Arabic Dioscorides, and of both to Paris ar. 2849, see Saliba and Komaroff, "Illustrated Books," 15.

84. Ettinghausen, *Arab Painting*, 70; Topkapi Dioscorides, fol. 252v.

85. See chapter 5.

86. Topkapi Dioscorides, fol. 96v; Toresella, "Il Dioscoride di Istanbul," 33–34.

87. Parchment Arabic Dioscorides, fol. 44r. The folio is currently mutilated, so only two of the garlic plants are visible.

88. Topkapi Dioscorides, fols. 143v and 144v. On these nature prints, see Roderick Cave, *Impressions of Nature: A History of Nature Printing* (New York: Mark Batty, 2010), 19 and 21.

89. Sergio Toresella has suggested that they represent parsley and coriander: Toresella, "Il Dioscoride," 35, 39. But the Arabic *karafs*, a loanword from Syriac, refers to parsley, celery, and other related plants, such as Alexanders and Angelica, but not coriander. The picture on fol. 143v likely illustrates *karafs al-bustānī* or cultivated celery, *Apium graveolens* L. The name below the illustration, *karafs jabalī*, is a calque of the Greek *oreoselinon*, i.e., either Macedonian, or mountain parsley, *Athamanta macedonica* (L.) Spreng. and *Peucedanum oreoselinum* Moench, respectively. See Marwān ibn Janāḥ, *Kitāb al-Talkhīṣ (On the Nomenclature of Medicinal Drugs)* in Gerrit Bos, Fabian Käs, Mailyn Lübke, and Guido Mensching, trans. and eds., *Marwān ibn Janāḥ, On the Nomenclature of Medicinal Drugs (Kitāb al-Talkhīṣ): Edition, Translation and Commentary with Special Reference to the Ibero-Romance Terminology*, 2 vols. (Leiden: Brill, 2020), nos. 152 [vol. 1:342–44], 447 [1:616–17], and 646 [2:807]; see also Albert Dietrich, *Die Dioskurides-Erklärung des Ibn al-Baiṭār: Ein Beitrag zur arabischen Pflanzensynonymik des Mittelalters* (Göttingen: Vandenhoeck and Ruprecht, 1991), 176–78.

The Arabic name *karafs al-ṣakhra* is similarly a calque of the Greek *petroselinon* and could refer to cultivated parsley, *Petroselinum crispum* (Mill.) Fuss. The name in the entry below the picture on fol. 143v, *karafs kabīr*, or "big *karafs*," seems to take the place of *karafs al-ʿaẓīm*, literally "huge *karafs*," in other Arabic herbals. This can be identified as Alexanders (*Smyrnium olusatrum* L.; see Dietrich, *Dioskurides-Erklärung*, 177–78; and Mohamed Nazir Sankary, *The Cilician Dioscorides' Plant Materia Medica as Appeared in Ibn al-Baitar, the Arab Herbalist of the 13th Century* [Aleppo: Aleppo University Publications, 1989], 93), or *Angelica* (see Nawal Nasrallah, trans. and ed., *Annals of the Caliphs' Kitchens: Ibn Sayyār al-Warrāq's Tenth Century Baghdadi Cookbook* [Leiden: Brill, 2007], 657).

90. I thank Heba Mostafa for our engaging conversations on *karafs*. On *karafs*, see, e.g., Dietrich, *Dioscurides Triumphans*, 2:412–14. See also Ibn al-Bayṭār, e.g., Dietrich, *Dioskurides-Erklärung*, 177–78; as well as Nasrallah, *Annals of the Caliphs' Kitchens*, 656–58. On the Syriac origin of the word *karafs*, see Bos et al., trans. and eds., *Marwān ibn Janāḥ, On the Nomenclature of Medicinal Drugs*, 1:344, citing Immanuel Löw, *Aramäische Pflanzennamen* (Hildesheim: Georg Olms Verlag, 1973), 222, no. 167; and Carl Brockelmann, *Lexicon Syriacum* (Eugene: Wipf and Stock, 2004), 349a.

91. Albert Atkin identifies five claims that Charles Sanders Peirce makes in identifying an index: "1) Indices use some physical contiguity with their object to direct attention to that object. 2) Indices have their characteristics independently of interpretation. 3) Indices refer to individuals. 4) Indices assert nothing. 5) Indices do not resemble, nor do they share any lawlike relation with, their objects." Nature prints, when used in the context of botanical illustration, cannot satisfy all of these criteria. See Albert Atkin, "Peirce on the Index and Indexical Reference," *Transactions of the Charles S. Peirce Society* 41, no. 1 (2005): 161–88, at 163–64.

92. Toresella, "Il dioscoride," 34.

93. On block printing, see Mark Muehlhaeusler, "Eight Arabic Block Prints from the Collection of Aziz S. Atiya," *Arabica* 55, nos. 5–6 (2008): 528–82; Richard W. Bulliet, "Medieval Arabic Tarsh: A Forgotten Chapter in the History of Printing," *Journal of the American Oriental Society* 107, no. 3 (1987): 427–38; Richard W. Bulliet, "Printing in the Medieval Islamic Underworld," *Columbia Library Columns* 36 (1987): 13–20; Karl R. Schaefer, "Eleven Medieval Arabic Block Prints in the Cambridge University Library," *Arabica* 48 (2001): 210–39; and Karl R. Schaefer, *Enigmatic Charms: Medieval Arabic Block Printed Amulets in American and European Libraries and Museums* (Leiden: Brill, 2006). On Hajj certificates, see Şule Aksoy and Rachel Milstein, "A Collection of Thirteenth-Century Illustrated Hajj Certificates," in *M. Uğur Derman armağani: altmışbeşinci yaşı münasebetiyle sunulmuş tebliğler / M. Uğur Derman Festschrift: Papers Presented on the Occasion of His Sixty-fifth Birthday*, ed. Irvin Cemil Schick (Istanbul: Sabancı Üniversitesi, 2000), 101–34.

94. E.g., an even earlier example of nature printing can be found in the negative leaf print in the dado zone of a wall from Room F in the Villa of P. Fannius Synistor at Boscoreale now in the Metropolitan Museum of Art, New York.

95. Ibn al-Bayṭār notes that aloe was commonly grown in homes. See Dietrich, *Die Dioskurides-Erklärung*, 3.22, pp. 159–60.

96. Ettinghausen, *Arab Painting*, 67; Sadek, *Arabic Materia Medica*, 47.

97. 'Abd al-Laṭīf al-Baghdādī, *A Physician on the Nile*, 62–105.

98. E.g., Suzan Yalman, "'Ala Al-Din Kayqubad Illuminated: A Rum Seljuq Sultan as Cosmic Ruler," *Muqarnas* 29 (2012): 151–86; Jill Meredith, "The Arch at Capua: The Strategic Use of Spolia and References to the Antique," in *Intellectual Life at the Court of Frederick II Hohenstaufen* (Washington DC: National Gallery of Art, 1994), 109–25.

99. Collins, *Medieval Herbals*, 256.

100. Givens, *Observation and Image-Making*, 73–74, 148–50; and Giulia Orofino, "Il rapporto con l'antico e l'osservazione della natura nell'illustrazione scientifica di età sveva in Italia meridionale," in *Intellectual Life at the Court of Frederick II Hohenstaufen* (Washington DC: National Gallery of Art, 1994), 129–50. The famous Manfred codex copy contains many illustrations of varying accuracy; see Vatican, Biblioteca Apostolica Vaticana, MS Pal. lat. 1071.

Chapter 8

1. Biblioteca Apostolica Vaticana, gr. 284, here: fol. 150r. See Touwaide, "Un recueil grec." On the original date of the manuscript, see Jean Irigoin, "Pour une étude de centres de copie byzantins, II. Quelques groupes de manuscrits," *Scriptorium* 13 (1959): 177–209, at 190–95.

2. Compare Touwaide, "Un recueil grec," 46–56; and Collins, *Medieval Herbals*, 70–71.

3. Milan, Biblioteca Ambrosiana, MS A 95 sup. ("the Ambrosiana Notebook"); and Venice, Biblioteca Nazionale Marciana, Cod. gr. XI, 2121 = coll. 453 ("the Marciana Handbook"). See Marie Cronier and Patrick Gautier Dalché, "A Map of Cyprus in Two Fourteenth-Century Byzantine Manuscripts," *Imago Mundi* 69, no. 2 (2017): 176–87. On the Ambrosiana Notebook, see Emilio Martini and Domenico Bassi, *Catalogus codicum Graecorum Bibliothecae Ambrosianae* (Milan: Hoepli, 1906), 1:23–28 (no. 24). On the Marciana Handbook, see Elpidio Mioni, *Bibliothecae Divi Marci Venetiarum codices Graeci manuscripti*, vol. 3: *Codices qui in classes IX, X, XI inclusos et supplementa duo continens* (Rome: Istituto poligrafo dello stato, 1972), 112–15. See also Touwaide, "Un recueil grec," 46, 52; and Collins, *Medieval Herbals*, 75.

4. At around 142 × 100 mm with 166 extant folios, the Marciana Handbook is a small, even hand-sized, cleaned-up copy of the Ambrosiana Notebook, which, at about 233 × 153 mm, is roughly two times larger and fourteen folios longer (180 folios total). Though larger, and probably the model for the Marciana codex, the Ambrosiana codex is likely a notebook: it is heterogeneous, with irregular formatting, different inks, and at least six different papers (based on watermarks), some of it clearly reused. On the designation of a heterogeneous notebook, see Peter Gumbert, "Codicological Units: Towards a Terminology for the Stratigraphy of the Non-homogeneous Codex," in *Il codice miscellaneo. Tipologie e funzioni. Atti del Convegno internazionale, Cassino 14–17 maggio 2003*, ed. Edoardo Crisci and Oronzo Pecere, *Segno e testo* 2 (2004): 17–42, at 18. Touwaide also calls the codex a notebook; see Touwaide, "Development," 196. By contrast, the Marciana Handbook is homogeneous. It has consistent formatting and rubrication, headpieces, and initials throughout. It has the bookplate of Bernardo Nani (1712–61).

5. These include an astrological text, brief works on weights and measures, a text on prognostics, a botanical lexicon, various prescriptions and remedies, a work on food and nutrition, antidotes, a lapidary, and excerpts of medical authors, Paul of Aegina, and Hippocrates, as well as *De opificio hominis*. Marie Cronier and Patrick Gautier Dalché think an intermediary unlikely. See Cronier and Dalché, "A Map of Cyprus," 177.

6. The plants here are labeled: ἀείζωον τὸ μίκρον, ἀείζωον τὸ λεπτοφύλλον, ἄλλον (*sic*), fol. 19v.

7. Ambrosiana Notebook, fol. 14r: τοῦ πρώτου σχήματος τῇ μιμήσει τῇ κατὰ πρόθεσιν—can be sufficient for an unskilled observer (τὸν ἄπειρον) to conjecture at its aim (διὰ τῶν φαινομένων τοῦ σκοποῦ τῆς τέχνης καταστοχάσασθαι).

8. Fol. 21r: ἀρτεμισία ἑτέρα πολύκλωνος: πρασῖνο ερά[νο]| ἠδὲ ρίζα ὀξό|χρους.

9. Ambrosiana Notebook, fol. 95v, at ἄλλον, i.e., ἀλόη: πρασινόχρουν μέχρι τῆς ριζ[ης] π . . . [ὁ]ξόχρουν ἡ ρί|ζα ὥσ|περ τρυ|γία | οἴνου. This longer note was abbreviated when copied into the Marciana Handbook (fol. 95v): Πρασινό . . . τῆς ρίζης π.

10. See Cronier and Gautier Dalché, "Map of Cyprus," 176–87.

11. This note appears on fol. 50r.

12. Also on fol. 50r.

13. The note including μαϊστρο τζιαννε appears on fol. 179v.

14. Cronier and Dalché, "A Map of Cyprus," 185n8.

15. Cronier and Dalché, "A Map of Cyprus," 176–87.

16. Cronier and Dalché, "A Map of Cyprus," 180.

17. See Barbara Zipser, *John the Physician's Therapeutics: A Medical Handbook in Vernacular Greek* (Leiden: Brill, 2009).

18. Alain Touwaide, "The 'Letter . . . to a Cypriot Physician,' Attributed to Johannes Argyropoulos (ca. 1448–1453)," *Medicina nei secoli arte e scienza* 11, no. 3 (1999): 585–601.

19. Munich, Bayerische Staatsbibliothek, Cod. Clm. 337, the "Munich Dioscorides." See Collins, *Medieval Herbals*, 149–54.

20. Collins, *Medieval Herbals*, 52. See also Salvatore Lilla, "Studio del Codice: A Study of the Manuscript," in *Dioscurides Neapolitanus: Biblioteca Nazionale di Napoli Codex ex Vindobonensis Graecus 1*, Commentarium, ed. Carlo Bertelli, Salvatore Lilla, and Giulia Orofino (Rome: Salerno editrice, 1992), 49–82, especially 72–74, on Latin hands, and 76–79, on later provenance.

21. Collins, *Medieval Herbals*, 168–79; and Grape-Albers, *Spätantike Bilder*, 7–14.

22. On Salerno, see Kristeller, *Studi sulla scuola*; Cuna, *Per una bibliografia della scuola medica salernitana*. More recently, see the contributions in *La scuola medica salernitana*, ed. Jacquart and Bagliani; and *Salerno nel XII secolo*, ed. Delogu and Peduto.

23. See Alain Touwaide, "Botany and Humanism in the Renaissance: Background, Interaction, Contradictions," in *The Art of Natural History: Illustrated Treatises and Botanical Paintings, 1400–1850*, ed. Therese O'Malley and Amy R. W. Meyers, Studies in the History of Art 69 (Washington, DC: National Gallery of Art, 2008), 32–61.

24. Gerstinger, *Dioscurides*, 3; and Otto Mazal, *Der Wiener Dioskurides*, 1:10.

25. Early Southern Gothic *rotunda* Latin inscriptions occur in at least two

hands on fols. 13r–v, 14r, 15r, 16v, 17v, 18v, 20r, 20v, 21v, 22v, 23v, 24v, 26v, 27v. Moreover, some of the ancient Latin synonyms were transliterated from Greek into Roman letters by a later Byzantine hand.

26. Copenhagen, Kongelige Bibliotek, MS Thott 190 4° (the "Greek Alphabetical Dioscorides"). On this manuscript, see Touwaide, "Latin Crusaders."

27. Touwaide compared the Thott manuscript to the Morgan and Vienna manuscripts but not to the Naples Dioscorides. In some crucial ways, the Thott manuscript follows the Morgan Dioscorides and the Naples Dioscorides when it comes to the *bouglosson* (*lingua bovina*, fol. 30r) as well as *nymphaia* (*ninofar mayor*, fol. 51r). The rose (*rosyer*, fol. 57v) is also closer to that in the Naples and Morgan Dioscorides manuscripts. But the Thott manuscript also includes an illustration of *bouglosson* (also labeled *lingua bovina*, fol. 83r) that is related to that found in the Vienna Dioscorides.

28. While the term "Western" risks flattening important distinctions between regions and schools of painting in Latinate Europe, the designation highlights broader stylistic differences between the Thott Codex and Byzantine and Arabic manuscripts.

29. Touwaide, "Latin Crusaders," 28.

30. On the emergence of the codex, see Touwaide, "Latin Crusaders," 44, 49.

31. Premerstein, "De codicis Dioscuridei," 54; Gerstinger, *Dioscurides*, 3; and Mazal, *Wiener Dioskurides*, 1:10; Touwaide, "Latin Crusaders," 27 and 45.

32. Minta Collins reasonably suggests that the manuscript need not have fallen into Crusader hands in order to acquire Latin annotations. See Collins, *Medieval Herbals*, 97n73. See also Elizabeth Fisher, "Monks, Monasteries and the Latin Language in Constantinople," in *Change in the Byzantine World in the Twelfth and Thirteenth Centuries*, ed. Ayla Ödekan, Engin Akyürek, and Nevra Necipoglu (Istanbul: Vehbi Koç Foundation, 2010), 390–95.

33. See "William of Moerbeke," in *The Oxford Dictionary of the Christian Church*, ed. Elizabeth A. Livingstone and Frank Leslie Cross (Oxford: Oxford University Press, 2005). For a list of translations and editions, see *Guillaume de Moerbeke. Recueil d'études à l'occasion du 700e anniversaire de sa mort (1286)*, ed. Jozef Brams and Willy Vanhamel (Leuven: Leuven University Press, 1989), 301–83.

34. Marshall Clagett, "William of Moerbeke: Translator of Archimedes," *Proceedings of the American Philosophical Society* 126, no. 5 (1982): 356–66, at 359. See also Martin Grabmann, *Guglielmo di Moerbeke, O.P., il traduttore delle opere di Aristotele*, Miscellanea historiae pontificiae 11, Collectionis totius 20 (Rome: Pontificia Università Gregoriana, 1946).

35. Fisher, "Monks," 393; Marie-Hélène Congourdeau, "Note sur les Dominicains de Constantinople au début du 14e siècle," *Revue des études byzantines* 45 (1987): 175–81; and Marie-Hélène Congourdeau, "Frère Simon le Constantinopolitain, O.P. (1235?–1325?)," *Revue des études byzantines* 45 (1987): 165–74, at 167–69.

36. On Pietro d'Abano, see Sante Ferrari, "Per la biografia e per gli scritti di Pietro d'Abano," *Atti della Reale accademia dei Lincei, anno CCCXII* 15 (1915): 629–725; Leo Norpoth, "Zur Bio-Bibliographie und Wissenschaftslehre des Pietro d'Abano, Mediziners, Philosophen, und Astronomen in Padua," *Kyklos* 3 (1930): 291–353; Lynn Thorndike, "Manuscripts of the Writings of Peter of Abano," *Bulletin of the History of Medicine* 15 (1944): 201–19.

37. J. Thomann, "Pietro D'Abano on Giotto," *Journal of the Warburg and Courtauld Institutes* 54 (1991): 238–44; Eva Frojmovič, "Giotto's Circumspection," *Art Bulletin* 89, no. 2 (2007): 195–210.

38. Paris, Bibliothèque nationale de France, MS gr. 54. See Kathleen Maxwell, *Between Constantinople and Rome: An Illuminated Byzantine Gospel Book (Paris gr. 54) and the Union of Churches* (London: Routledge, 2014), 145–74.

39. Maxwell, *Between Constantinople and Rome*, 175–216.

40. Fisher, "Monks," 394; and Elizabeth Fisher, "Manuel Holobolos, Alfred Sareshal, and the Greek Translator of ps.-Aristotle's *De plantis*," *Classica et mediaevalia* 57 (2006): 189–211.

41. Venice, Biblioteca Nazionale Marciana, Cod. gr. Z 516 [=904]. See Francesco Lovino, "Un miniatore nella bottega degli Astrapas? Alcune osservazioni attorno alle immagini del Tolomeo Marciano gr. Z. 516 (904)," *Hortus artium medievalium* 22 (2016): 384–98, at 397.

42. Givens, *Observation and Image-Making*, 73–74, 148–50. Frederick II both experimented with classical imagery and undertook quasi-scientific experiments, as reported in his *De arte venandi cum avibus*. A famous example of this text in an illustrated manuscript is Vatican, Biblioteca Apostolica Vaticana, MS Pal. lat. 1071.

43. London, British Library, MS Egerton 2020. On the Carrara Herbal, most recently, see Kyle, *Medicine and Humanism*.

44. Venice, Biblioteca Marciana, Cod. lat. VI, 59=2548.

45. London, British Library, Add. Ms. 41623.

46. Annaclara Cataldi Palau, "Learning Greek in Fifteenth-Century Constaninople," in *Studies in Greek Manuscripts* (Spoleto: Centro italiano di studi sull'alto medioevo, 2008), 1:219–34.

47. *Ambrosii Traversarii generalis Camaldulensium aliorumque ad ipsum epistulas*, ed. L. Mehus (Florence: Ex typographio Caesareo, 1759), lib. 24, ep. 58, col. 1033. Giovanni Aurispa, *Carteggio di Giovanni Aurispa*, ed. Remigio Sabbadini (Rome: Tipografia del senato, 1931), 67–68: "mirae antiquitatis in quo depictae sunt et herbae et radices et quaedam animalia."

48. See *Giovanni Tortelli, "Della medicina e die medici." Gian Giacomo Bartolotti, "Dell'antica medicina." Due storie della medicina del XV secolo*, ed. and trans. Luigi Belloni and Dorothy M. Schullian (Milan: Industrie grafiche italiane stucchi, 1954), 14, and translation here adapted from translation given at 85: "Vidi ego Constantinopoli eiusdem auctoris [i.e., Dioscoridis] codicem litteris graecis antiquissimisque exaratum, in quo non solum herbarum effigies, sed volatilium, quadrupedum et reptilium tanto artificio et proprietate effictae erant, quanto natura ipsa, ut puto, producere potuit."

49. These include Paris, Bibliothèque nationale de France, MS gr. 2180; Salamanca, University Library, MS 2659; Cambridge, University Library, MS Ee. 5.7; Milan, Biblioteca Ambrosiana, MS C 102 sup.; and Vienna, Österreichische Nationalbibliothek, Cod. 2277. On these manuscripts, see Francesca Marchetti, "La trasmissione delle illustrazioni del Dioscoride di Vienna negli anni intorno alla caduta di Costantinopoli (Cod. Banks Coll. Dio. 1, Natural History Museum, Londra; Ee. V. 7, Cambridge University Library, Cambridge; e C 102 sup., Biblioteca Ambrosiana, Milano)," *Jahrbuch der Österreichischen Byzantinistik* 66 (2016): 153–78. See also Collins, *Medieval Herbals*, 82–84; Pächt, "Die

früheste abendländische Kopie," 201–14; Brigitte Mondrain, "Lettrés et copistes à Corfou au XVᵉ et au XVIᵉ siècle," in *Puer Apuliae. Mélanges offerts à Jean-Marie Martin, Centre de recherche d'histoire et civilisation de Byzance*, ed. E. Cuozzo (Paris: ACHCByz, 2008), 463–76; Alain Touwaide, "Une note sur la thériaque attribuée à Galien," *Byzantion* 67 (1997): 439–82; and Touwaide, "The Salamanca Dioscorides (Salamanca, University Library, 2659)," *Erytheia. Revista de estudios bizantinos y neogriegos* 24 (2003): 125–58; and Teresa Martínez Manzano, "De Corfú a Venecia. El itinerario primero del *Dioscórides* de Salamanca," *Medioevo greco* 12 (2012): 133–54.

50. Niccolò Leoniceno, *De Plinii et aliorum medicorum in medicina erroribus* (1492). On Renaissance critiques of Pliny, see Brian Ogilvie, *The Science of Describing: Natural History in Renaissance Europe* (Chicago: University of Chicago Press, 2006), 122–33; Karen Reeds, "Renaissance Humanism and Botany," *Annals of Science* 33 (1976): 519–42, at 523–24.

51. Ogilvie, *Science of Describing*, 30–33. Lynn Thorndike, *A History of Magic and Experimental Science* (New York: Columbia University Press, 1934), 4:593–610.

52. Ermolao Barbaro, *Castigationes plinianae* (Venice: D. Barbarus, 1493).

53. Reeds, "Renaissance Humanism," 525, see especially note 23.

54. Pietro Andrea Mattioli, *Commentarii in libros sex Pedacii Dioscorides Anazarbei, de materia medica* (Venice: Apud Vicentium Valgrisium, 1554).

55. London, Natural History Museum, MS Banks Coll. Dio. 1.

56. Marchetti, "La trasmissioni," 155. She compares the watermark to Harlfinger Ciseaux 30. In a letter now pasted to the inside cover of the codex, however, Robert Farquharson Sharp dates the watermarks anywhere from 1458 to 1477.

57. Andrew Griebeler, "*Aeizōon to amaranton*," 20.

58. E.g., fols. 25v, 344v, 370v, 385v, 391v.

59. Fol. 25v: περὶ τούτου παρὰ πλυνίῳ ἐν τῷ Κ ὅπου περὶ στρύχνου φαρμάκων ἁλικάκαβον ὀνομάζει καὶ ἁλικάλλιον. ἡμεῖς δὲ vesicariam. A second note switches between Greek and Latin mid-sentence. See fol. 391v: καμαιδαφνην πλυνη[ος] vincampervincam nominat ἤ φανεται ἄλλη τις εἶναι· ἐν τῷ Κ. The word καμαιδαφνην is normally spelled χαμαιδάφνην. That a *kappa* is used instead of a *chi* could indicate that the author of the notes is a speaker of a language lacking the /χ/ sound. I thank Maria Mavroudi for this observation.

60. This training is especially evident in the depiction of Poseidon on fol. 399r; see Griebeler, "Aeizōon," 24, 26.

61. Marchetti, "Trasmissione," 156.

62. Fol. 1r, ἀείζωον τὸ ἀμάραντον. For more on this illustration, see Griebeler, "*Aeizōon to amaranton*." (Note the name change from *Sedum* to *Petrosedum*.) The plant is either *Petrosedum rupestre* (L.) P. V. Heath or *Petrosedum sediforme* (Jacq.) Grulich. I thank Urs Eggli for our engaging correspondence on succulents and for pointing out the change of *Sedum* to *Petrosedum*. I also thank Andrew S. Doran and Dean G. Kelch at the University and Jepson Herbaria at the University of California, Berkeley, for their help in identifying this picture. On sedums in the Mediterranean, generally, see Henk 't Hart, *Sedums of Europe—Stonecrops and Wallpeppers*, ed. Urs Eggli (Lisse: A. A. Balkema, 2003).

63. The name only appears in late Byzantine botanical lexica. On Greek botanical lexica, see Delatte, "Glossaires de botanique," 277–454; and Thomson,

Textes grecs inédits, especially 125–77. More recently, see Touwaide, "Lexica medico-botanica byzantina." See also Jerry Stannard, "Byzantine Botanical Lexicography," *Episteme* 5 (1971): 168–87.

64. I thank Henrike Lange and C. Jean Campbell for our conversations about Italian style.

65. E.g., Paris, Bibliothèque nationale de France, MS gr. 2224, fols. 70–71. Alternatively, the word *amaranton* may have resonated with the commissioner or recipient of the book. It appears in a Marian epithet and icon-type *to rhodon to amaranton* (τὸ ῥόδον τὸ ἀμάραντον), ultimately descended from the epithets given in the Akathist hymn. See Dēmḗtrios I. Pallas, "Η Θεοτόκος Ρόδον τό Ἀμάραντον. Εἰκονογραφική ἀνάλυση καὶ καταγωγή τοῦ τύπου," *Archaiologikón Deltíon* 26 (1971): 225–38. I thank Anna Kartsonis for this recommendation.

66. Marie Cronier, "La production de manuscrits scientifiques dans l'atelier de Michel Apostolis. L'exemple du *De materia medica* de Dioscoride," in *The Legacy of Bernard de Montfaucon: Three Hundred Years of Studies on Greek Handwriting: Proceedings of the Seventh International Colloquium of Greek Palaeography (Madrid—Salamanca, 15–20 September 2008)*, ed. Antonio Bravo García, Inmaculada Pérez Martín, and Juan Signes Codoñer (Turnhout: Brepols, 2010), 1:463–72. On the life of Michael Apostolis, see Deno John Geanakoplos, *Greek Scholars in Venice: Studies in the Dissemination of Greek Learning from Byzantium to Western Europe* (Cambridge: Cambridge University Press, 1962), 73–110.

67. Mondrain, "Lettrés et copistes à Corfou," 463–76; see also Brigitte Mondrain, "Les manuscrits grecs de médecine," in *Colloque. La médecine grecque antique. Actes, Cahiers de la villa Kérylos 15*, ed. J. Jouanna (Paris: De Boccard, 2004), 267–85.

68. In 1481, the scribe Demetrios Tribolis signed his name into Paris, Bibliothèque nationale de France, MS gr. 2182, an apograph of Paris, Bibliothèque nationale de France, MS gr. 2183. Both codices, together with Paris, Bibliothèque nationale de France, MS gr. 2286, would eventually become part of a cache of manuscripts that Antonio Eparchos, the grandson of Andronikos Eparchos, sold to François I in 1538. Teresa Martínez Manzano has further argued, largely on the basis of shared watermarks, that during the same year that Demetrios Tribolis copied Paris, Bibliothèque nationale de France, MS gr. 2183, it was also copied by John Moschos into yet another codex, now Salamanca, University Library, MS 2659. Mondrain, "Lettrés et copistes à Corfou," 472–62; Touwaide, "Une note," 451–52; Touwaide, "Salamanca Dioscorides"; Marie Cronier, "Comment Dioscoride est-il arrivé en Occident? A propos d'un manuscrit byzantin, de Constantinople à Fontainebleau," *Néa Ῥώμη* 10 (2013): 185–209; Martínez Manzano, "De Corfú a Venecia," especially 140. Teresa Martínez Manzano argues against Touwaide's hypothesis that the Salamanca Dioscorides was produced in Venice.

69. According to Touwaide, the Corfiot manuscript Salamanca, University Library, MS 2659, served as the model for Vatican, Biblioteca Apostolica Vaticana, MS Pal. gr. 48, which was in turn the model for the Aldine, "without being . . . the editor's manuscript itself." See Touwaide, "Salamanca Dioscorides," 128.

70. According to Touwaide, Demetrios Moschos copied Paris, Bibliothèque nationale de France, MS gr. 2183 around 1480 into Milan, Biblioteca Ambrosiana, L 119 sup., which was used by Ermolao Barbaro. See Touwaide, "Salamanca Dioscorides," 128.

71. Vienna, Österreichische Nationalbibliothek, cod. 2277, the "Vienna Atlas." See Pächt, "Die früheste abendländische Kopie," at 205. For an overview of the Vienna Atlas, including contents, see Petra Hudler, "Die Pflanzenbilder in den Codices 187 und 2277 der Österreichischen Nationalbibliothek," *Codices manuscripti* 66 (2008): 1–54, especially 25–43.

72. Three are only partly colored paintings, and four remain as underdrawings without any color.

73. For a recent study of varieties of *Cytinus*, see Clara de Vega, Regina Berjano, Montserrat Arista, Pedro L. Ortiz, Salvador Talavera, and Tod F. Stuessy, "Genetic Races Associated with the Genera and Sections of Host Species in the Holoparasitic Plant Cytinus (Cytinaceae) in the Western Mediterranean Basin," *New Phytologist* 178, no. 4 (2008): 875–87.

74. Vienna Atlas, fol. 36r. I thank Clara de Vega for answering my questions about *Cistus* and *Cytinus*.

75. On the identification of the hand, see Mantuani, "Die Miniaturen," 479.

76. *Inepta icon.*

77. The first is Cambridge, University Library, Ee 5.7 (the "Cambridge Dioscorides"), and the second is Milan, Biblioteca Ambrosiana MS C 102 sup. On these manuscripts, see Marchetti, "Trasmissione."

78. Marchetti, "Trasmissione," 159, 164. On Manuel Grēgoropoulos, see Stephanos Kaklamanis and Stelios Lampakis, eds., Μανουήλ Γρηγορόπουλος, νοτάριος Χάνδακα 1506–1532: Διαθήκες, απογραφές, εκτιμήσεις (Iraklion: Vikelaia Dēmotikē Vivliothēkē, 2003).

79. Bologna, Biblioteca Universitaria, Cod. 3632. See Marchetti, "Trasmissione," 165.

80. For example, on fol. 56r (fol. 61r in Marchetti, "Trasmissione," 169), the Hebrew note for the plant darnel (*aira*) reads: *eira r(ozeh) l(omar) lolium min meḥitah,* that is, "*aira* means to say *lolium,* a kind of wheat." I thank Maʾayan Sela and Philip Hollander for helping me with these notes.

81. See Stefan C. Reif, *Hebrew Manuscripts at Cambridge University Library: A Description and Introduction* (Cambridge: Cambridge University Press, 1997), 356.

82. Doukas, *Historia Turcobyzantina,* chap. 39, sec. 15, in *Decline and Fall of Byzantium to the Ottoman Turks by Doukas,* trans. Harry Magoulias (Detroit: Wayne State University Press, 1975), 225.

83. See Théoharis Stavrides, *The Sultan of Vezirs: The Life and Times of the Ottoman Grand Vezir Mahmud Pasha Angelović (1453–1474)* (Leiden: Brill, 2001), 92–93. According to Janin, however, Pierre Gilles observes that the monastery had largely fallen into ruin around 1537–40. I have not been able to confirm this report. Raymond Janin, *La géographie ecclésiastique de l'empire byzantin,* part. 1, vol. 3, *Les églises et les monastères* (Paris: Institut français d'études byzantines, 1953), 439.

84. On Moses Hamon, see Uriel Heyd, "Moses Hamon, Chief Jewish Physician to Sultan Süleymān the Magnificent," *Oriens* 16 (1963): 152–70, at 164–65. See also Henri Gross, "La famille juive des Hamon," *Revue des études juives* 56 (1908): 1–26, and 57 (1909): 55–78.

85. Some scholars have thought that the Hebrew notes on fols. 1v and 2r apparently say *Mosheh ben Mosheh* (Moses, son of Moses) and indicate ownership. Uriel Heyd has, however, argued that the inscription is to be read as *Mazzeh ben*

Mazzeh, with *mazzeh* designating "sprinkling," which dictionaries of Talmudic Hebrew define as the priest who purifies by sprinkling, that is, by extension, the "holder of a high office, a man of distinction." See Heyd, "Moses Hamon," 168.

86. Heyd cites the Spanish *Viaje de Turquía* for this figure. See Heyd, "Moses Hamon," 166."

87. B. Lewis, "The Privilege Granted by Mehmed II to His Physician," *Bulletin of the School of Oriental and African Studies* 14 (1952): 550–63; Eleazar Birnbaum, "Hekim Ya'qub, Physician to Sultan Mehemmed the Conqueror," *Harofé Haivri* 34 (1961): 222–50.

88. On these earlier models, see chapter 5.

89. Russell, "Physicians at the Ottoman Court."

90. Russell, "Physicians at the Ottoman Court."

91. Cronier, "The Manuscript Tradition," 147–51.

92. Collins, *Medieval Herbals*, 279; and Vera Segre Rutz, *Il giardino magico degli alchimisti. Un erbario illustrato trecentesco della Biblioteca universitaria di Padova e su tradizione* (Milan: Edizioni il Polifilo, 2000).

93. See discussion in Dominic Olariu, "The Misfortune of Philippus de Lignamine's Herbal or New Research Perspectives in Herbal Illustrations from an Iconological Point of View," in *Early Modern Print Culture in Central Europe*, ed. Stefan Kierdroń and Anna-Maria Rim (Wrocław: Wydawnictwo Uniwersytetu Wrocławskiego, 2014), 39–62.

94. Olariu, "Misfortune of Philippus," 51–52; on the dating, see Vito Capialbi, *Notizie circa la vita, le opere, e le edizioni di Messer Giovan Filippo La Legname Cavaliere Messinese e Tipografo del secolo XV raccolte dal Conte Vito Capialbi Napoli* (Naples: Porcelli, 1853), 43.

95. Montecassino, Archivio della Badia, Cod. 97; F. W. T. Hunger, *The Herbal of Pseudo-Apuleius from the Ninth-Century Manuscript in the Abbey of Monte Cassino (Codex Casinensis 97) Together with the first printed edition of Joh. Phil. de Lignamine* (Leiden: E. J. Brill, 1835), xxxv.

96. Olariu, "Misfortune of Philippus," 53.

97. For example, see Karen M. Reeds, "Leonardo da Vinci and Botanical Illustration: Drawing, Nature Printing, and Printing circa 1500," in *Visualizing Medieval Medicine*, 205–37, at 236–37.

98. Reeds, "Renaissance Humanism," 527–39, especially 534.

99. Olariu, "Misfortune of Philippus," 54–56.

100. Kusukawa, *Picturing the Book*, 19.

101. The subtitle of Brunfels's work also emphasizes the revival of antique botany. See Reeds, "Renaissance Humanism," 520n2.

102. David Landau and Peter Parshall, *The Renaissance Print: 1470–1550* (New Haven: Yale University Press, 1994), 252; Kusukawa, *Picturing the Book*, 16–20.

103. Kusukawa, *Picturing the Book*, 13–18.

104. Kusukawa, *Picturing the Book*, 19.

105. Translation as given in Kusukawa, *Picturing the Book*, 111.

106. Kusukawa, *Picturing the Book*, 114–19.

107. Kusukawa, *Picturing the Book*, 109.

108. Kusukawa, *Picturing the Book*, 131.

109. Pächt, "Die früheste abendländische Kopie," 210: "Kurz, was die antike Herbarillustration bietet, ist in der Regel manipulierte, vordemonstrierte Natur,

eine auf Wiedererkennbarkeit und objektive Bestimmbarkeit gerichtete Empirie, nie der subjektive Eindruck des in spontaner Wahrnehmung Erschauten."

110. On aspects of manuscript culture in printed books, e.g., the survival of glossing and marginal annotations, see, for example, Byron Ellsworth Hamann, *Translations of Nebrija: Language, Culture, and Circulation in the Early Modern World* (Amherst: University of Massachusetts Press, 2015), 108–20.

111. Agnes Arber, *Herbals: Their Origin and Evolution* (Cambridge: Cambridge University Press, 1912), 74 and 190.

112. The picture appears to show a birdsfoot trefoil in the genus *Lotus*, such as *Lotus ornithopodioides* L. It could also be orange birdsfoot, *Ornithopus pinnatus* (Mill.) Druce. Note that the term *Coronopus* can be associated with other plants, too, such as *Plantago coronopus* L. and *Lepidium coronopus* (L.) Al-Shehbaz.

113. I thank Joshua Allbright for checking my translation. Rembert Dodoens, *Stirpium historiae pemptades sex sive libri triginta* (Antwerp: Ex officina Christophori Plantini, 1583), 110: "Quod si vero (et) fides iconi habenda, quae in Codice Caesareo reperitur, longissime differens Herba Stella a Coronopo est; Depingitur enim in hoc Coronopus tenuibus ac viticulosis cauliculis, (et) recurvis corvorum pedum digitos referentibus siliquis, ut ipsa figura ostendit. Caesareani autem veteris manuscripti codicis fidem nobis asservit doctissimus ac diligentissimus Bernardus Paludanus, qui huiuscemodi Coronopum, qualem icon exprimit, a se visum ac repertum haud longe a Tripoli (et) ad radices montis Libani nobis coram narravit."

114. Dodoens, *Stirpium historiae pemptades sex sive libri triginta*, 123: "Extat autem Stoebes spinosae imago in vetere Caesareae bibliothecae codice, quam hoc loco adiiciendam putauimus, ut vel ex hac cognoscatur, quam multum Stoebe (et) Scabiosa different. Atque huiuscemodi spinosam Stoeben nostra aetate reperiri est in Cypro atque Peloponneso, vulgo Morea, planis ac apricis locis: unde aridam secum attulit doctissimus Bernardus Paludanus stirpium diligentissimus observator." Translation: "Moreover, an image of spiny Stoebe stands out in the old codex from the imperial library, which we thought should be put here, so that from this it can be seen how much the Stoebe and Scabiosa differ. This kind of spiny Stoebe can be found today in Cyprus and the Peloponnesus, commonly called Morea, in sunny and flat places. The most learned Bernard Paludanus, a very diligent observer of plants, has brought with him a dried plant from here." I thank Joshua Allbright for his suggestions on this translation.

115. Italics in the original. Bruno Latour, "Visualization and Cognition: Drawing Things Together," in *Knowledge and Society Studies in the Sociology of Culture Past and Present*, ed. H. Kuklick, vol. 6 (Greenwich, CT: Jai Press, 1979), 1–40, at 7. I thank Annabel Wharton for recommending this essay.

116. Italics in original. Latour, "Visualization and Cognition," 7.

117. John Sibthorp to John Hawkins, quoted in Hans Walter Lack and David J. Mabberley, *The Flora Graeca Story: Sibthorp, Bauer, and Hawkins in the Levant* (Oxford: Oxford University Press, 1999), 33.

118. On this project, see Hans Walter Lack, "Die Kupferstiche von frühbyzantinischen Pflanzenabbildungen im Besitz von Linné, Sibthorp und Kollár," *Annalen des Naturhistorischen Museums in Wien. Serie B für Botanik und Zoologie* 100 (1998): 613–55.

119. Jacquin had written to Carl von Linné to tell him of the ancient

illustrations. See Linné's response to Jacquin, written in Uppsala, January 28, 1763, quoted in Lack, "Die Kupferstiche," 649: "Stupefactus legi, quas narras de antiquissimo auctore Botanico cum figuris, adjectis nominibus graecis. Non capio quomodo mei oculi in hac mortalitate hunc aureum viderent librum." And see letter of Linné to Jacquin, written in Uppsala, April 13, 1763, quoted in Lack, "Die Kupferstiche," 649: "Quam antiquum crederes Mss. Dioscoridis vestrum esse ? an scriptum sit literis graecis uncialibus s. initialibus ? . . . Exspecto avidissimo videre figuras antiquissimas Dioscoridis."

120. The letter read aloud to the Linnean Society in London on November 1, 1808. See James Edward Smith, "Remarks on the *Sedum ochroleucum*, or Αειζωον το μικρον of Dioscorides; In a Letter to Alexander MacLeay, Esq. Sec. Linn. Society," *Transactions of the Linnean Society of London* 10 (1810): 6–9: "Ἀειζωον ἕτερον [*sic*] [in Dioscorides] . . . seems to be *Sedum acre* as Matthiolus and Clusius judged, though Dr. Sibthorp took it for a *Sedum ochroleucum*, on the authority of a figure in the celebrated imperial MS of Dioscorides at Vienna, which he considered as of great authority. The qualities, however, recorded of this third Αειζωον [*sic*] are quite at variance with those which Dr. Sibthorp himself attributes to the *S. ochroleucum*, and which agree with those ascribed by Dioscorides to his second species."

121. The British botanist Charles Daubeny (d. 1867), for example, studied the engravings as part of his work on ancient Roman agriculture. E.g., Charles Daubeny, *Lectures on Roman Husbandry* (Oxford: J. Wright, 1857). In 1933, Daubeny's identifications of the engraved plants and his characterizations of their quality were published in Robert T. Gunther's adaptation of John Goodyer's English translation of Dioscorides from 1655.

Conclusion

1. Reported in William Bertram Turrill, "A Contribution to the Botany of Athos Peninsula," *Bulletin of Miscellaneous Information (Royal Botanic Gardens, Kew)* 4 (1937): 197–273.

Acknowledgments

1. Claudia Sobrevila, *The Role of Indigenous Peoples in Biodiversity Conservation: The Natural but Often Forgotten Partners*, report no. 44300 (World Bank, 2008).

2. At present, the app Native Land (native-land.ca) provides a good, though still developing, starting place with links to additional resources.

SELECT BIBLIOGRAPHY

Aletta, Alessia A. "Per una puntualizzazione cronologica del Morgan 652 (Dioscoride)." In *Praktika tou 6' Diethnous Symposiou Ellenikes Palaiographias (Drama, 21–27 Septembriou 2003)*, ed. Basiles Atsalos and Nike Tsirone, 2:771–87. 3 vols. Athens: Société hellénique de reliure, 2008.

Allen, Terry. *A Classical Revival in Islamic Architecture.* Wiesbaden: Reichert, 1986.

Amar, Zohar, Efraim Lev, and Yaron Serri. "On Ibn Juljul and the Meaning and Importance of the List of Medicinal Substances Not Mentioned by Dioscorides." *Royal Asiatic Society, Series 3*, 24, no. 4 (2014): 529–55.

Amigues, Suzanne, ed. and trans. *Théophraste. Recherches sur les plantes. Tome V. Livre IX. Texte établi et traduit par S. Amigues.* Paris: Les Belles Lettres, 2006.

Angelova, Diliana. *Sacred Founders: Women, Men, and Gods in the Discourse of Imperial Founding, Rome through Early Byzantium.* Oakland: University of California Press, 2015.

Arber, Agnes. *Herbals: Their Origin and Evolution.* Cambridge: Cambridge University Press, 1912.

Balme, D. "Aristotle's Use of Division and Differentiae." In *Philosophical Issues in Aristotle's Biology*, ed. Allan Gotthelf, 69–89. Cambridge: Cambridge University Press, 1987.

Baumann, Felix. *Das Erbario Carrarese und die Bildtradition des Tractatus de herbis. Ein Beitrag zur Geschichte der Pflanzendarstellung im Übergang von Spätmittelalter zu Frührenaissance.* Bern: Benteli, 1974.

Beck, Lily, trans. *Pedanius Dioscorides of Anazarbus, De materia medica.* Hildesheim: Olms-Weidmann, 2005.

Bekker, Immanuel, ed. *Theophanes Continuatus, Ioannes Cameniata, Symeon Magister, Georgius Monachus.* Corpus scriptorum historiae Byzantinae. Bonn: Weber, 1838.

Bethe, Erich. *Buch und Bild in Altertum*. Leipzig: Harrassowitz, 1945.

Bleichmar, Daniela. *Visible Empire: Botanical Expeditions and Visual Culture in the Hispanic Enlightenment*. Chicago: University of Chicago Press, 2012.

Bogen, Steffen, and Felix Thürlemann. "Jenseits der Opposition von Text und Bild: Überlegungen zu einer Theorie des Diagramms und des Diagrammatischen." In *Die Bildwelt der Diagramme Joachims von Fiore: Zur Medialität religiös-politischer Programme im Mittelalter*, ed. Alexander Patschovsky, 1–22. Ostfildern: Thorbecke, 2003.

Bonnet, Edmond. "Étude sur les figures de plantes et d'animaux peintes dans une version arabe manuscrite de la *Matière médicale* de Dioscoride conservée à la BN de Paris." *Janus* 14 (1909): 294–303.

Boudon, Véronique. "Galen's *On My Own Books*: Material from Meshed, Rida, Tibb. 5223." *Bulletin of the Institute of Classical Studies*, suppl. 77 (2002): 9–18.

Bouras-Valliantos, Petros. *Innovation in Byzantine Medicine: The Writings of John Zacharias Aktouarios (c. 1275–c. 1330)*. Oxford: Oxford University Press, 2020.

Brubaker, Leslie. "The Vienna Dioskorides and Anicia Juliana." In *Byzantine Garden Culture*, ed. Antony Robert Littlewood, Henry Maguire, and Joachim Wolschke-Bulmahn, 189–214. Washington, DC: Dumbarton Oaks, 2002.

Cataldi Palau, Annaclara. "Legature costantinopolitane del monastero di Prodromo Petra tra i manoscritti di Giovanni di Ragusa (t 1443)." *Codices manuscripti* 37–38 (2001): 11–50.

Cataldi Palau, Annaclara. "The Manuscript Production in the Monastery of Prodromos Petra (twelfth–fifteenth centuries)." In *Studies in Greek Manuscripts*, 197–208. Spoleto: Fondazione Centro Italiano di Studi sull'alto Medioevo, 2008.

Cavallo, Guglielmo. "Introduction," trans. Salvatore Lilla. In *Dioscorides Neapolitanus. Biblioteca Nazionale di Napoli. Codex ex Vindobonensis Graecus 1. Commentarium*, ed. Carlo Bertelli, Salvatore Lilla, and Giulia Orofino, 9–13. Rome: Salerno Editrice, 1992.

Cave, Roderick. *Impressions of Nature: A History of Nature Printing*. New York: Mark Batty, 2010.

Christodoulou, Geōrgios A. Σύμμικτα Κριτικά. Athens: By the author, 1986.

Collins, Minta. *Medieval Herbals: The Illustrative Traditions*. Toronto: University of Toronto Press, 2000.

Cronier, Marie. "L'herbier alphabétique grec de Dioscoride. Quelques remarques sur sa genèse et ses sources textuelles." In *Fito-zooterapia antigua y altomedieval. Textos y doctrinas*. Ed. Arsenio Ferraces Rodriguez, 33–59. Coruña: Universidade da Coruña, 2009.

Cronier, Marie. "Un manuscrit méconnu du Περὶ ὕλης ἰατρικῆς de Dioscoride: New York, Pierpont Morgan Library, M. 652." *Revue des études grecques* 125, no. 1 (2012): 95–130.

Cronier, Marie. "Transcrire l'arabe en grec. À propos des annotations du Parisinus gr. 2179." In *Manuscripta graeca et orientalia. Mélanges monastiques et patristiques en l'honneur de Paul Géhin*, 247–65. Leuven: Peeters, 2016.

Cronier, Marie, and Patrick Gautier Dalché. "A Map of Cyprus in Two Fourteenth-Century Byzantine Manuscripts." *Imago Mundi* 69, no. 2 (2017): 176–87.

Dalby, Andrew. *Geoponica: Farm Work: A Modern Translation of the Roman and Byzantine Farming Handbook*. Totnes, Devon, UK: Prospect Books, 2011.

Daly, Lloyd W. *Contributions to a History of Alphabetization in Antiquity and the Middle Ages*. Brussels: Latomus, 1967.

Deichgräber, Karl. *Die griechische Empirikerschule: Sammlung der Fragmente und Darstellung der Lehre*. Berlin: Weidmann, 1965.

Delatte, Armand. "Glossaires de botanique." In *Anecdota Atheniensia et alia*, 2:277–454. Paris: Droz, 1939.

Dietrich, Albert. *Die Dioskurides-Erklärung des Ibn al-Baiṭār: Ein Beitrag zur arabischen Pflanzensynonymik des Mittelalters*. Göttingen: Vandenhoeck and Ruprecht, 1991.

Dietrich, Albert. *Die Ergänzung Ibn Ǧulǧul's zur "Materia Medica" des Dioskurides*. Göttingen: Vandenhoeck and Ruprecht, 1993.

Dietrich, Albert. *Dioscurides Triumphans: Ein anonymer arabischer Kommentar (Ende 12. Jahrh. n. Chr.) zur Materia medica*. 2 vols. Göttingen: Vandenhoeck and Ruprecht, 1988.

Dodoens, Rembert. *Stirpium historiae pemptades sex sive libri triginta*. Antwerp: Ex officina Christophori Plantini, 1583.

Dols, Michael. "The Origins of the Islamic Hospital: Myth and Reality." *Bulletin of the History of Medicine* 61, no. 3 (1987): 367–90.

Ettinghausen, Richard. *Arab Painting*. Geneva: Skira, 1962.

Fähndrich, Hartmut, ed. and trans. *Treatise to Ṣalāḥ ad-Dīn on the Revival of the Art of Medicine by Ibn Jumayʿ*. Wiesbaden: Franz Steiner, 1983.

Fausti, Daniela. "Erbari illustrati su papiro e tradizione iconografica botanica." In *Testi medici su papiro. Atti del Seminario di studio (Firenze, 3–4 giugno 2002)*, 131–50. Florence: Istituto papirologico G. Vitelli, 2004.

Fisher, Elizabeth. "Monks, Monasteries and the Latin Language in Constantinople." In *Change in the Byzantine World in the Twelfth and Thirteenth Centuries*, ed. Ayla Ödekan, Engin Akyürek, and Nevra Necipoglu, 390–95. Istanbul: Vehbi Koç Foundation, 2010.

Flemming, Rebecca. "Empires of Knowledge: Medicine and Health in the Hellenistic World." In *A Companion to the Hellenistic World*, ed. Andrew Erskine, 449–63. Oxford: Blackwell, 2003.

Fowler, Henry North, trans. *Plutarch. Moralia X*. Loeb Classical Library 321. Cambridge, MA: Harvard University Press, 1936.

Gerstinger, Hans, ed. *Dioscurides. Codex Vindobonensis Med. Gr. 1, Der Österreichischen Nationalbibliothek*. Kommentarband zu der Faksimileausgabe. Graz: Akademische Druck-u. Verlagsanstalt, 1970.

Givens, Jean. *Observation and Image-Making in Gothic Art*. Cambridge: Cambridge University Press, 2005.

Givens, Jean. "Reading and Writing the Illustrated *Tractatus de herbis*, 1280–1526." *Visualizing Medieval Medicine and Natural History, 1200–1550*. Ed. Jean Givens and Alain Touwaide, 115–45. Aldershot, UK: Ashgate, 2006.

Gombrich, Ernst H. *Art and Illusion: A Study in the Psychology of Pictorial Representation*. London: Phaidon Press, 1984 [1960].

González Manjarrés, Miguel Ángel, and María Cruz Herrero Ingelmo. *El Dioscórides Grecolatino del Papa Alejandro VII. Manuscrito Vat. Chigi 53 (F. VII 159)*. Madrid: Testimonio, 2001.

Goodman, Nelson. *Languages of Art: An Approach to a Theory of Symbols*. Indianapolis: Hackett, 1976.

Grape-Albers, Heide. *Spätantike Bilder aus der Welt des Arztes. Medizinische Bilderhandschriften der Spätantike und ihre mittelalterliche Überlieferung.* Wiesbaden: Guido Pressler, 1977.

Griebeler, Andrew. "*Aeizōon to amaranton*: Intercultural Collaboration in a Late Byzantine Nature Study." *Convivium: Exchanges and Interactions in the Arts of Medieval Europe, Byzantium, and the Mediterranean: Seminarium Kondakovianum,* n.s. 2, no. 6 (2019): 16–29.

Griebeler, Andrew. "Botanical Illustration and Byzantine Visual Inquiry in the Morgan Dioscorides." *Art Bulletin 105, no. 1* (2023): 93–116.

Griebeler, Andrew. "How to Illustrate a Scientific Treatise in the Palaiologan Period." In *Late Byzantium Reconsidered: The Arts of the Palaiologan Era in the Mediterranean,* ed. Maria Alessia Rossi and Andrea Mattiello, 85–103. London: Routledge, 2019.

Griebeler, Andrew. "Production and Design of Early Illustrated Herbals." *Word & Image: A Journal of Verbal/Visual Enquiry* 38, no. 2 (2022): 104–22.

Gruendler, Beatrice. *The Rise of the Arabic Book.* Cambridge, MA: Harvard University Press, 2020.

Guasparri, Andrea. "Explicit Nomenclature and Classification in Pliny's *Natural History* XXXII." *Studies in History and Philosophy of Science Part A* 44, no. 3 (2013): 347–53.

Gutas, Dimitri. "Arabic into Byzantine Greek: Introducing a Survey of Translation." In *Knotenpunkt Byzanz: Wissensformen und kulturelle Wechselbezeihungen,* ed. Philipp Steinkruger and Andreas Speer, 246–62. Berlin: De Gruyter, 2012.

Gutas, Dimitri. *Greek Thought, Arabic Culture: The Graeco-Arabic Translation Movement in Baghdad and Early Abbasid Society, 2nd–4th/8th–10th Centuries.* New York: Routledge, 1998.

Hallé, Francis. "A Life Drawing Trees: Interview with Emmanuele Coccia." Trans. Emma Lingwood. In *Trees,* ed. Pierre-Édouard Couton, 32–47. Paris: Fondation Cartier, 2019.

Halporn, James W., trans. *Cassiodorus: "Institutions of Divine and Secular Learning" and "On the Soul".* Liverpool: Liverpool University Press, 2004.

Hämeen-Antilla, Jaako. *The Last Pagans of Iraq: Ibn Waḥshiyya and His Nabatean Agriculture.* Leiden: Brill, 2006.

Hankinson, R. J., trans. *Galen, On the Therapeutic Method: Books I and II.* New York: Oxford University Press, 1991.

Hankinson, R. J. "Usage and Abusage: Galen on Language." In *Language: Volume 3 of Companions of Ancient Thought,* 166–87. Cambridge: Cambridge University Press, 1994.

Hanson, Ann Ellis. "Greek Medical Papyri from the Fayum Village of Tebtunis: Patient Involvement in a Local Health-Care System?" In *Hippocrates in Context, Papers Read at the XIth International Hippocrates Colloquium, University of Newcastle upon Tyne, 27–31 August 2002,* ed. Philip van der Eijk, 387–402. Studies in Ancient Medicine 31. Leiden: Brill, 2005.

Hanson, Ann Ellis. "Text and Context for the Illustrated Herbal from Tebtunis." In *Atti del XXII Congresso di Papirologia, Firenze, 23–29, 1998,* ed. Isabella Andorlini, Guido Bastianini, Manfredo Manfredi, and Giovanna Menci, 585–604. Florence: Istituto papirologico G. Vitelli, 2001.

Hardy, Gavin, and Laurence Totelin. *Ancient Botany*. New York: Routledge, 2016.

Henry, René, ed. and trans. *Photius Bibliothèque*. Vol. 2: *Codices 84–185*. Paris: Les belles lettres, 1960.

Heurgon, Jacques. "L'agronome carthaginois Magon et ses traducteurs en latin et en grec." *Comptes rendus des séances de l'Académie des inscriptions et belles-lettres* 120 (1976): 441–56.

Heyd, Uriel. "Moses Hamon, Chief Jewish Physician to Sultan Süleymān the Magnificent." *Oriens* 16 (1963): 152–70.

Hoffman, Eva R. "The Author Portrait in Thirteenth-Century Arabic Manuscripts: A New Islamic Context for a Late Antique Tradition." *Muqarnas* 10 (1993): 6–17.

Hoffman, Eva R. "The Beginnings of the Illustrated Arabic Book: An Intersection between Art and Scholarship." *Muqarnas* 17 (2000): 37–52.

Holler, Theresa. "Naturmaß, künstlerisches Maß und die Maßlosigkeit Ihrer Anwendung. Simplicia in zwei 'Tractatus de herbis'-Handschriften des 13. und 14. Jahrhunderts." *Das Mittelalter* 13, no. 1 (2018): 1–25.

Howald, E., and H. Sigerist, eds. *Antonii Musae De herba vettonica liber. Pseudo-Apulei Herbarius. Anonymi De taxone liber. Sexti Placiti Liber medicinae ex animalibus etc.* Corpus medicorum latinorum 4. Leipzig: Teubner, 1927.

Hunger, Herbert. *Die hochsprachliche profane Literatur der Byzantiner*. Munich: Beck, 1978.

Hutter, Irmgard. "Decorative Systems in Byzantine Manuscripts, and the Scribe as Artist: Evidence from Manuscripts in Oxford." *Word & Image* 12, no. 1 (1996): 4–22.

Issa, Ahmed. *Histoire des Bimaristans (hopitaux) à l'époque islamique*. Cairo: Paul Barbey, 1928.

Jacques, Jean-Marie, ed. and trans. *Nicandre: Oeuvres. Tome II: Les Thériaques. Fragments iologiques antérieurs à Nicandre*. Paris: Les belles lettres, 2002.

Johnson, J. de M. "A Botanical Papyrus with Illustrations." *Archiv für die Geschichte der Naturwissenschaften und der Technik* 4 (1912–13): 403–8.

Jones, Peter Murray. *Medieval Medicine in Illuminated Manuscripts*. London: British Library, 1998.

Jones, W. H. S., trans. *Pliny the Elder, Natural History, Volume VII: Books 24–27*. Loeb Classical Library 393. Cambridge, MA: Harvard University Press, 1956.

Julius Africanus. *Cesti: The Extant Fragments*. Ed. Martin Wallraff et al. Berlin: De Gruyter, 2012.

Kavrus-Hoffmann, Nadezhda. "Catalogue of Greek Medieval and Renaissance Manuscripts in the Collections of the United States of America, Part IV.2: The Morgan Library and Museum." *Manuscripta* 52, no. 2 (2008): 207–324.

Kerner, Jaclynne J. "The Illustrated *Herbal* of al-Ghāfiqī: An Art Historical Introduction." In *The Herbal of al-Ghāfiqī: A Facsimile Edition with Critical Essays*, ed. F. Jamil Ragep and Faith Wallis with Pamela Miller and Adam Gacek, 121–56. Montreal: McGill-Queen's University Press, 2014.

Kimmerer, Robin Wall. *Gathering Moss: A Natural and Cultural History of Mosses*. Corvallis: Oregon State University Press, 2003.

Kind, Friedrich Ernst. "Krateuas." In *Paulys Realencyclopädie der classischen Altertumswissenschaft*, 11, 2:1644–46. Stuttgart: Alfred Druckenmüller, 1922.

Kotsifou, Chrysi. "Books and Book Production in the Monastic Communities of Byzantine Egypt." In *The Early Christian Book*, ed. William E. Klingshirn and Linda Safran, 48–66. Washington, DC: Catholic University of America Press, 2007.

Kristeller, Paul Oskar. "The School of Salerno: Its Development and Its Contribution to the History of Learning." *Bulletin of the History of Medicine* 17 (1945): 138–94.

Kubler, George. *The Shape of Time: Remarks on the History of Things*. New Haven: Yale University Press, 1962.

Kuhn, Carl Gottlieb, ed. *Clavdii Galeni opera omnia*. 20 vols. Hildesheim: Olms, 1964, 1965 [1821–33].

Kusukawa, Sachiko. *Picturing the Book of Nature: Image, Text, and Argument in Sixteenth-Century Human Anatomy and Medical Botany*. Chicago: University of Chicago Press, 2011.

Kuttner, Anne. "Looking Outside Inside: Ancient Roman Garden Rooms." *Studies in the History of Gardens & Designed Landscapes* 19, no. 1 (1999): 7–35.

Kyle, Sarah R. *Medicine and Humanism in Late Medieval Italy: The Carrara Herbal in Padua*. Abingdon, UK: Routledge, 2017.

Lack, Hans Walter. "Die Kupferstiche von frühbyzantinischen Pflanzenabbildungen im Besitz von Linné, Sibthorp und Kollár." *Annalen des Naturhistorischen Museums in Wien. Serie B für Botanik und Zoologie* 100 (1998): 613–55.

Laourdas, V., and L. G. Westerink, eds. *Photii Patriarchae Constantinopolitani Epistulae et Amphilochia*. Leipzig: Teubner, 1984.

Latour, Bruno. *Pandora's Hope: Essays on the Reality of Science Studies*. Cambridge, MA: Harvard University Press, 1999.

Latour, Bruno. "Visualization and Cognition: Drawing Things Together." In *Knowledge and Society: Studies in the Sociology of Culture Past and Present*, ed. H. Kuklick, 6:1–40. Greenwich, CT: Jai Press, 1979.

Lazaris, Stavros. "L'illustration des disciplines médicales dans l'antiquité. Hypothèses, enjeux, nouvelles interprétations." In *La collezione di testi chirurgici di Niceta. Firenze, Biblioteca medicea laurenziana, Plut. 74.7. Tradizione medica classica a Bisanzio*, ed. Massimo Bernabò, 99–109. Rome: Edizioni di storia e letteratura, 2010.

Lazaris, Stavros. "L'image paradigmatique. Des *Schémas anatomiques* d'Aristote au *De materia medica* de Dioscuride." *Pallas* 93 (2013): 131–64.

Leith, David. "The Antinoopolis Illustrated Herbal (PJohnson + PAntin. 3.214 = MP3 2095)." *Zeitschrift für Papyrologie und Epigraphik* 156 (2006): 141–56.

Lewis, Charlton T., William Freund, E. A. Andrews, and Charles Short. *A Latin Dictionary: Founded on Andrews' Edition of Freund's Latin Dictionary*. New York: Oxford University Press, 1995.

Liddell, Henry George, Robert Scott, Henry Stuart Jones, and Roderick McKenzie. *A Greek-English Lexicon*. 9th ed. Oxford: Clarendon Press, 2007.

Lilla, Salvatore. "Studio del Codice: A Study of the Manuscript." In *Dioscurides Neapolitanus: Biblioteca Nazionale di Napoli Codex ex Vindobonensis Graecus 1, Commentarium*, ed. Carlo Bertelli, Salvatore Lilla, and Giulia Orofino, 49–82. Rome: Salerno Editrice, 1992.

Lopes, Dominic. *Understanding Pictures*. Oxford: Clarendon Press, 2004 [1996].

Lowden, John. "The Transmission of 'Visual Knowledge' in Byzantium through Illuminated Manuscripts: Approaches and Conjectures." In *Literacy, Education and Manuscript Transmission in Byzantium and Beyond*, ed. Catherine Holmes and Judith Waring, 59–80. Leiden: Brill, 2002.

Maguire, Henry. *The Icons of Their Bodies: Saints and Their Images in Byzantium*. Princeton: Princeton University Press, 2000.

Maguire, Henry. *Nectar and Illusion: Nature in Byzantine Art and Literature*. Oxford: Oxford University Press, 2012.

Mango, Cyril. *Art of the Byzantine Empire, 312–1453*. Sources and Documents in the History of Art. Englewood Cliffs, NJ: Prentence-Hall, 1972.

Mantuani, Joseph [Josip Mantuani]. "Die Miniaturen im Wiener Kodex Med. Graecus I." In *De codicis Dioscuridei Aniciae Iulianae, nunc Vindobonensis Med. Gr. 1*, ed. Josef Karabacek, 353–490. Leiden: Sijthoff, 1906.

Marasco, Gabriele. "L'introduction de la médecine grecque à Rome. Une dissension politique et idéologique." In *Ancient Medicine in Its Socio-Cultural Context: Papers Read at the Congress Held at Leiden University, 13–15 April 1992*, ed. Philip van der Eijk, Herman F. J. Horstmanshoff, and Piet H. Schrijvers, 1:35–48. Amsterdam: Rodopi, 1995.

Marchetti, Francesca. "La trasmissione delle illustrazioni del Dioscoride di Vienna negli anni intorno alla caduta di Costantinopoli (Cod. Banks Coll. Dio. 1, Natural History Museum, Londra; Ee. V. 7, Cambridge University Library, Cambridge; e C 102 sup., Biblioteca Ambrosiana, Milano)." *Jahrbuch der Österreichischen Byzantinistik* 66 (2016): 153–78.

Marganne, Marie-Hélène. *Le livre médical dans le monde gréco-romain*, Cahiers du CeDoPaL 3. Liège: Les Éditions de l'Université de Liège, 2004.

Martínez Manzano, Teresa. "De Corfú a Venecia. El itinerario primero del *Dioscórides* de Salamanca." *Medioevo greco* 12 (2012): 133–54.

Massar, Natacha. *Soigner et servir. Histoire sociale et culturelle de la médecine grecque à l'époque hellénistique*. Paris: De Boccard, 2005.

Mattern, Susan P. *Galen and the Rhetoric of Healing*. Baltimore: Johns Hopkins University Press, 2008.

Mauch, Ute. "Pflanzenabbildungen des Wiener Dioskurides und das Habituskonzept: Ein Beitrag zur botanischen Charakterisierung von antiken Pflanzen durch den Habitus." *Antike Naturwissenschaft und ihre Rezeption* 16 (2006): 125–38.

Mavroudi, Maria. *A Byzantine Book on Dream Interpretation: The Oneirocriticon of Achmet and Its Arabic Source*. Leiden: Brill, 2002.

Mavroudi, Maria. "The Naples Dioscorides." In *Byzantium and Islam: Age of Transition: Catalogue of the Exhibition at the Metropolitan Museum of Art*, ed. Helen Evans and Brandie Ratliff, 22–26. New Haven: Yale University Press, 2012.

Maxwell, Kathleen. *Between Constantinople and Rome: An Illuminated Byzantine Gospel Book (Paris gr. 54) and the Union of Churches*. London: Routledge, 2014.

Mayhoff, Karl, ed. *C. Plini Secundi Naturalis historiae libri XXXVII*. Stuttgart: Teubner, 1892–1909.

Mazal, Otto. *Der Wiener Dioskurides: Codex medicus graecus 1 der Österreichischen Nationalbibliothek*. 2 vols. Graz: Akademische Druck- u. Verlagsanstalt, 1998.

Mioni, Elpidio. "Un ignoto Dioscoride miniatio." In *Libri e stampatori in Padova. Miscellanea di studi storici in onore di Mons. G. Bellini*, 345–76. Padua: Tipografia Antoniana, 1959.

Mondrain, Brigitte. "Lettrés et copistes à Corfou au XVe et au XVIe siècle." In *Puer Apuliae. Mélanges offerts à Jean-Marie Martin*, ed. E. Cuozzo, 463–76. Centre de recherche d'histoire et civilisation de Byzance, Monographies 30. Paris: ACHCByz, 2008.

Müller, A. E. "Ein vermeintlich fester Anker. Das Jahr 512 als zeitlicher Ansatz des 'Wiener Dioskurides.'" *Jahrbuch der österreichischen Byzantinistik* 62 (2012): 103–9.

Munitiz, Joseph A. "Dedicating a Volume: Apokaukos and Hippocrates (Paris. gr. 2144)." In *Φιλέλλην, Studies in Honour of Robert Browning*, ed. Costas N. Constantinides, Nikolaos M. Panagiotakis, Elizabeth Jeffreys, and Athanasios D. Angelou, 267–80. Istituto Ellenico di Studi Bizantini e Postbizantini di Venezia, Bibliotheke 17. Venice: Istituto Ellenico, 1996.

Mynors, R. A. B., ed. *Cassiodori Senatoris Institutiones*. 2nd ed. Oxford: Clarendon Press, 1963.

Nasrallah, Nawal, trans. and ed. *Annals of the Caliphs' Kitchens: Ibn Sayyār al-Warrāq's Tenth Century Baghdadi Cookbook*. Leiden: Brill, 2007.

Nicholls, Matthew. "Parchment Codices in a New Text of Galen." *Greece & Rome*, 2nd ser., 57, no. 2 (2010): 378–86.

Nijhuis, Karin. "Greek Doctors and Roman Patients: A Medical Anthropological Approach." In *Ancient Medicine in Its Socio-Cultural Context*, ed. Philip van der Eijk, Herman F. J. Horstmanshoff, and Piet H. Schrijvers, 49–67. Amsterdam: Rodopi, 1995.

Nutton, Vivian. *Ancient Medicine*. 2nd ed. London: Routledge, 2013.

Odorico, Paolo. "La cultura della *Συλλογή*." *Byzantinische Zeitschrift* 83 (1990): 1–21.

Ogilvie, Brian. *The Science of Describing: Natural History in Renaissance Europe*. Chicago: University of Chicago Press, 2006.

Olariu, Dominic. "The Misfortune of Philippus de Lignamine's Herbal or New Research Perspectives in Herbal Illustrations from an Iconological Point of View." In *Early Modern Print Culture in Central Europe*, ed. Stefan Kierdroń and Anna-Maria Rim, 39–62. Wrocław: Wydawnictwo Uniwersytetu Wrocławskiego, 2014.

Olson, Alexander. *Environment and Society in Byzantium, 650–1150: Between the Oak and the Olive*. Cham, Switzerland: Palgrave Macmillan, 2020.

Orofino, Giulia. "The Dioscorides of the Biblioteca Nazionale of Naples: The Miniatures." Trans. Linda Lappin. In *Dioscorides Neapolitanus: Biblioteca Nazionale di Napoli Codex ex Vindobonensis Graecus 1, Commentarium*, ed. Carlo Bertelli, Salvatore Lilla, and Giulia Orofino, 99–113. Rome: Salerno editrice, 1992.

Pächt, Otto. "Early Italian Nature Studies and the Early Calendar Landscape." *Journal of the Warburg and Courtauld Institutes* 13, nos. 1–2 (1950): 13–47.

Pächt, Otto. "Die früheste abendländische Kopie der Illustrationen des Wiener Dioskurides." *Zeitschrift für Kunstgeschichte* 38, nos. 3–4 (1975): 201–14.

Pancaroğlu, Oya. "Socializing Medicine: Illustrations of the *Kitāb al-diryāq*." *Muqarnas* 18 (2001): 155–72.

Papaioannou, Stratis, ed. *Michael Psellus, Epistulae*. 2 vols. Berlin: De Gruyter, 2019.

Park, Katharine. "Observations in the Margins, 500–1500." In *Histories of Scientific Observation*, ed. Lorraine Daston, 15–44. Chicago: University of Chicago Press, 2011.

Philippides, Marios, and Walter K. Hanak. *Cardinal Isidore, c. 1390–1462: A Late Byzantine Scholar, Warlord, and Prelate*. Abingdon, UK: Routledge, 2008.

Philipsborn, Alexandre. "Der Fortschritt in der Entwicklung des byzantinischen Krankenhauswesens." *Byzantinischen Zeitschrift* 54, no. 2 (1961): 338–65.

Pollard, Elizabeth Ann. "Pliny's Natural History and the Flavian Templum Pacis: Botanical Imperialism in First-Century C.E. Rome." *Journal of World History* 20, no. 3 (2009): 309–88.

Pomata, Gianna. "A Word of the Empirics: The Ancient Concept of Observation and Its Recovery in Early Modern Medicine." *Annals of Science* 68, no. 1 (2011): 1–25.

Premerstein, Anton von. "Anicia Juliana im Wiener Dioskorides-Kodex." *Jahrbuch der kunsthistorischen Sammlungen des allerhöchsten Kaiserhauses* 24 (1903): 105–24.

Premerstein, Anton von [Antonius de Premerstein]. "De codicis Dioscuridei Aniciae Iulianae, nunc Vindobonensis Med. Gr. I: Historia, Forma, Argumento." In *De codicis Dioscuridei Aniciae Iulianae, nunc Vindobonensis Med. Gr. 1: Historia, Forma, Scriptura, Picturis*, ed. Josef Karabacek, 3–228. Leiden: Sijthoff, 1906.

Ragab, Ahmed. *Medieval Islamic Hospital: Medicine, Religion, and Charity*. Cambridge: Cambridge University Press, 2015.

Reeds, Karen. *Botany in Medieval and Renaissance Universities*. New York: Garland, 1991.

Reeds, Karen. "Renaissance Humanism and Botany." *Annals of Science* 33 (1976): 519–42.

Rheinberger, Hans-Jörg. "Experimental Systems: Historiality, Narration, and Deconstruction." *Science in Context* 7, no. 1 (1994): 65–81.

Riddle, John. "Byzantine Commentaries on Dioscorides." *Dumbarton Oaks Papers* 38 (1984): 95–102.

Riddle, John. *Dioscorides on Pharmacy and Medicine*. Austin: University of Texas Press, 1985.

Roberts, Colin H., and T. C. Skeat. *Birth of the Codex*. London: Oxford University Press, 1983.

Rosenthal, Franz. *The Classical Heritage of Islam*. Berkeley: University of California Press, 1975.

Ryholt, Kim. "The Illustrated Herbal from Tebtunis: New Fragments and Archaeological Context." *Zeitschrift für Papyrologie und Epigraphik* 187 (2013): 233–38.

Ryholt, Kim. "Libraries from Late Period and Graeco-Roman Egypt, c. 800 BCE–250 CE." In *Libraries before Alexandria: Ancient Near Eastern Traditions*, ed. Kim Ryholt and Gojko Barjamovic, 388–472. Oxford: Oxford University Press, 2019.

Sabbadini, Remigio, ed. *Carteggio di Giovanni Aurispa*. Rome: Tipografia del senato, 1931.

Sadek, Mahmoud M. *The Arabic Materia Medica of Dioscorides*. Quebec: Éditions du Sphinx, 1983.

Sadek, Mahmoud M. "Notes on the Introduction and Colophon of the Leiden Manuscript of Dioscorides' 'De Materia Medica.'" *International Journal of Middle East Studies* 10, no. 3 (1979): 345–54.

Saliba, George, and Linda Komaroff. "Illustrated Books May Be Hazardous to Your Health: A New Reading of the Arabic Reception and Rendition of the 'Materia Medica' of Dioscorides." *Ars Orientalis* 35 (2008): 6–65.

Savage-Smith, E., S. Swain, G. J. van Gelder, ed. and trans. *A Literary History of Medicine—The 'Uyūn al-anbā' fī ṭabaqāt al-aṭibbā' of Ibn Abī Uṣaybi'ah Online.* Leiden: Brill, n.d. Open access edition: https://brill.com/view/db/lhom.

Scarborough, John, and Vivian Nutton. "The Preface of Dioscorides' *De materia medica*: Introduction, Translation and Commentary." *Transactions and Studies of the College of Physicians of Philadelphia* 4, no. 3 (1982): 187–227.

Singer, Charles. "The Herbal in Antiquity and Its Transmission to Later Ages." *Journal of Hellenic Studies* 47, no. 1 (1927): 1–52.

Smith, James Edward. "Remarks on the *Sedum ochroleucum*, or Αειζωον το μικρον of Dioscorides; In a Letter to Alexander MacLeay, Esq. Sec. Linn. Society." *Transactions of the Linnean Society of London* 10 (1810): 6–9.

Staden, Heinrich von. "Liminal Perils: Early Roman Receptions of Greek Medicine." In *Tradition, Transmission, Transformation: Proceedings of Two Conferences on Pre-Modern Science Held at the University of Oklahoma*, ed. F. Jamil Ragep and Sally P. Ragep, 369–418. Leiden: Brill, 1996.

Stannard, Jerry. "Byzantine Botanical Lexicography." *Episteme* 5 (1971): 168–87.

Strasser, Bruno. "Collecting Nature: Practices, Styles, and Narratives." *Osiris* 27, no. 1 (2012): 303–40.

Strömberg, Reinhold. *Griechische Pflanzennamen.* Göteborgs Högskolas Årsskrift 46, no. 1. Goteborg: Elanders Boktryckeri Aktiebolag, 1940.

Stückelberger, Alfred. *Bild und Wort: Das illustrierte Fachbuch in der antiken Naturwissenschaft, Medizin und Technik.* Munich: P. von Zabern, 1994.

Tait, W. J. *Papyri from Tebtunis in Egyptian and in Greek.* London: Egyptian Exploration Society, 1977.

Talbot, Alice Mary. "Byzantine Monastic Horticulture: The Textual Evidence." In *Byzantine Garden Culture*, ed. Antony Littlewood, Henry Maguire, and Joachim Wolschke-Bulmahn, 37–67. Washington, DC: Dumbarton Oaks Research Library, 2002.

Thomas, John, and Angela Constantinides Hero, eds. and trans. *Byzantine Monastic Foundation Documents: A Complete Translation of the Surviving Founders' Typika and Testaments.* Washington, DC: Dumbarton Oaks Library and Collection, 2000.

Thomas, Joshua. "The Illustrated Dioskourides Codices and the Transmission of Images during Antiquity." *Journal of Roman Studies* 109 (2019): 241–73.

Thomson, Margaret H. *Textes grecs inédits relatifs aux plantes.* Paris: Les belles lettres, 1955.

Toresella, Sergio. "Il Dioscoride di Istanbul e le prime figurazioni naturalistiche botaniche." *Atti e memorie della Accademia italiana di storia della farmacia* 13, no. 1 (1996): 21–40.

Totelin, Laurence. "Botanizing Rulers and Their Herbal Subjects: Plants and Political Power in Greek and Roman Literature." *Phoenix* 66, nos. 1–2 (2012): 122–44.

Totelin, Laurence. "Mithridates' Antidote: A Pharmacological Ghost." *Early Science and Medicine* 9, no. 1 (2004): 1–19.

Touwaide, Alain. "Art and Science: Private Gardens and Botany in the Early Roman Empire." In *Botanical Progress, Horticultural Innovation and Cultural Change*, ed. Michel Conan and W. John Kress, 37–49. Washington, DC: Dumbarton Oaks Research Library and Collection, 2004.

Touwaide, Alain. *A Census of Greek Medical Manuscripts: From Byzantium to the Renaissance*. London: Routledge, 2016.

Touwaide, Alain. "Crateuas." In *Brill's New Pauly: Encyclopedia of the Ancient World*, ed. Hubert Cancik and Helmuth Schneider, 3:920–21. Leiden: Brill, 2003.

Touwaide, Alain. "Latin Crusaders, Byzantine Herbals." In *Visualizing Medieval Medicine and Natural History, 1200–1550*, ed. Jean A. Givens, Karen M. Reeds, and Alain Touwaide, 25–50. New York: Routledge, 2016.

Touwaide, Alain. "Lexica medico-botanica byzantina. Prolégomènes à une étude." In *Tês filiês ta'de dôra: Miscelánea léxica en memoria de Conchita Serrano*, 211–28. Manuales y Anejos de Emerita 41. Madrid: Consejo superior de investigaciones científicas, 1999.

Touwaide, Alain. "Un recueil grec de pharmacologie du Xe siècle illustré au XIVe siècle. Le Vaticanus gr. 284." *Scriptorium* 39, no. 1 (1985): 13–56.

Touwaide, Alain. "The Salamanca Dioscorides (Salamanca, University Library, 2659)." *Erytheia. Revista de estudios bizantinos y neogriegos* 24 (2003): 125–58.

Tucci, Pier Luigi. *The Temple of Peace in Rome*. Cambridge: Cambridge University Press, 2017.

Turrill, William Bertram. "A Contribution to the Botany of Athos Peninsula." *Bulletin of Miscellaneous Information (Royal Botanic Gardens, Kew)* 4 (1937): 197–273.

Uexküll, Jacob von. *A Foray into the Worlds of Animals and Humans with a Theory of Meaning*. Trans. Joseph D. O'Neil. Minneapolis: University of Minnesota Press, 2010.

Uexküll, Jacob von. *Streifzüge durch die Umwelt von Tieren und Menschen: Ein Bilderbuch unsichtbarer Welten*. Hamburg: Rowohlt, 1956 [1934].

Ullmann, Manfred. *Untersuchungen zur arabischen Überlieferung der* Materia medica *des Dioskurides*. Wiesbaden: Harrassowitz, 2009.

Van Minnen, Peter. "Boorish or Bookish? Literature in Egyptian Villages in the Fayum in the Graeco-Roman Period." *Journal of Juristic Papyrology* 28 (1998): 99–184.

Ventura, Iolanda. "Il *Circa Instans* attribuito a Platearius. Trasmissione manoscritta, redazioni, criteri di costruzione di un'edizione critica." *Revue d'histoire des textes* 10 (2015): 249–362.

Walzer, Richard, and Michael Frede, trans. *Galen, Three Treatises on the Nature of Science: On the Sects for Beginners, An Outline of Empiricism, On Medical Experience*. Indianapolis: Hackett, 1985.

Webb, D. A. "Flora of European Turkey." *Proceedings of the Royal Irish Academy*. Section B: Biological, Geological, and Chemical Science, 65 (1966/1967): 1–100.

Weitzmann, Kurt. *Ancient Book Illumination*. Cambridge, MA: Harvard University Press, 1959.

Weitzmann, Kurt. "The Greek Sources of Islamic Scientific Illustrations." In

Archaeologica Orientalia in Memoriam Ernst Herzfeld, 244–66. Locust Valley, NY: J. Augustin, 1952.

Weitzmann, Kurt. *Illustrations in Roll and Codex: A Study of the Origin and Method of Text Illustration.* Princeton: Princeton University Press, 1975.

Wellmann, Max. "Krateuas." *Abhandlungen der königlichen Gesellschaft der Wissenschaften zu Göttingen. Philologisch-historische Klasse*, n.s. 2, 1 (1897): 2–32.

Wellmann, Max, ed. *Pedanii Dioscuridis Anazarbei De materia medica.* 3 vols. Berlin: Weidmann, 1907–14.

Wellmann, Max. "Die Pflanzennamen des Dioskurides." *Hermes* 33 (1898): 360–422.

Wilson, Nigel G. "Two Notes on Byzantine Scholarship: I. The Vienna Dioscorides and the History of Scholia." *Greek, Roman and Byzantine Studies* 12 (1971): 557–58.

GENERAL INDEX

Page numbers in italics refer to figures.

'Abd al-Raḥmān I (emir), 112
'Abd al-Raḥmān III (caliph),
    113–14
abrotonon (Greek). *See*
    southernwood
abstraction, 53–54, 57, 258n31
Abū 'Abd Allāh al-Ṣiqillī, 113–14
Abū Bakr al-Rāzī, 106, 185, 273n28
Abū Ḥanīfa al-Dīnāwarī, 106,
    272n20
Abū Sālim al-Malṭī, 106, 272n22
Abū Yūsuf Bihnām ibn Mūsā al-
    Mawṣilī, 123, 193, 279n106
acanthus, *Acanthus* sp., 23
accidents, 39
accuracy, 3, 48, 50, 57, 81, 102, 109,
    143, 150, 156, 163, 165, 167–69,
    171, 175–76, 178–80, 189, 192,
    202, 209, 214–15, 221, 228,
    257n17, 289n13, 290n29, 293n60
aconite, monkshoods, *Aconitum*
    sp., 247n78, 247n81
aeizōon (Greek), *Sempervivum* sp.,
    85, 86, 91, *198*
aeizōon to amaranton (Greek),
    Petrosedum sp. *See* pale
    stonecrop

aerial perspective. *See* perspective
Aetius of Amida, 78, 103
Africa/Africans, 84. *See also* Con-
    stantine the African
aglaophotis (Greek), *Paeonia* sp.,
    241n21
Aleppo, 278n99
Alexander of Aphrodisias, 201
Alexander of Macedon, 22,
    279n106
Alexanders, *Smyrnium olusatrum*
    L., 296n89
Alexandria, 22, 247n86, 262n3
aloe, *Aloe vera* L., *94*, 180, *181*,
    189, 197, *198*, 267n81, 293n62,
    296n95
Alphabetical Herbarium. See *De
    materia medica*: Alphabetical
    Herbarium version of
alphabetization, 78, 82, 84, 104,
    262n3
Amadio, Andrea, 202
amaranton (Greek), 253n28. *See
    also* pale stonecrop
Amerias of Macedon, 19, 245n60
Anatolia, 178, 180, 184, 193, 230,
    291n41, 293n60

Andronikos II (emperor), 201
anemone, *Anemone coronaria* L.,
    67, *68*, 69, 70, 86, 103, 157,
    242n29, 260n59, 265n59
Angelica sp., 295n89
Anicia Juliana, *16*, 25, 26, 28, 29,
    128, 135, 240nn16–17,
    249n108
annotations: in Arabic letters, *64,
    66, 68, 73, 74, 83, 88, 89, 92, 93,
    94, 96, 98, 115, 116, 117, 123, 132,
    140, 141, 143, 145, 146, 147, 149,
    152, 157, 170, 172, 173, 174, 177,
    181, 212*; bilingual Greco-Latin
    notes, 205, 301n59; in Greek
    letters, *12, 14, 15, 16, 64, 66, 68,
    69, 73, 74, 83, 89, 92, 96, 111,
    117, 128, 139, 147, 148, 150, 152,
    153, 158, 196, 197, 198*, 275n56,
    281n119, 286n27, 298nn6–9,
    298n13; in Hebrew letters, *34,
    64, 66, 68, 73, 74, 75, 83, 88,
    152*, 211–13, 303n80, 303n85; in
    Roman letters, *34, 70, 72, 82,
    87, 90, 92, 93, 94, 95, 96, 98, 118,
    119, 152, 155, 156, 158*, 200, *209,
    211*, 298n25, 303n76

318 / 319

Lignamine, Giovanni Filippo de (Johannes Philippus de Lignamine), 214

Linnaean botany, 53; non-Linnean botany, 53, 54

Linné, Carol von, 53, 223, 226, 305n119

Livia (empress), 23, 24, 26, 29, 249nn103–104

lonchitis (Greek), *Serapias lingua* L., 97, *98*, 268n89, 268n91

Lopes, Dominic, 51, 257n25

lotus, *Nelumbo nucifera* Gaertn., 146, *147*, 286n31, 289n24

Lucius Licinius Lucullus, 23

lupine, *Lupinus* sp., 114, *116*, *117*, *120*, 277n81

Macer Floridus, 108

MacLeay, Alexander, 226

Madonna lily (*krinon basilikon*), *Lilium candidum* L., 284n8

madrasa, 123, 131, 282n147

Magdalino, Paul, 109

Mago the Carthaginian, 19

Magi, 84

magic and the occult, 42, 123, 279n103. *See also* eryngo; mandrake

Maguire, Henry, 109

male fern, *Dryopteris filix-mas* (L.) Schott, 88

Malik al-Mu'azzam, al- (sultan), 182, 193

mandrake, *Mandragora* sp., 13, *14*, *15*, 52, 53, 71, 123, *125*, 126, 128, *224*, *225*, 241n21, 280n117, 281n120

Manfredus of Monte Imperiale, 128, 131, 132, 281n128, 282n142

Manuzio, Aldo (Aldus Manutius), 208

Marcellus Empiricus, 107

Marchetti, Francesca, 205, 301n56, 303n80

mare's tail, *Hippuris vulgaris* L., 103

marginalia. *See* annotations

Maria Theresa (empress), 223

marshmallow, *Althaea officinalis* L., 123, 126

mastic, *Pistacia lentiscus* L., 293n52

materia medica (medicinal substances), 21, 28, 41, 42, 104, 117, 204; storage of, 27, 42

Matthaeus Platearius, 107, 164

Mattioli, Pietro Andrea, 205, 306n120

Mauch, Ute, 53

Mayhoff, Karl Friedrich Theodor, 10

Medea, 17

medical sects, 21, 36

medicine, study of, 102, 107, 108, 123, *125*, 126, 128, 129, 200, 228, 278n95

medicines. See *materia medica*; *mithridatium*

Medicinii Plinii, 107

medlar, *Mespilus germanica* L., 293nn52–54

Medusa. *See* Gorgon

Mehmed II (sultan), 213

memory, 36, 37, 60, 199, 247n86. *See also* mnemonic devices; search image

mental rotation, 62, 259n49

Mesokēpion (Constantinople), 112, 179

Mesopotamia, 1, 105

Metrodorus, 9, 17, 19, 28, 243n50

Meyer, Albrecht, 217

Michael VIII Palaiologos (emperor), 201

Mikion, 19, 245n60

milk thistle, *Sillybum marianum* (L.) Gaertn., 67, *70*

minerals, 44, 78–79, 84, 132, 243n42

Mithridates VI Eupator (king), 18, 20, 21, 22, 23, 28, 39, 40, 41, 42, 45, 46, 243n47, 247n81

mithridatium, 21, 247n85

mnemonic devices, 47, 70, 71, 117, 214

monasteries, 6, 28, 101, 122, 126, 130, 226, 228, 291n41, 293n55; Kyr Meletios Monastery (Attica), 201; Montecassino, 107, 214; Pantokrator Monastery (Constantinople), 126,

288n57, 288n59; St. Catherine's Monastery (Sinai) (*see index of manuscripts*); St. Demetrios Monastery (Constantinople), 285n19; St. John the Forerunner Monastery (Petra, Constantinople), 126, 161, 162, 199, 200, 204, 208, 212, 213, 280nn116–117, 285n19, 303n83; Stoudios Monastery (Constantinople), 130; Vivarium Monastery (Calabria), 102, 270n2

Montpellier, 108

mosaics, 112, *113*, 240n19

Moschos, Demetrios (copyist), 302n70

Moschos, John (copyist), 302n68

Mount Athos, 232. *See also index of manuscripts*

Mount Lebanon, 182, 184, 221, 222

Mount Vesuvius, 9

mouseion (Alexandria), 21

mugwort. *See* artemisia (multiple species)

mulberry, *Morus nigra* L., *173*, *175*, 258n38, 291nn32–33, 291nn40–41

Mulk, Niẓām al-, 131

mullein, *Verbascum* sp., *61*, 81

Müller, Carl Friedrich Wilhelm, 10

Munya al-Ruṣāfa (Córdoba), 112

mushrooms, 267n79, 289n24

Mutawakkil, al- (caliph), 105

myos ōta (Greek), unidentified, *96*, 99

myrtle, *Myrtus communis* L., 24, 42, 290n29, 291n32, 292n41, 292n50

Nabāṭī, al-, Abū l-'Abbās (also al-Jayyānī), 184, 295n73

Najm al-Dīn (ruler), 106, 272n22

naming of plants, 5, 6, 7, 18, 23, 24, 27, 31–38, 47, 70–71, 76, 86, 88–90, 91, 97, 100, 103, 104, 105, 110, 111, 114, 115, 132, 144, 154, 188, 205, 208, 211, 214, 217, 228, 229, 243n47, 252n15, 264n47, 266n68, 275n56, 285n19, 288n5, 294n68, 295n89,

Ten Broeke, Berend (Bernard Paludanus), 219, 221–22

terebinth, *Pistacia terebinthus* L., 179, 293n52

Tertullian, 20

testing, 21, 36, 42, 128, 161, 247n81. See also *peirai*

Thalassa, 75, 261n77, 261n78

Theodore the Studite, 109

Theophrastus, 2, 17, 39, 54, 65, 173, 205, 231, 235, 254n52, 255n64, 262n6, 291n35

time, 65, 67, 70. *See also* seasons

Titus (emperor), 9

Tortelli, Giovanni, 204

Touwaide, Alain, 18, 99, 116, 299n27, 302n68, 302n69, 302n70

toxicology, 20, 105, 135

Tractatus de herbis, 4

tragos (Greek)/*ṭrāghūs* (Arabic), grain product and fictional plant, 92, *93*, 97, 114, *115*, 268n93

translations: Arabic translations of Greek, 106; Arabic translations of Syriac, 272n22; in general, 105–107; Greek translations of Arabic, 106, 272n28; Greek translations of Latin, 106, 202; Greek translation of Punic, 19, 245n56; Hebrew translation, 211; Latin translation of Arabic, 106, 107; Latin translation of Greek, 22, 102, 108, 201, 273n38; Latin translation of Punic, 22; Paduan translation, 108, 202; Syriac translations of Greek, 105, 106, 271n17,

272n22, 273n31, 278n95. See also *De materia medica*

transliteration, 7, 105, 110, 114, 126, 158, 211, 212, 265n53, 266n68, 271n19, 299n25

Traversari, Ambrogio, 204

tree heath, *Erica arborea* L., 169, *170*, 290n29

trees, 25, 44, 53, 78, 79, 84, *111*, 117, 132, *165*, 167, 168, 169, *170*, *171*, *172*, *173*, *174*, 175, *176*, *177*, *178*, 179, 180, 255n73, 289n24, 292n42, 293nn52–54

typikon, 126, 280n113, 288n59

Uexküll, Jacob von, 56, 258n42

van Swieten, Gerard, 223

variants, 138–44; defined, 138

Venice, 197, 202, 234, 302n68

vetch, *Vicia sativa* L., 150, 287n41

viburnum, *Viburnum* sp., 24

vine. *See* grapevine

violet, *Viola odorata* L., 24, *64*, 65, *92*, 94, 219, *220*, 259n53, 267n84

visual knowledge, 3–4; defined, 2

wall-paintings, 23, 24, *25*, 201

walnut, *Juglans regia* L., 175, 291n32, 291n41

water lettuce, *Pistia stratiotes* L., 154, *155*

waterlily (*nymphaia* [Greek]), *Nymphaea* sp., *89*, *90*, 263n25, 266n66, 287n46, 299n27

watermarks, 205, 297n4, 301n56, 302n68

Weiditz, Hans, 215

Wellmann, Max, 17, 18, 19, 32, 78, 80, 82, 84, 91, 177, 238n18, 242n28, 244n50, 244n54, 245n62, 264n37, 284n16, 287n48

white bryony, *Bryonia alba* L., 165, *166*

white hellebore, *Veratrum album* L., *120*

white poplar, *Populus alba* L., 291n32

William of Moerbeke, 201

willow, *Salix alba* L., 291n32

wine, 44, 78, 79, 84, 168, 197

Wisdom, personified, 128, 135

woad, *Isatis tinctoria* L., *96*, 97, 142, *143*, 144, 157, 285n21

women: healthcare, 26, 249n111

Xenōn of the Kral (Constantinople), 161, 281n119. *See also* monasteries: St. John the Forerunner Monastery

Zachalias of Babylon, 21

Zacharias of Chalcedon, 103, 104, 110, 144, 161

Zakynthos. *See* Corfu

zoomorphism, 92, *93*, 97, 114, *115*

zōonychon (Greek), unidentified, 265n48

Zoroaster, 84

Zsámboky, János (Johannes Sambucus), 7, 211

INDEX OF MANUSCRIPTS

Listed by shelf mark; names in parenthesis according to the author's usage.